EQUILIBRIUM IN SOLUTIONS

SURFACE AND COLLOID CHEMISTRY

This volume is published as part of a long-standing cooperative program between Harvard University Press and the Commonwealth Fund, a philanthropic foundation, to encourage the publication of significant scholarly books in medicine and health.

George Scatchard in 1954

Equilibrium in Solutions

Surface and Colloid Chemistry

GEORGE SCATCHARD

A Commonwealth Fund Book

HARVARD UNIVERSITY PRESS

Cambridge, Massachusetts, and London, England   1976

Library of Congress Cataloging in Publication Data

Scatchard, George, 1892–1973.
   Equilibrium in solutions *and* Surface and colloid
chemistry.

   "A Commonwealth Fund book."
   Bibliography: p.
   Includes index.
   1. Chemical equilibrium. 2. Solution (Chemistry)
3. Surface chemistry. 4. Colloids. I. Title.
QD503.S28 1975   541'.345   75-15753
ISBN 0-674-26025-2

Contents

Introduction

by I. Herbert Scheinberg

In the late 1940s George Scatchard's laboratory at the Massachusetts Institute of Technology consisted of one rather large room flanked by two smaller rooms, with a cold room in one of the latter. Across the hall in Room 6-221 was his office, which in those years looked out on a pleasant expanse of grass and the Charles River. Doris Currier, his secretary, worked in a nearby office. During this period Scatchard's associates at M.I.T. included S. Howard Armstrong, Jr., Jeanette Weeks, Allen Gee, Robert W. Bridgforth, Judith Bregman, James Coleman, Harriet Anderson, Ernestine Rolls, and myself.

Only the offices had a view; the laboratories looked out on a courtyard. Scatchard's desk was next to a window, so that by lifting his eyes he could see the river. On entering his office one was likely to find him writing in pencil on lined white paper, sitting, with his back to the door, at his calculating machine (a "Millionaire"), or gazing pensively out of the window.

Scatchard's office was about 12 by 25 feet, with a window occupying the entire far wall. There were two or three bookcases against the left wall. The other was lined with cabinets containing unused equipment, and filing cabinets; between these and the window were a blackboard and a coatrack. A long, linoleum-covered table in the center of the room was laden with journals, data notebooks, and manuscripts. The desk was placed with its long axis perpendicular to the window. The "Millionaire" was on Scatchard's left, so with a push of his hands and rotation of his castered chair, it took no more than a second or two to turn from writing to calculating. A quarter-turn to the right brought him to a typewriter. The desk always had four to five sharpened yellow pencils, his preferred writing tool.

The office was where Scatchard thought, read, calculated, wrote, and did a good deal of his talking. He made forays into the laboratory

—with some reluctance—and always in a Brooks Brothers suit; occasionally in shirtsleeves; never in a laboratory coat. With very rare exceptions the only laboratory work he performed was to remedy the consequences of a student's ineptness or repair a balky instrument. He was faintly contemptuous of apparatus, and once described his extremely sensitive equipment for measuring conductivity as "a box where a lot of things go on inside that I really don't care about as long as I can get the numbers that come out."

What he did care about were ideas. Not long after joining his laboratory I realized what pleasure he got from proving, or disproving, a hypothesis, using the published results of someone else. His pleasure was greatest if the author had not thought of the ideas that Scatchard succeeded in developing from his data.

The lunch hour was fixed. About 12:30—to miss the noon rush of students—we would walk across the wide lawn to the Walker Memorial. Scatchard's lunch was always shredded wheat, milk, and chocolate ice cream. Weather permitting, there was a leisurely postprandial walk and talk along the Charles. This occasionally extended to crossing to Boston over the Longfellow Bridge, strolling along the Embankment to the Massachusetts Avenue Bridge, and back to M.I.T.

Of his M.I.T. colleagues he was closest to Walter H. Stockmayer, whose office was several doors away from his, and to Isadore Amdur, a floor below. Stockmayer and Scatchard had a close scientific relationship and, in their discussions of theoretical or applied thermodynamics, almost every word that was not essential to the ideas they were pursuing was eliminated. Very few people could follow one of these conversations. Both wrote in a similar spare, understated way. James S. Johnson, Jr., of Oak Ridge, recalls, "I hope I am not alone in sometimes finding his writing difficult—he is in my book the world's unchallenged master of the indefinite antecedent." Although he shared fewer scientific interests with Amdur, Scatchard was much involved with him in affairs of the American Academy of Arts and Sciences. There was a certain aloofness between Scatchard and Louis Harris and James Beattie, the other two professors of physical chemistry whose offices were off the same corridor, although Beattie's interest in Gibbsian thermodynamics was at least as great as his own.

In any scientific conversation Scatchard assumed at the outset that he was talking with someone who was just as quick and as well-informed as he. If this was not in fact the case he de-escalated the sophistication of what he was saying and added necessary explanation so that his—by then—listener rarely left unenlightened. This intuitive ability to sense the maximal intellectual level at which he could communicate with an individual made him a unique teacher and an extraordinarily valuable consultant. Johnson has pointed out how his scientific group at Oak Ridge "particularly benefited, because of several common interests—ultracentrifugation, ion exchange, thermodynamics of solutions—(and because) he has a trait oddly rare, in my experience, in consultants . . . he discusses with you your problems, rather than his . . . His patience in explicating his ideas to us has been immense. Once one understands what he is about, intriguing subtleties frequently become apparent in his style. There are other occasions, particularly in his M.I.T. lecture notes, when he has produced about the only worthwhile discussion of many topics. And there is no ally I'd rather have in a scientific controversy."

Scatchard's ability to understand the scientific problems of others, his superb knowledge of thermodynamics, and an imagination which permitted flexibility in his thinking made him uniquely valuable to the work on proteins carried out in the laboratory of Edwin J. Cohn at the Harvard Medical School. His association with Cohn, which began in their undergraduate days at Amherst, brought Scatchard to thinking about proteins long before this was considered a respectable activity for physical chemists. In 1948 Herbert Harned, a more orthodox physical chemist from Yale, said, at a meeting of the New York Academy of Sciences, "George, I don't see how you possibly can spend time working on proteins when there's so much we don't know about potassium chloride!"

Actually he probably was induced to begin thinking about proteins 20 years earlier. Around 1926 James Conant told Cohn and Scatchard that a number of European chemists he had just visited had indicated that proteins had molecular weights of only several hundred, but agglomerated in water to larger aggregates. Cohn proceeded to test this experimentally by measuring the vapor pressure or freezing-point depression of various protein solutions. The results differed wildly from day to day. Scatchard recalls meeting Cohn on the days that the molecular weights seemed to be in the hundreds of thousands and Conant on the days when they were apparently in the hundreds. It occurred to him that such discrepant results could only mean that the activity of the water was not constant from day to day. He suggested Cohn make it so by adding to his solutions a mixture of anhydrous and hydrated sodium sulfate which, in equilibrium with water, would establish its activity as unvarying. Constant values for the molecular weights were immediately obtained.

At about the same time Scatchard, Cohn, and Conant came to recognize the importance of constant solubility of a protein, in the presence of increasing amounts of the solid phase, as a prime criterion of a protein's purity. Discussion about these and other problems gradually led Scatchard to spend more time at Harvard Medical School. Thomas L. McMeekin described Scatchard's involvement with Cohn's laboratory. "I am glad to record my impressions of the effect of Dr. Scatchard's talks at the Department of Physical Chemistry weekly seminar, which he attended regularly. His and Kirkwood's discussions served as a stimulus and a guide for our experimental [work] on amino acids and proteins and in interpreting the results . . . The work that I was involved in . . . was largely the effect of structural relationships on the properties of amino acids and proteins, including related substances, which of course Dr. Scatchard was also much interested in. I am particularly indebted . . . for a chance remark of his which led to one of the few of my scientific accomplishments that I am proud of. In discussing the properties of amino acids, Dr. Scatchard remarked that what we needed was to compare the properties of amino acids with the properties of related uncharged molecules. This led to (previously unacknowledged) comparative studies by Cohn, Edsall and me on the amino acids and their uncharged isomers." James L. Tullis referred to these same weekly seminars. "I have no pictures except the vivid mental one of George quietly sitting at that big table in Edwin's office—appearing to be asleep—then abruptly awaking to utter the best, and sometimes the only, sage remark of the day."

Hubert Vickery of the Connecticut Agricultural Experiment Station remembers—that "George was the ultimate court of appeal from the Cohn group on any problem in physical chemistry that came up. It was a strong team with Edsall and the junior members of the group, and there was little in the literature that someone could not recall when arguments developed." Jeffries Wyman recollects seeing "a great deal of both him and Cohn during those years of their close collaboration, [when I myself was working in Cambridge in collaboration with McMeekin and later with Jesse Greenstein, both of whom were members of Edwin's department.] It was apparent to all of us how much Edwin relied on George's deep background in physical chemistry in all his work up to the time when it took a more practical turn in the war years, and even then the influence continued. George's impact was of course particularly strong in the case of Cohn's elaborate solubility studies. Also, it was through George, if I am not greatly mistaken, that Jack Kirkwood was drawn into the picture, and . . . his influence in developing a model for the electrostatic effects to be attributed to proteins as a result of their dipolar ionic properties was very great. . . . I think it is only fair to say that although George and Jack Kirkwood both contributed so much in furnishing . . . ideas to Cohn, *his* influence in confronting them with provocative problems was very considerable, and led them into paths which they would probably not otherwise have attempted to explore . . . I was well aware throughout the whole period of the very close personal dependence, quite reciprocal, of the two characters on one another."

Because of the depth of his knowledge and the quiet, unemotional way in which he talked, Scatchard could evoke awe—perhaps even terror—in those with whom he had scientific contact. Charles Tanford, who worked in Cohn's laboratory around 1950, writes "I learned what thermodynamics I happen to know indirectly from GS. I was in continual fear of his disapproval—his frown of disapproval at seminars was deadly—and I tried valiantly and often unsuccessfully to avoid it. One of my pleasantest memories is of the paper I gave at a meeting of the Electrochemical Society in 1953 (on serum albumin expansion), which GS actually approved of! But my success didn't last. I had a talk with GS in 1966 or 1967, where he said 'I have read papers both by you and Harold Scheraga on this subject (thermodynamics of denaturation), and I think one of you is much better than the other, but I've forgotten which one.'"

John T. Edsall, who had close scientific relationships with both Cohn and Scatchard, wrote about joining Cohn's laboratory in 1926. "Edwin had just initiated some research on muscle proteins, and suggested that I take up work on the problem in the Department of Physical Chemistry at Harvard Medical School under his guidance. Departmental seminars were held frequently, usually in the small room next to the Bowditch Library, one floor below Physical Chemistry, adjoining Walter Cannon's office and the Physiology Department. It was in these seminars that I first came to know George Scatchard, and to realize the pervading influence that he had on the thinking of everyone in the Department. Cohn was of course deeply concerned, from the beginning of his research, with the separation of pure proteins from the complex mixtures found in nature and therefore was eager to learn as much as he could about the conditions governing the solubilization and precipitation of proteins. He naturally sought a rational basis for describing these phenomena, partly to provide a theoretical basis for protein fractionation, partly to understand the properties of proteins in relation to their size, shape and distribution of electric charge. As I listened to the discussions of these matters, I soon came to realize the constant interplay of ideas and experimental plans between Edwin and George. George's deep knowledge of the physical chemistry of solutions was indispensable, not only to Edwin, but to all of us who were concerned with these problems of protein chemistry. He knew, for instance, that the phenomenon of salting out was not at all confined to molecules like proteins, but was manifest in all sorts of simpler systems such as ethyl acetate-water mixtures, where the addition of simple salts, like sodium chloride or sodium sulfate, caused the solubility of the ethyl acetate to decrease.

"Debye and McAulay had proposed a simple theory of salting out . . . and George had elaborated and refined this picture in various ways. I am sure that his influence proved to be important when Edwin wrote his first major review on proteins (Physiological Reviews, 1925) in which he proposed a linear equation to describe the salting out of proteins:

$$\text{Log } S = \beta - K_s \mu$$

Here S is the solubility, μ is the ionic strength of the salt solution and β is a constant obtained by extrapolating the linear part of the solubility curve, obtained at high ionic strength, back to $\mu = 0$. It is of the same form as the linear salting out equation for simple molecules like ethyl acetate in water, only in those cases the curve is linear all the way from ionic strength zero on up, so that $\beta = \log S_0$, where S_0 is the solubility in the absence of salt. The big difference is that K_s for a protein and a given salt is higher by an order of magnitude than K_s for a small molecule and the same salt. In a logarithmic expression this makes a huge difference. As George put it [Chem. Revs. 4, 383–402 (1927)] 'while the solubility of ethyl acetate is reduced one half, that of pseudoglobulin is reduced to one thousandth.' (p. 397) Obviously George drew on Edwin's calculations for proteins in discussing the general significance of the salting out effect. Conversely I feel sure that the background in the general physical chemistry of solutions, that Edwin had acquired from George, greatly influenced him in looking for a simple relation to describe salting out of proteins.

"In 1923 Debye and Hückel published their famous paper on interionic forces in solution . . . It was certainly George Scatchard who guided Edwin in his use of the Debye-Hückel equations in interpreting the solubility of proteins . . . For example, Arda Alden Green had extended earlier work of Cohn and Prentiss so as to include both salting in and salting out of horse hemoglobin, by a variety of salts, and over a considerable range of pH. Dr. Green was a superb experimentalist, and her resulting papers rank as classics in the study of protein solubility, but the interpretation of the data in terms of Debye-Hückel theory was inevitably rather empirical. It became gradually clear to us, and to George and Edwin first of all, that there was a great need for the study of simple model substances of known structure, especially amino acids and peptides . . . There were important implications of the 'zwitterion' (or dipolar ion) structure of

amino acids that we gradually came to perceive. . . . For one thing
we realized early—and I suspect that George Scatchard was the first
to perceive it—that dipolar ions, with their substantial separation of
positive and negative charges . . . must have dipole moments in water
which are much larger than those of ordinary polar molecules. . . .
Edwin induced Jeffries Wyman to take up the problem, and . . .
Wyman's work gave a tremendous impetus to the study of the dielectric constants of polar liquids, and led to the later theories of polar
liquids that Lars Onsager and John G. Kirkwood developed.

"Edwin had taken sabbatical leave from Harvard in 1931–32 and
spent much of that year in Munich. George and Jack Kirkwood, who
had just taken his Ph.D. at M.I.T., were abroad also and saw a great
deal of Edwin. Agreeing that there was a need for a better theory to
provide a model for calculating the activity coefficients, and other
properties, of dipolar ions, George and Jack set out to provide one.
Their first model, described in a paper in the 'Physikalische Zeitschrift'
(1932, GS paper No. 32) consisted of two charged spheres separated
by a thin rigid connection. Edwin called it the 'dumbbell model,' and
George had suggested that the term was intended, by implication, to
refer to the authors as well as the model.

"It was not a very satisfactory model and Kirkwood was dis-
satisfied with it, soon coming up with a more realistic treatment, using
several different models. One was a sphere with a dipole inside it; two
others were ellipsoidal, with charges at the foci or with a point dipole
at one of the foci. After George, Jack, and Edwin had returned from
Europe, we had a whole series of seminars, discussing these models and
the experiments that they suggested.

"T. L. McMeekin, who had come to Edwin's laboratory in 1928, had
already embarked on his solubility studies of amino acids and peptides and their derivatives, and as mentioned above by McMeekin,
the interplay of theory and experiment led to a tremendous range of
studies on these compounds in solvents of varying dielectric constant
over a wide range of ionic strengths. Kirkwood's theory turned out
to explain a wide range of phenomena in a very elegant way.

"George was thoroughly at home with non-electrolyte as well as
electrolyte systems; indeed there were few people in the world who
understood either as well as he did. This became important when
McMeekin's studies showed the great importance of the nonpolar
side chains of amino acids and their derivatives on their solubility
and other properties.

"These explorations—studies of partial molal volumes of dipolar
ions, the electrostriction produced by their charged groups, studies
of partial molal heat capacities and compressibilities, which Frank

Gucker in Evanston undertook with stimulus from Edwin and George, and the studies of dielectric dispersion in dipolar ion solutions that Larry Oncley undertook—are all discussed in the book 'Proteins, Amino Acids and Peptides,' the writing of which was finished just before the United States became involved in World War II. The book carried on its cover only the names of Cohn and Edsall, but the title page indicated that we had enlisted several other authors who wrote a number of essential chapters. The most comprehensive of these was the one by George Scatchard on 'Thermodynamics and Simple Electrostatic Theory.' This packed into a little over 50 pages a tremendous sweep of fundamental ideas that were basic to everything that followed. The word 'simple' in the title had different connotations for different people. Compared to Jack Kirkwood's chapter on dipolar ions, later in the book, George's mathematics was undoubtedly a good deal easier; but quite a few people, myself included, found it hard going in places. (Hubert B. Vickery—'Vic'—said he found George's chapter very useful as a bedside book at night: a few pages were enough, for him, to produce a highly soporific effect.) Looking over that chapter again, as I write this, I marvel at its power and insight; it says so much of fundamental importance in so small a space.

"George read every chapter in the book, at an early stage; he may indeed have read every draft of every chapter. He saved me from serious weaknesses that he found in an early draft of Chapter 8, on solubility in water and organic solvents. I had made a number of misleading statements, and perhaps downright errors also. I do not now remember just what the trouble was; at any rate, George straightened me out, and the chapter emerged under our joint authorship. He caught some of my mistakes in a later chapter also, on solubility of proteins, and got it right before the book went to the printers. His suggestions helped us all out in numerous other places.

"Of course the Department of Physical Chemistry had been put upon a war footing long before the book appeared—for all practical purposes by the summer of 1941. We had become completely mobilized for the study of fractionation of blood plasma, both human and bovine, to produce albumin, gamma globulin, fibrin products, and other plasma products that might be of clinical value. Edwin Cohn has told the story in *Advances in Military Medicine*, Volume I, Chapter 28 (1948). Here there is no need to retell it, except to point out that albumin solution for the treatment of shock was the first and foremost need of the Armed Forces, and the effectiveness of albumin was dependent on its osmotic action in causing liquid to be drawn into the blood from the surrounding tissues. It was George who understood the principles of osmotic pressure far more deeply than anyone else in the group, and he who got the measurements going, along with Alan Batchelder and Alexander Brown, that determined the molecular weight of albumin, and gave reliable values for its capacity to expand blood volume. He also became deeply involved in studying the other striking property of albumin, which we recognized only gradually as the work went on—namely, its capacity to bind all sorts of anions and many neutral molecules. This had some major practical results, such as the use of sodium caprylate, which binds very strongly to albumin, to stabilize the albumin solutions prepared for clinical use against heat denaturation, so that they could be heated for 10 hours at 60° to destroy the hepatitis virus. George was a major figure in the whole enterprise, and we got wise advice from him on all kinds of problems. Yet he was, at the same time, commuting to New York every week, where he was heavily involved in one section of the Manhattan Project, and he still had his Department at M.I.T., which had to keep running through the war. Just how he managed this three-ring circus I do not know; I do know that he always seemed unhurried and was always available to give advice and help.

"After the war, George went more deeply into the osmotic pressure of serum albumin and other proteins, and into the binding of ions and neutral molecules by proteins. His basic theoretical paper on osmotic pressure made very tough reading for me—it still does, if I turn to it again—but it was indispensible for me when I later became involved in studies of light scattering in protein solutions. As to ion binding in albumin solutions, George's great achievements in that area are too well known to need remark from me here: the 'Scatchard plot' for analysis of binding data is to be found in hundreds of papers from laboratories all over the world.

"There is another 'Scatchard plot' also, less widely used, but one that proved a most valuable tool in some of our own researches. When Frank Gurd and I, about twenty years ago, were trying to analyze our data for the binding of imidazole to copper and zinc ions, George suggested to us an elegant way of analyzing our binding data, so as to calculate the first and last binding constants. It worked beautifully, and I have used it on many occasions since, including the analysis of Roughton's data on the binding of oxygen by hemoglobin. I owe it entirely to George; I doubt if I would have had the wit to think of it on my own, though it is simple enough when the equations are stated.

"As George himself has said in his Cohn Memorial Lecture, he and Edwin collaborated in the closest fashion for about thirty years, although they never published a paper together. It was, in my experience, a unique relationship between two close friends whose gifts largely supplemented each other—Edwin with his colossal energy and drive, his capacity to envisage a large enterprise in broad qualitative terms, and George with his deep and subtle power of analysis, and his creative and yet resolutely critical outlook. The combination led to some great and inspiring results, and it was a rare privilege for some of the younger people like myself to work close to them, for many of those years."

Scatchard was one of the first scientists—if not the first—to apply the fruits of Gibbs's and Debye's genius to the study of proteins as well as to more orthodox problems of inorganic physical chemistry. He was Debye's friend and, particularly in the 1920s, when there were some doubters among the members of the chemical establishment about the correctness and power of the Debye-Hückel theory, one of his apostles. Ironically, Scatchard's success in shedding light on the interactions of proteins with themselves, and with smaller molecules and ions, was probably considerably greater than Debye's less deeply going light-scattering and dielectrophoresis experiments with proteins.

Scatchard's research on the binding of small molecules, or ions, to proteins, is perhaps the best example of his efforts to teach the importance and usefulness of physical and chemical theory in solving a variety of biological problems. Up to 1949, work in this field was very largely limited to discovering new protein-small molecule compounds. Scatchard was the first to call attention to the theoretical and practical importance of asking "How many? How tightly? Where? Why?" for each specific small molecule-protein combination (G. Scatchard, The Attractions of Proteins for Small Molecules and Ions, Ann. N.Y. Acad. of Sci. 51:660–672 (1949)). "Scatchard plots" were the means he devised to relate experimental results to theory since the intercepts of the curve drawn through the points answer the first two of these questions quantitatively. Thus here, as in his usage of graphs generally, the curve did not just connect experimental points but possessed theoretical significance. He never drew a curve merely as a visual aid.

It seems a safe bet that Scatchard plots will serve biological chemists for a long time. At a Symposium of the American Chemical Society in Chicago in the summer of 1973, three papers made use of this procedure, and it is difficult to find a current issue of the *Journal of Biological Chemistry* that does not have at least one article in which they are mentioned.

That much of Scatchard's work was devoted to proteins was in part a result of his life-long friendship with Edwin Cohn, through which he was exposed to the problems of protein chemistry every week of the academic year. Wyman and Edsall, in the passages quoted above, have pointed this out. But there was also his desire to see just how far he could extend the theory and rigorous techniques of physical chemistry to these complex, unstable molecules that play such a central role in the phenomena of life. He did indeed go well beyond the domain of the pure physical chemists, as Harned noted, and in so doing, made fundamental and far-reaching advances in biochemistry.

George Scatchard was not unaware of his exceptional powers of analysis and understanding of chemical phenomena, but he was a modest man. Once Cohn, in discussing a monograph by another chemist, said, "He has certainly written a good book. What a pity that he felt he had to try to save the world in the last chapter." Scatchard wrote many superb papers, and the two remarkable treatises reprinted in this volume, but he was never led to overestimate his immense talents in an attempt to save the world.

ACKNOWLEDGMENTS

In 1971 several former students of George Scatchard made tentative arrangements to reprint his collected publications and the mimeographed notes he had prepared for two of his courses at M.I.T. On learning of this I wrote to Robert J. Glaser, then Vice-President of The Commonwealth Fund, to determine if the Fund would be interested in sponsoring publication of such a volume as part of its cooperative program with The Harvard University Press. In awarding a grant-in-aid for the project Quigg Newton, President of The Commonwealth Fund, expressed his belief that the volume, to include retrospective annotations of Scatchard, would be "an important contribution to the scholarly literature of American science."

Because of rapidly rising costs and the availability in journals or books of his published papers this volume has been limited to Scatchard's autobiographical memoir, incorporating comments on his publications, and his two unpublished texts, *Equilibrium in Solutions* and *Surface and Colloid Chemistry*. Fortunately, he was able to correct the typed manuscripts of both before his death in 1973. Proofs have been read critically by Joseph Elder, John T. Edsall, Walter H. Stockmayer, and myself. Stockmayer has written the following two paragraphs:

"Publication of any text 20 years after its last revision implies that it is unique and in some measure timeless. The two Scatchard texts are judged to pass this test. *Equilibrium in Solutions* is not just another book on chemical thermodynamics. It was written for a graduate course (offered to physical chemists at M.I.T.) variously entitled 'Theory of Solutions' or 'Chemical Thermodynamics III.' It conveys, in the same generally compressed style that makes his major papers difficult, Scatchard's methodology in dealing with solutions of any complexity from noble-gas mixtures to multicomponent electrolyte or protein systems. It is this breadth and the continuous subtle interplay between strict thermodynamic reasoning and intuitive molecular interpretation which are unique. Although the modern student of solutions will not find the extensive statistical-mechanical developments of the past two decades, he will find in Scatchard's chapters many guides and examples of procedures that are still viable and valuable. It is not irrelevant (or irreverent) to suggest a qualitative parallel with the papers of Gibbs, and indeed this suggestion has already been made by others.

"*Surface and Colloid Chemistry* was written for senior undergraduates, to follow two semesters of standard classical physical chemistry. It too is less than modern in some spots, but it offers a strong and continuous demonstration that the subject is a completely legitimate part of physical chemistry. As Scatchard recalled in his charming Kendall Award Address entitled 'Half a Century as a Part-time Colloid Chemist,' the colloid-chemical textbooks of his student days seemed 'mostly cook-book or nonsense,' and he resolved to remedy the situation. If the book is read in company with a modern text, the deliberate omission of literature references will not be a handicap."

Autobiographical Note

by George Scatchard

I was born March 19, 1892, in Oneonta, New York, the second son and fourth child of Elmer Ellsworth and Fanny Lavinia Harmer Scatchard. My older brother was named after our father and I after my paternal grandfather, who jealously protected me from any middle name because he himself was plain George Scatchard. (My father wanted Van Rensselaer, after our family doctor.)

Oneonta was a village of about 10,000, on the Susquehanna River in the northern foothills of the Catskills, about 20 miles south of the much better known Cooperstown. The largest village for 50 miles in any direction, its most important industry was the shops of the Delaware and Hudson Railroad. It was also the site of the Oneonta Normal School, which had about one boy student to ten girls in my time, and an excellent High School Department when I was ready for it.

When I was starting first grade, my mother went to Philadelphia to help care for my grandfather, who was ill. She took me and my youngest sister, aged five, with her. The necessity of being quiet and entertaining ourselves was solved by my teaching my sister to read. I can remember no texts other than the daily paper and the Bible, but when we returned home she went into the second grade and I into the third. I was an avid reader, often for the almost mechanical pleasure of reading. Among the books I remember were *The Autobiography of Ulysses S. Grant* and some publisher's *History of the United States*. When I was seven, there were two exceptions to reading for the sake of reading: Kipling's *Just So Stories*, then being serialized in the *Ladies' Home Journal*, and Chaucer's *Canterbury Tales*. I heard my father tell my mother to keep that book away from the children. It did not take me long to find her hiding place and his reasons, and not much longer to pick out the spicy places, in spite of Chaucer's language. I did help protect the rest of the family.

Except for reading, I had the normal childhood of a small-town boy—hiking, bird-watching, swimming, skating, coasting, and later bicycling. I helped organize and played on the "Oneonta Juniors" basketball team and later on the Oneonta Normal School team. I was manager then and officiated more than I played. (At Amherst basketball was not an intercollegiate sport; I played a little on my class team.)

I was a very shy child; my shyness was cured only when I worked in my brother's drugstore during my high-school years. I still remember the agony of waiting on a customer at the beginning, but it could not last. At first I opened the store in the morning, swept and mopped the floor, and snaked the 150-pound cakes of ice from the sidewalk to the icebox behind the store. On ice-cream days I fetched the cream about three blocks from the creamery, added the sugar and flavoring, and turned the crank of the five-gallon freezer. Later I graduated from the early morning and menial work. My brother had the theory that in the evenings, when our neighbors from the hardware and clothing stores had gone home, we should do no work not directly connected with waiting on trade. My reading branched out into drug journals, the pharmacopoeia, materia medica, and prescriptions—the last with difficulty. Most physicians had terrible handwriting, and prescriptions were written with English or Latin abbreviations. There was one street in Oneonta with fifteen or twenty houses numbered chronologically, not geographically. One physician cared for the people in most of them, so I had to learn the numbers too. I did make myself into a good practical pharmacist, and managed the store when my brother was absent.

The first college I visited was Hamilton, where I was pledged to a fraternity by Alexander Woollcott. After a visit to Amherst, I decided to go there instead. I could have gone prepledged to Woollcott's fraternity or to either of two others, but I joined a fourth fraternity, Phi Gamma Delta, in my junior year. I joined the mandolin club as guitar player in the same year because I was the only eligible player who could read music, and this proved useful in my senior year when they wanted a large group close to Amherst.

I took no physics at Amherst because the senior professor, Arthur Lalanne Kimball, forgot some elementary practical physics: a backing automobile, when unrestrained, curves more and more. His driveway was parallel to the railroad track and ended in a road with a bridge over the tracks. He backed onto the bridge, then fell onto the track, and he could not teach again for some years.

(As I confirmed later when I went back to teach, there was a great difference between him and Professor Joseph Osgood Thompson, the next senior professor. I served on a committee to assign freshmen to faculty advisers of which Professor Thompson was chairman. I suggested that the system was not working well and perhaps we should try having the student pick his own adviser. I still remember the reply almost verbatim: "Why, Professor Scatchard, sometimes a single word from a member of the faculty will change a boy's whole career in college. I remember one boy. It was before midterm, and I did not need to say anything, but I did tell him that if he did not do better he would flunk the course and flunk out of college. He did not do better, he did flunk the course, and he did flunk out of college. I tell you, Professor Scatchard, sometimes a single word from a member of the faculty will change a boy's whole career in college." So I subsided and merely tried to limit the chairman to one story to every five freshmen assigned.)

At the end of my junior year I went to Greece on a University Travel Tour with my three sisters and some friends from home. We started in Olympia, then toured the Aegean on the University Travel ship, the *Athena*. We were escorted by the professor of archaeology at Cornell and the dean of the Graduate School at the University of Colorado. We visited Athens, then took the Orient Express, stopping at the important cities. We then went through Austria, Germany, Belgium, and Holland, most of which I saw, literally, through jaundiced eyes. I recovered from jaundice the day we crossed to England and Scotland. After a further stay in France, Switzerland, and Italy, I got back to Amherst about Thanksgiving and worked in the organic chemistry laboratory without course credit the rest of the first term. I graduated at the end of the year with the highest marks in my class.

Although all my chemistry teachers at Amherst were Johns Hopkins alumni, Howard W. Doughty sent me to Columbia. He was one of the first people in the country, and probably in the world, to use simple physical chemistry in analytic chemistry. At Amherst I assisted him in the laboratory, but neither of us thought of any subject for my thesis other than synthetic organic chemistry.

At Columbia I worked on quinazolines with Professor Marston T. Bogert. I purified 2,5-dinitroquinazoline by dissolving it in hot water in a porcelain casserole. The crystals were colorless but turned yellow when the dish was washed with tap water. We used Harold A. Fales's buffers and Sörensen charts to determine the pK of the indicator but were given normal NaOH instead of tenth normal, so we went completely through two color changes instead of partly through one. This convinced us even more that we had an exceptionally sensitive indicator. Before we published we knew that the sensitivity comes from the fact that the pK of the first change is 6.8.

Like almost all Columbia organic majors, I took my second minor in physiological chemistry because it was the easiest. Professor Gies always asked chemists on their orals about the ductless glands, so I prepared for mine from Bayliss' *Physiology*. When I saw his associate, P. E. Howe, I was so shocked because I was sure he did not know the rules that I said, "What are you doing here?" He asked me instead about precipitating agents for proteins. The one he could not extract from me was ethanol. We met next many years later at Edwin Cohn's to discuss precipitation of proteins by regulated alcohol concentration, ionic strength, and pH.

In 1916 while at Columbia I met my future wife, Willian Watson Beaumont, who was studying music in New York and taking her B.S. degree in education at Columbia to add to her two degrees in music from the college of Montana. We were married on July 28, 1928.

I finished my Ph.D. dissertation in the fall of 1916 and worked the next year as research assistant to Alexander Smith. My examining committee was G. B. Pegram, chairman, and Professors Bogert, Smith, and Howe. The chemists asked me no questions, the physiological chemist just one, and Professor Pegram was intrigued by the yellow-green color of my indicator. His only question was how it would work with a blue filter. I could have flunked by admitting that I did not know because I had never tried. For the rest of my examination I stood aside while he quizzed Professor Bogert, who knew no more than I but could not say so.

The first part of my dissertation (2) was printed from the type of the journal discussion of the indicator action (1). The second part (4) did not appear in the journal until three years later because Professor Bogert and I were both busy with war work.*

My paper with Professor Smith (3) illustrates the great difference then and now in publication procedures. He had the reputation of requiring repeated rewriting from his junior colleagues, so I did not send a copy of the first draft to the third coauthor. I sent Professor Smith a copy with at least fifteen heating or cooking curves starting from the center of one sheet, asking which ones he wanted published and whether I had been too rough on Wegscheider, who had published an explanation that seemed to me absurd. I got it back with the statement that the editor wanted the drawing simplified. In that paper we showed that the quantity of dry ammonium chloride in the cell could be more than that which corresponds to NH_4Cl, but we missed the conclusion that much of it was adsorbed on the dried walls.

*The numbers given here and in subsequent paragraphs correspond to the numbers in the bibliography of George Scatchard's papers; pages 283–293.

During 1917–18 I was teaching organic chemistry, mostly in the laboratory, doing a little war research at Columbia under the direction of W. K. Lewis of M.I.T. When I wanted to quit to join ROTC, they told me how important my work was and persuaded me to stay. However, when I was drafted, Professors Bogert and Lewis thought they should set an example and ask exemption only for essential workers, which I was not. But I was also doing nonwar research for Professor Smith, which he wanted continued. So he applied for my exemption from his summer home in the South, and I did the leg work for him. He said I knew more about my work than anyone else, which seemed more persuasive to politicians than to scientists.

Bogert and Lewis had not forgotten me, however. That afternoon I found in my mailbox a commission as First Lieutenant in the Sanitary Corps, the only place a chemist could be commissioned before the days of the Chemical Warfare Service. I waited until I had my uniform before I reported to the District Board that I had given up the work for which they had exempted me. The chairman, Charles Evans Hughes, said "The Army got you into this, they'll have to get you out of it." The chairman of my local board, proprietor of the little news store near Columbia, saw no problem but wanted to be sure that they sent back word from Washington when I reported for duty, so that I would be included in his quota. I received orders on the same day to report to Fort Upton and the Army Medical School. There my orders were "to proceed to France and on arrival at that Capital to report to the American Ambassador for duty under the French Interministerial Commission on War Research." No one had heard of an officer reporting to a civilian. The easiest way to pass the buck was to send me on, so I made a record trip. France had had several ministries since they asked for me, and it took several days for the ambassador's office to find the Commission—who no longer knew why they had asked for me. They showed off their Paris laboratories, which were not too secret, and let me choose where I wanted to work. I naturally chose to work with Victor Grignard, who was in the laboratory of Georges Urbain because his laboratory in Nancy was occupied by the Germans. The equipment of a research laboratory on rare earths in those days was a good spectroscope and lots of evaporating dishes, very poorly adapted to defensive gas warfare. The method of the laboratory was for everybody to work on analysis when a batch of samples arrived and to turn to research when they were all analyzed. My first contribution was to keep the best men from routine analysis. It took a little time before I graduated to this class.

Grignard and Urbain were as different as their fields. Grignard was the only person I have known who disliked all music, and it would not have helped him if jazz and rock had been invented. Urbain was an able sculptor, and had the attitude of an artist. Grignard was extremely patient with my poor French and would always listen to my discussions with my coworker until he understood me. He was responsible for the ideas in our paper (5). Perhaps the most important result of this work for me was burning my throat with mustard gas and phosgene.

I started home about the first of the year by way of St. Aignan, a camp for casuals returning to the States. By summer St. Aignan was a showpiece, but in January it was a mud hole. They had been sending out a medical officer with each company of 150 men until they were all used up, and anybody with a caduceus was accepted. When I protested they said, "You can read a clinical thermometer, can't you? You're better than nobody. If they don't have you, they'll have nobody, and you never will get out of here." So we—a dentist and a chemist—went to Brest, as medical officers for 1500 men unfit for further duty. At Brest there were plenty of physicians, but I had to take responsibility for my five companies. I put myself on the list of the first company to sail, but I had to return to Paris where my oldest sister died of meningitis. I arrived back at Brest just in time to join my company. We sailed on the *Orizaba*, one of the transports to the Philippines in 1898, whose sanitary conditions caused a scandal. The sanitary conditions had been improved, but the only other significant change was that a lot of iron had been piled around the rudder post. So the ship had to be steered by hand from signals wigwagged from the bridge. The crew had plenty of time to practice since we were 21 days from Brest to Newport News—mostly because of the weather. The captain of my company had forgotten to list me on his roster, but the authorities admitted I was there and sent me on to New York where I arrived on my twenty-seventh birthday, March 19, 1919.

I received one of the first group of National Research Council fellowships, but I gave it up to return to Amherst to teach. They were exciting years for a young teacher there. Alexander Meiklejohn believed that to teach well one must be learning. He also believed that chemistry and physics are so simple that to learn about them a man must do research. He brought me to Amherst to prove that research in the physical sciences could be done in a small liberal arts college with the right atmosphere. He brought Niels Bohr to lecture for a term, principally to stimulate me. But by the time Bohr arrived, Meiklejohn and I were both gone—he to New York on his way to Wisconsin to found the Experimental College, and I to M.I.T. The National

Research Council generously renewed my fellowship. This was followed by appointments to the staff.

The first papers that contained my own ideas were the three published from Amherst. The first idea was that in ideal solutions the activities are proportional to the mole fractions rather than to the volume concentrations and that for sugar solutions the mole fractions might be changed by hydration, which could therefore be calculated from the vapor pressure.

Some years later, when John J. McCloy (class of 1916) was chairman of the board of trustees, a student protest against compulsory chapel induced him to show them a chapel service such as he had experienced with Meiklejohn, at which a quotation from Epictetus was more likely than one from the Bible. So he invited Meiklejohn to conduct a chapel service. Some of the Meiklejohn supporters returned for it, and the next day we had brunch at the home of Eli (A. W.) Marsh, Professor of Physical Education, who was the only ardent Meiklejohn supporter remaining at Amherst in 1923. When I came in, Meiklejohn put his arm around my shoulders and said, "This is the man who drove me out of Amherst." This paradox was strictly true. After the Alumni banquet in 1923 I drove him and his daughter Annaletta to Keene, New Hampshire, to the home of his chief supporter in Brown, who was the father of the future, second Mrs. Meiklejohn.

The Amherst Faculty Club invited me to a reception for Bohr. At the instigation of a mutual friend, I obtained an invitation for Edwin Cohn. We stayed with another friend in Amherst, and I rode back to Cambridge with Cohn. I had known him since he entered Amherst in 1910, a year behind me. I thought of him as an esthete who fraternized with the faculty—which I disapproved of then. I found on our way to Cambridge that he thought more like me than anyone else I had known. I became almost a member of the Cohn family. Marianne Cohn asked if I minded their son calling me "George," because Mr. Scatchard was a big mouthful for a three-year-old. When his younger brother was three, he was calling me "Georgie Scatchard." This tickled his father, who called me "Georgie" until almost his last words. He was seized by a fatal stroke while we were talking on the telephone. The phone dropped to his chest and his last words to me were, "Georgie, I can't hear you." He lost consciousness very soon afterward.

Those were exciting years scientifically, too, for a young chemist interested in solutions and rates of reaction. Bjerrum, Brönsted, A. A. Noyes, and G. N. Lewis were at their peaks. Brönsted's critical complex appeared in 1922; Lewis' ideas were summarized in Lewis and Randall's *Thermodynamics*, published in 1923. My chief contributions were the meticulous survey of the assumptions underlying the various treatments and the definition of ideal solutions in terms of mole fractions instead of volume concentrations. I assumed that carbohydrates were so much like water that their solutions could be considered ideal except for the chemical combination of the carbohydrate with water—hydration. I called such solutions physically ideal or semi-ideal.

My first paper from Amherst (6) was on the rates of reaction, my third (8) was the same, but using Brönsted's critical complex which had appeared meanwhile. My second Amherst paper (7) calculated the hydration of sucrose from vapor pressure measurements, and the fourth (9) was a study of the electromotive force of concentration cells with transference and with a saturated potassium chloride bridge.

My first paper from the Massachusetts Institute of Technology (10) was a polemic. Duncan MacInnes and I had become fed up with the quality of some of the papers published in the *Journal of the American Chemical Society* and had decided to submit criticisms for publication in the *Journal*. My target was two papers on the effect of gelatine on transference numbers—of sulfuric acid by Ferguson and France and of hydrochloric acid by France and Moran—both from Ph.D. theses. They posed some interesting questions and probable answers to some of them, which were not related to their own experimental work. The second paper used for the univalent chloride ion the same equation as the first did for the bivalent sulfate ion.

The next paper (11) describes my measurements of the cell with hydrogen and silver chloride electrodes in aqueous hydrochloric acid. When it was first submitted, Merle Randall wrote that his freezing-point measurements with Vanselow made all others obsolete. I waited. Then he wrote that he should be permitted to discuss his measurements before anyone else. I fought.

At a meeting of the American Chemical Society in Pasadena in 1925, the first thing that William Bray of Berkeley said to me was, "Scatchard, you don't look nearly as belligerent as I expected." He had seen all the correspondence. It was easy for us to become friends because his sister in Amherst was a very good friend. I rode to Berkeley with him and stayed at his house while there, and he showed me around the laboratory. When we came to Randall, I suggested that a good way to check the Debye-Hückel equation would be to measure the freezing points of potassium nitrate. He said he did not want to "because the nitrate ion is thermodynamically unstable." I left it to his older colleague to settle that.

The next paper (12) also discusses the measurements of others with aqueous alkali halides, for the next (13) which measures the electromotive force of the cell Hg/HgCl, KCl (sat) // HCl (C_1), AgCl/Ag with flowing junction and after flow is stopped.

The comparison of the Debye theory with measurements in mixtures of water and alcohol for HCl, NaCl, and KCl is discussed in (14), and (15) discusses my own measurements and those of Lewis and his coworkers with aqueous sucrose solutions of HCl and of KCl.

The very accurate measurements of the rate of inversion of sucrose by Pennycurck in Australia are treated more accurately in (16), and the apparent change of the rate constant with time is attributed to incomplete mixing at the start.

Another polemic is started in (17). The calculations of Milner and the limiting law of Debye and Hückel are compared by Pike and Nonhebel and by Nonhebel and Hartley. I showed that the complete Debye-Hückel equation fitted better than the calculations of Milner, and I discussed some other erroneous statements. Hartley ended the polemic by asking Milner's opinion. Milner replied that of course Debye's result was right and he was sorry that he had not thought of it. Hartley also invited me to a meeting of the Faraday Society and sat me next to a man who immediately began talking economics, so I caught his name as "Lord Barclay" and associated him with the bank. Only when he had left did I realize that he was the Earl of Berkeley, whose papers on the vapor pressure of aqueous sugar solutions I had used a decade earlier.

Hartley also organized a small meeting in his office—Hartley, Debye, Bjerrum, Fajans, Scatchard, and Bury. Debye automatically took the chair and silenced the discussion between Bjerrum and Fajans.

About that time the physicists held a meeting in Zurich at the University and the Technical High School for students in physics. Einstein, Schrödinger, and others stressed the nobility of theoretical physics and said that those who would have entered the ministry a generation earlier were now becoming physicists. Debye's contribution was, "Ich bin Physiker weil es mir Spass macht" (I'm a physicist because it's fun for me—except that "Spass" is a much rougher word than "fun").

Numbers 18, 19, and 20 discuss electrolyte-non-electrolyte mixtures. Number 21 is a tabulation of the density and thermal expansion under atmospheric pressure of aqueous solution of strong electrolytes, with J. A. Beattie, L. J. Gillespie, W. C. Schumb, and R. F. Tefft. It was my share of the International Critical Tables and, though it took most of my time for a year, I accepted it out of gratitude for my

National Research Council Fellowship. Number 22 discusses the then new Moore Laboratory at Amherst College. Number 23 discusses the effect of changing environments on the rate of reaction. Numbers 24 and 25, with R. F. Tefft, discuss the electromotive force of cells with calcium chloride and with zinc chloride. Number 26 discusses the equation of state explicit in the volume. Number 27, with T. F. Buehrer, discusses the effect of breadth of liquid junction on the electromotive force of a concentration cell.

Number 28, my first very important paper, presents a simple theory of nonelectrolyte solutions. Number 29 discusses the fact that measurements of heats of dilution are almost always made at constant pressure and most of the calculations are at constant volume. Number 30 presents a comparison of the Langmuir and Van Laar theories of alloys. Number 31, of which Number 37 is an abstract, extends the Debye theory to concentrated solutions. Number 32 discusses the rates of reactions in concentrated solutions. Number 33, with J. G. Kirkwood, discusses an extreme model of a Zwitterion, which we called the dumbbell model. Numbers 34, 35, and 42, with P. T. Jones and S. S. Prentiss, and numbers 36, 41, 43, 45, 46, and 47, with S. S. Prentiss, and number 50, with Marjorie A. Benedict, describe our measurements of the freezing points of aqueous solutions, mostly of electrolytes.

I had a Guggenheim Fellowship in 1931–32, and my wife and I spent those two summers and the intervening year in Europe. We started at the Anglo-American Conference of Music Teachers in Lausanne, and attended several other music meetings as we worked our way to London. Then we went directly to Leipzig, where I worked a little with Debye but spent most of my time writing. When I introduced Kirkwood to Debye, he was apparently very busy and did not invite us into his office. He asked what we were working on. I said we were correcting Bjerrum's statement that molecules with widely separated charges behave like two independent ions even in the limiting law. Debye said, "Bjerrum must be right." I said, "Do you forget that at infinite dilution the ion atmosphere is at an infinite distance?" Debye said, "That's right. Bjerrum must be wrong."

We went to Munich the Saturday before Lent. On Sunday we went to Kitzbühl with the Cohns, who were skiing to recover from the beginning of Fasching. Again I worked largely on my own. Later we took a Rhine steamer to Frankfurt for a meeting of the Bunsen Gesellschaft. From there we went to Copenhagen, where I gave a talk before the Danish Chemical Society. The abstract (38), in Danish, in the *Kemisk Maanesblad*, is the only paper I have ever written that I cannot read. We had planned to go on to Norway and Sweden, but were having such a good time in Copenhagen that we stayed there. Everyone paired Niels Bohr and Niels Bjerrum, who were classmates through school, but my wife was the only one who included Niels Bohr in the group of Niels B's who were friendly to her. Bjerrum and Bohr shared a sailboat, which Bohr was allowed to sail, once clear of all obstacles. About the first of July we went to the Isle of Man to visit my wife's family for the rest of the summer.

Number 39 is the introduction to a symposium which notes that 1933 is the twenty-first anniversary of Milner's great paper and the tenth anniversary of the Debye-Hückel theory. Number 40, with S. S. Prentiss, describes our least square study of our freezing-point measurements in dilute solutions (up to $0.1m$) and the measurements of others which are sufficiently accurate in this concentration range. Number 44 is a polemic with J. H. Hildebrand et al. on their discussion of number 28. Numbers 48 and 49, with W. J. Hamer, discuss our analytical equations for the equilibrium between partially miscible liquids and between liquid and solid solutions.

Number 51 discusses the treatment of concentrated solutions, including the extension of the Debye equations for charge-charge and charge-molecule effects and a simple treatment of molecule-molecule effects. Number 52 discusses the effect of change of volume on the properties of solutions. Number 53 is a book review of R. W. Gurney's *Ions in Solution*.

Number 54, with C. L. Raymond and H. H. Gilmann, number 55, with C. L. Raymond, and numbers 58, 60, and 61, with S. E. Wood and J. M. Mochel, describe our apparatus for studying vapor-liquid equilibrium and measurements with it. Number 56, with W. J. Hamer and S. E. Wood, is a description of our measurements on isotonic (isopiestic) solutions. Number 57 is the introduction to a symposium on intermolecular action. Number 59 is a discussion of the critical complex.

Number 62 is a brief polemic with Laidler and Eyring on the effect of solvents on reaction rates. Number 63 describes the determination of the equilibrium between two phases by plotting the activity of one component against that of the other with the same standard state for each.

Number 64 is the extension of the Debye theory of the interaction between ions and neutral molecules over the whole composition range for water-ethanol mixtures. It is suggested that the poor agreement with experiment may be due to the failure to take into account the discrete structure of the solvents. Number 65, with L. F. Epstein, extends the Debye theory over the whole concentration range of the measurements and illustrates with aqueous sodium chloride and sulfuric acid.

Number 66 is my address as retiring vice president and chairman of the Section on Chemistry of the American Association for the Advancement of Science. Number 67 is a review of the book, *Elementary Physical Chemistry* by H. S. Taylor and H. A. Taylor. Number 68 is the application of the revised constants of Birge to our equation for aqueous electrolyte solutions.

Number 69 is my contributions to the book by E. J. Cohn and J. T. Edsall: *Proteins, Amino Acids and Peptides.*

Number 70, with S. T. Gibson, L. M. Woodruff, A. C. Batchelder and A. Brown, number 71, with A. C. Batchelder and A. Brown, and number 75, with L. E. Strong, W. L. Hughes, Jr., J. N. Ashworth, and A. H. Sparrow, are parts of the studies from the Harvard Medical School of plasma proteins and the fractionation products. Number 72, with J. L. Oncley, J. W. Williams, and A. Brown, is a preliminary study of the size distribution in gelatin solutions.

Number 73 is a review of *The Physical Chemistry of Electrolytic Solutions* by H. S. Harned and B. B. Owen. Number 74 is an obituary of Louis John Gillespie. Numbers 76 and 77, with S. E. Wood and J. M. Mochel, report measurements of vapor-liquid equilibrium with one polar and one non-polar component.

In number 78 I derive the equations for osmotic pressure. Number 79, with A. C. Batchelder and A. Brown; number 80, with A. C. Batchelder, A. Brown, and Mary Zosa; number 81, with J. L. Oncley and A. Brown; number 84, with S. S. Gellis, J. R. Neefe, J. Stokes, Jr., L. E. Strong, and C. A. Janeway; and number 86, with Elizabeth S. Black, evaluate the study of plasma proteins.

On July 1, 1946, I flew to Germany for six months as scientific adviser to General Lucius D. Clay, Deputy Military Governor of the United States Office of Military Government. Flying across the ocean we wore parachutes and Mae Wests. There was practically no one in Berlin when I arrived. I learned that the Russians were reopening the Prussian Academy of Science as the German Academy of Science and was told that the Americans were about to confiscate the homes of three Nobel Prize winners. So I had both fists flying when I met General Clay and told him that that must not be. He did not wait fifteen seconds before stopping the one in Berlin, and not more than a minute before he called headquarters in Frankfurt of General McCartney, Commander of the American Zone, and stopped the other two. I soon learned that I was wrong about one. Bothe had not yet received the Nobel Prize; he probably would have, had it not been for my error.

FIAT (Field Intelligence Agency Technical) adopted me, for no reason I can remember except kindness of heart. My second job was as Chief of the Research and Development Branch of the Ecoin Solutions and on Surface and Colloid Chemistry. Number 122 is chapter 2, "The Interpretation of Activity and Osmotic Coefficients," and number 123, with B. A. Soldano, R. W. Stoughton, and R. J. Fox, is chapter 14, "A High Temperature Isopiestic Unit," in *The Structure of Electrolytic Solutions*, W. J. Hamer, Editor. Number 124, with J. S. Johnson and K. A. Kraus, is a refinement of number 110. Number 125, with Judith Bregman, describes the effect of temperature on the osmotic coefficients of human serum albumin solutions in aqueous sodium chloride. Number 126, with S. Zaromb, is a discussion of measurements of the osmotic coefficient of solutions of human serum albumin in aqueous solutions of sodium trichloroacetate. Number 127, with Y. V. Wu and Amy L. Shen, is a study of the binding to serum albumin of chloride, fluoride, thiocyanate and trichloroacetate ions from solutions of their sodium salts by measurements of pH and the membrane potentials and, in the more concentrated solutions, of the osmotic pressures.

Number 128 is an obituary of John Gamble Kirkwood.

Number 129, with B. Vonnegut and D. W. Beaumont, is the description of a new type of freezing-point apparatus and the measurement of the freezing points of lanthanum chloride solutions. Number 130, with J. S. Johnson and K. A. Kraus, discusses measurements of the distribution of silicotungstic acid of sodium silicotungstate in the ultracentrifuge. Number 131 is a review of the osmotic and activity coefficients in solutions. Number 132, with J. N. Anderson, is a study of the water content of ion exchange resin beads in equilibrium with aqueous solutions of HCl and of NaCl. Number 133 is a review of *Electrolytic Dissociation*, by C. B. Monk. Number 134, with R. M. Rush, is a study of the molal volumes and refractive indices of aqueous solutions of $BaCl_2$ and HCl. Number 135, with T. P. Lin, is a study of the equilibrium with zinc and cadmium vapors with ternary alloys of those metals with silver. Number 136, with J. Pigliacampi, is a study of the osmotic pressures of serum albumin, carboxyhemoglobin and their mixtures.

Number 137 is a study of the Gibbs adsorption isotherm. Derived simply and illustrated for ethanol-water mixtures, it is defended from the misunderstanding of Guggenheim. Number 138 is a review of E. A. Moelwyn-Hughes' *States of Matter*. Number 139 is a review of K. S. Pitzer and L. Brewer's revision of G. N. Lewis and M. Randall's *Thermodynamics*. Number 140 is my address on receiving the Kendall Award in Colloid Chemistry. Number 141 is the review of electrolyte solutions for the 1963 Annual Review of Physical Chemistry. Number 142 is a discussion, in the Conference on the Ultracentrifuge, of the basic equations for equilibrium in the ultracentrifuge. This is based on the extension of Gibbs' equation for the gravitational field. Number 143, with G. M. Wilson and F. G. Satkiewicz, is a description of our apparatus for measuring static equilibrium between vapor and liquid mixtures. Number 144, with F. G. Satkiewicz, and number 145 with G. M. Wilson, are examples of the use of this apparatus with mixtures with very large differences in composition of liquid and of vapor. Numbers 146 and 149, with D. H. Freeman, are studies of ion-exchange resin beads by microscopy. Number 147 is a study of the effect of dielectric constant on ultrafiltration. Number 148, with W. T. Yap, is a study of the effect of temperature and of hydroxide ion on the binding of small anions to serum albumin. Number 150 uses the tables of Loeb, Overbeek, and Wiersema to compute the activity coefficients of small ions in solutions without colloids. Number 151 is chapter 16, "The Osmotic Pressure, Light Scattering and Ultracentrifuge Equilibrium of Polyelectrolyte Solutions," of *Chemical Physics of Ionic Solutions*, edited by B. E. Conway and R. G. Barradas. Number 152 is an introduction to a conference of the Federation of Societies of Biochemistry. Number 153 is a discussion of osmotic pressure at the same conference. Number 154, with W. H. Orttung, is a comparison of the freezing points of aqueous NH_4Br and KBr which explains the abnormalities of the former by the adsorption of Br^- on the surface due to the solubility of NH_4^+ in ice. Number 155, also with W. H. Orttung, is a study of the bi-ionic potential with HBr and $NaBr$ solutions separated by an ion exchanger membrane. Number 156, with S. Y. Tyree, Jr., R. L. Angstadt, F. C. Hentz, Jr., and R. L. Yoest, describes the osmotic coefficients and related properties of aqueous 12-tungstosilicic acid. Number 157 is a discussion of the excess free energies and related properties of electrolyte solutions. Number 158 gives corrections to 157. Number 159 is the Edwin J. Cohn Lecture: Edwin J. Cohn and Protein Chemistry. Numbers 160 and 161, with Y. C. Wu and R. M. Rush, and number 162, with R. M. Rush and J. S. Johnson, discuss the osmotic and activity coefficients of mixtures of sodium chloride, sodium sulfate, and magnesium chloride. Number 163, with H. F. Gibbard, Jr., describes the comparison in this apparatus of synthetic sea salt with sodium chloride. Number 164, also with H. F. Gibbard, Jr., discusses the liquid—vapor equilibrium of aqueous lithium chloride. Number 165, with H. F. Gibbard, Jr., R. A. Rousseau, and J. L. Creek, describes improvements in the apparatus for the measurement of liquid-vapor equilibrium, and the use of the apparatus to study aqueous sodium chloride.

At a Symposium on Solutions of Electrolytes, New Haven, Connecticut, June 16–18, 1954. (*left to right, standing:* P. Debye, George Scatchard, J. J. Hermans, J. W. Williams; *seated:* H. S. Harned, J. G. Kirkwood)

EQUILIBRIUM IN SOLUTIONS

Contents

Models *154*   8-9. Electromotive Force and Chemical
Potentials. Liquid-Junction Potentials *157*   8-10. Electrode
Potentials and Single-Ion Activities *160*   8-11 Other
Thermodynamic Functions *163*

1. Introduction

1-1. General

The foundation of physical chemistry as a separate division of the science of chemistry, in about 1890, was due to the increase in interest in solutions, and especially to the interplay of the ideas on this subject of Raoult, van't Hoff, Arrhenius, Ostwald, Planck, Nernst, and other less well-known men. The chief reason for the importance to the chemist of an understanding of solutions is that every chemical reaction involves at least one reactant and one product, and there are very few reactions in which each reactant or product forms a separate phase, even when there are no foreign substances present. So almost every chemical reaction involves one or more solutions. The understanding of solutions is also very useful in the separation and purification of substances.

Two criticisms made some 50 years ago have always impressed me: "The solutions you fellows deal with are so dilute that they're really only slightly polluted water" and "You professors pat each other on the back when you learn something about two-component solutions, but we practical chemists have to deal with four- and five-component solutions." Whatever basis there is for these criticisms comes from the fact that solution chemists have followed the usual scientific procedure of handling a complicated problem by dividing it up into a series of small problems and treating the simpler ones first, but too often in textbooks, and sometimes even in research, have forgotten to put the structure together again.

1-2. A Practical Example

Let us suppose that the practical chemist needs to know the relation of volume to mass for a liquid solution at a given temperature and pressure. First of all we advise him to work with the intensive quantities: the specific volume, or ratio of volume to mass, $\tilde{V} = V/W$, and the weight fraction, or ratio of the mass of a component to the total mass, $\tilde{W}_1 = W_1/W$.

This notation is an extension in two ways of the generally accepted Lewis and Randall symbolism. They use Φ_k to denote the partial derivative of any thermodynamic property Φ with respect to the number of moles of component k at constant temperature, pressure, and quantities of all components other than k. We use it in this sense except for $\overline{N}_k$, which by this definition would be unity, but which we define as $\overline{N}_k = N_k/\Sigma_i N_i$, the mole fraction of component k. Without the subscript, $\bar{\Phi} = \Phi/\Sigma_i N_i$. For the corresponding specific quantities we have

$$\Phi_k = (\partial\Phi/\partial W_k)_{T,p,w} = \bar{\Phi}_k/\tilde{W}_k,$$
$$\tilde{W}_k = W_k/W = W_k/\Sigma_i W_i,$$
$$\tilde{\Phi} = \Phi/W.$$

We find that for a binary system measurements of $\tilde{V}$ for the two components and for eight mixtures, with $\tilde{W}_2$ varying in steps of about 1/9, are enough to enable us to plot $\tilde{V}$ against $\tilde{W}_2$ so that a smooth curve may be drawn through them to give sufficiently accurate values of $\tilde{V}$ by interpolation, and of $d\tilde{V}/d\tilde{W}_2$ by interpolation and differentiation. The same density of measurements for a ternary system requires 55 measurements, which can be plotted as a surface, preferably with an equilateral-triangular base representing the composition. The construction of a surface is much more difficult than that of a line, and reading it or using it for differentiation is even more difficult. This same density of measurements requires 220 measurements for a four-component system and 715 measurements for a five-component system, and there is no way to plot either one graphically. The general formula is

$$N = (c + n - 2)!/(n - 1)!(c - 1)!,$$

if N is the total number of measurements required for a system of c components, and n is the number of measurements needed for a two-component system.

It is not surprising that the practical chemist thinks he should have more help. It is apparent that for four- or five-component systems he must use analytic expressions rather than graphs, that he will find such expressions much more convenient for three-component systems, and that he will probably prefer them for two-component solutions. Many workers believe, however, that, since most physical relations cannot be represented exactly by analytic expressions, the experimental measurements cannot be reproduced with sufficient precision by moderately simple analytic expressions. Even they will find it advantageous to express most of the function analytically and to plot only a small deviation function.

The practical chemist would also like us to do something about the number of measurements. If it takes him a day to make the 10 measurements for a two-component system, he will need more than 14 weeks for the five-component system. There are three ways in which theory may be brought to help him. The one that has interested theoretical-minded physical chemists longest is to calculate the volume of the pure substance from its chemical formula. This would save five measurements. The one that has been usually considered the province of the theory of solutions is to calculate the volume of binary mixtures from the properties of the components, including their volumes, which would save 80 measurements. We can also calculate the volume of polycomponent systems from those of the component two-component systems, and thereby save 630 of the 715 measurements. We would probably want to check the adequacy of our method by measuring at least a few of the ten 1–1–1 mixtures of the three-component systems. The first two calculations have the greater theoretical interest, but the third, which is the easiest, has by far the greatest practical importance.

1-3. Equilibrium

The study of solutions may be divided into three parts: (1) the study of equilibrium; (2) the study of the approach to equilibrium; and (3) the study of transport problems. As its title indicates, the present work is concerned mainly with the first of these. However, we shall study the others whenever it seems useful, just as we shall often consider one-component systems.

Our study of equilibrium is usually based on two findings of Willard Gibbs:[1] (1) that, under certain conditions, a system is in equilibrium with respect to a given change of state when the free energy of the system, expressed as a function of the extent of that change of state, is a minimum; and (2) that the change in free energy of the system produced by any change of state may be expressed by the sum of the products of the chemical potential, or partial free energy, of unit quantity of each component and the quantity of that substance

[1]J. W. Gibbs, *Trans. Connecticut Acad. 3*, 108–248 (1875–76), 343–524 (1877–78); *The Collected Works of J. Willard Gibbs* (Longmans, Green, New York, 1906, 1928), I, 55–371.

produced by the change of state. Quantities of substances that disappear are counted as negative quantities produced. The study of a very large number of reactions is thus reduced to the study of the potentials of a very much smaller number of substances. This study is systematized and simplified by expressing the potentials in real systems as differences from those in an idealized standard state.

These two findings may be expressed symbolically as

$$\Sigma_i \nu_i \bar{G}_i = 0, \tag{1-1}$$

in which Σ_i stands for the sum over all the z reactants and products, $i = 1, \ldots, z$, ν_i is the number of units of component i produced in the reaction, and $\bar{G}_i$ is the potential (for the same unit) of component i. If the change of state is a physical change of a single component from one phase to another, the potential of that component must be the same in the two phases at equilibrium. If the change of state is a chemical reaction in which a moles of A react with b moles of B to form e moles of E and f moles of F, or

$$aA + bB = eE + fF,$$

and the $\bar{G}_i$ are chemical potentials (for 1 mole), then

$$e\bar{G}_E + f\bar{G}_F - a\bar{G}_A - b\bar{G}_B = 0.$$

The conditions for the first finding are: (1) no heat or matter is transferred through a temperature gradient; (2) no matter is transferred through a pressure gradient; (3) no matter is transferred through a gravitational potential gradient; (4) no uncompensated electric charge is transferred through an electric potential gradient due to external fields; (5) no uncompensated magnetic pole is transferred through a magnetic potential gradient due to external fields; (6) there is no change in the area of any interface between two phases; and (7) there is no change in the state of strain of any solid body. Conditions (1), (2), (4), and (5) may be met exactly by having the whole system at the same temperature, without a diaphragm that is impermeable to any component and without any external electric or magnetic field. Conditions (3), (6), and (7) may be met approximately. The controlled deviations from conditions (2), (3)—with a centrifugal rather than a gravitational field—, (4), (5), and (6) give important methods of studying chemical potentials.

The two findings of Gibbs as stated here also depend upon the extrathermodynamic assumption that the potential of any substance is never greater in a phase in which that substance is absent than it is in a second phase in which that substance is present and which is in equilibrium with the first phase. Without this assumption,

Eq. (1–1) must be replaced by an inequality for an equilibrium that can be displaced in only one direction. Gibbs himself noted that the condition that the potential of any substance approaches a linear function of the logarithm of its concentration as the concentration becomes very small is sufficient to validate this assumption. We need not worry about this assumption, but we shall discuss the condition later.

1–4. Some Definitions

A *system* is any section or group of sections of the physical world that is under consideration, and generally of a more or less idealized world. For example, unless a definite statement to the contrary is made, we shall assume the seven conditions of the preceding section.

An *isolated system* is one that cannot exchange matter or heat with the rest of the world and can do no work, positive or negative, on the outside world. A *closed system* is one that cannot exchange matter with the outside world, and an *open system* is one that may gain or lose matter.

A *phase* is a homogeneous material, whether it is continuous or discontinuous. Since material made up of molecules cannot be strictly homogeneous if parts of the order of molecular dimensions are considered, the definition of phase must depend upon the definition of molecules. For gases this leads to no difficulty, and it is found in practice that two gaseous phases cannot exist in contact. For liquids and solids the molecules are so close together that they cannot be distinguished sharply. In the range of colloidal dimensions there is difficulty in determining the number of phases in a system. Many mixtures may be treated as single-phase or as polyphase according to which method is the more convenient, but any treatment will be more complicated than for those cases in which the distinction is clear cut. If the system is treated as single-phase, it will be necessary to consider many components, and if the system is treated as polyphase, the variation in interfacial area cannot be ignored. If a mixture can be separated into two parts by an ordinary filter, or if it appears inhomogeneous under an ordinary microscope, we may say that it is made up of more than one phase. If it cannot be separated by an ultrafilter or if it appears homogeneous under an ultramicroscope, we may say that it is a single phase. Mixtures that can be separated by an ultrafilter but not by an ordinary filter, or that appear heterogeneous in an ultramicroscope but homogeneous in an ordinary microscope, probably fall in the intermediate range discussed above.

There are many definitions of this kind in science which give con-

venient classifications in many cases, but do not divide all phenomena
into categories. If the conclusions drawn concerning things so defined
are clearly understood not to apply when the distinction is not clear,
there is much gain and little loss in the use of such a definition.

The *components* of a system are the molecular species that make
it up if each species can be added to, or subtracted from, any one of
the phases of the system, though not necessarily from each phase. If
the ratio of two or more species is always the same, or is fixed by the
external conditions in every phase of the system, it may be convenient
to treat those species as a single component. It is possible, however,
to treat them as two or more components with one less condition than
the number of molecular species whose ratios are fixed. For example,
if a mixture of isotopes is nowhere separated even partially, it is
conveniently considered as a single component, but each isotope may
be considered a component if the ratio of the quantity of one to that
of each of the others is taken as a condition. The same is true of a
salt, even though it is considered a mixture of ions. We may consider
each salt in a mixture as a component, or we may consider each ion
as a component with the condition that the number of anions must
be so related to the number of cations that the mixture is electrically
neutral. It is often convenient to consider a mixture such as that of
nitrogen dioxide and nitrogen tetroxide, which exist only in the
equilibrium mixture, as a single component. It is possible to consider
them as two components with the condition that they are in equilibrium. If it were possible to obtain a nonequilibrium mixture, it would
be necessary to consider them as two components except when at
equilibrium, or when satisfying some other condition. If we wish to
study the relative quantities of the two species, we must do the
equivalent of conceiving some operation by which one of them may
be added or subtracted independently of the other, and treat them as
two components.

A *solution* is a single phase that contains more than one component.
We may also look at a solution as a mixture of two kinds of molecules
that is homogeneous down to molecular dimensions, with the understanding that not all the molecules need be as simple as those that
exist in very dilute gases.

1–5. Quantity, Concentration and Composition

Among the things we need to know about a system are its pressure,
its volume, its temperature, and the quantity of each of its components. The definitions of the first two require no comments. For
temperature we shall use the thermodynamic scale, either celsius
or absolute. This agrees exactly with the ideal-gas scale (see Appendix

B-4) and very closely with the current International Practical Temperature Scale based mainly on the platinum resistance thermometer.

The quantity of a component may be measured in several ways. The mass W_i may be used, for we shall never be interested in conditions under which the theory of relativity tells us that mass is not an accurate measure of quantity. We may also use the volume V_i, measured at some specific temperature and pressure. For many purposes in physical chemistry it is much more convenient to use a measure more closely related to the number of atoms of each element. We shall use the number of formula weights corresponding to some specified formula that expresses the composition of the component. If that formula corresponds to the molecular weight as a dilute gas or in dilute solution, we call the corresponding formula weight the molecular weight $\bar{W}_i$, and we call the number of molecular weights, $W_i/\bar{W}_i$, the number of moles N_i. We shall also use the number of molecules, $N_i N_A = n_i$, in which N_A is Avogadro's number.

For many purposes only the relative quantities of the components, and not their absolute quantities, are important. If the quantities are expressed relative to the volume V of the system, they are known as densities or concentrations. If there are N_i moles in the volume V we shall say that the concentration is C_i [$= N_i/V$] moles per liter, or that the solution is C_i volume molal, or that its volume molality is C_i. If we are talking of formula weights, we shall say that its concentration is C_i formula weights per liter, or that it is C_i volume formal, or that its volume formality is C_i.

If we express the quantities of the components relative to one another, we may use any one of the measures discussed above, and it is not necessary to use the same measure for every component. We may use as our reference the sum of the quantities of the separate components, or we may use the quantity of any one component. The latter system is usually limited to the cases in which one component is in large excess. This component is called the *solvent*, the other components are called the *solutes*, and the quantity of the solvent is taken as the reference. In aqueous solutions the compositions are often expressed as the number of formula weights of the solutes per kilogram of water, and this is the commonest example of the use of mixed units.

The following expressions will often be found to be convenient. If the quantity of the whole system is used, the composition may be expressed as the *weight fraction* $\tilde{W}_i$ [$= W_i/\Sigma_j W_j$], the *volume fraction* ϕ_i [$= V_i/\Sigma_j V_j$] with the V_j measured at some standard temperature and pressure (which is usually somewhat different from the more frequently used fraction by volume V_i/V), or the mole fraction $\bar{N}_i$ [$= N_i/\Sigma_j N_j$]. If the quantity of one component is used, the composition may be expressed as the mole ratio N_i/N_s or the *weight molality* m_i $[= N_i/W_s]$ if the unit of W_s is the kilogram. Then we shall say that the solution is m_i weight molal. If we are talking of formula weights, we shall say that the weight formality is m_i or that the solution is m_i weight formal.

1-6.   Extensive and Intensive Properties

Some properties of a system, such as its mass, may be taken as measures of its quantity; if temperature and pressure are kept constant, the volume is such a measure, otherwise it is not. Other properties, such as the composition, or the density if the temperature and pressure are kept constant, are independent of the quantity. Those properties that are proportional to the quantity of the system when the temperature and the pressure are constant we shall call *extensive properties*; those that are independent of the quantity when the temperature and the pressure are constant we shall call *intensive properties*. It is convenient to consider the temperature and the pressure as intensive properties even though they are fixed by the definition. Most of the properties in which we are interested, although not all properties, are either intensive or extensive.

The following propositions are obvious when once stated: (1) the product or the ratio of two intensive quantities, such as TC or C/p, is intensive; (2) the product of an extensive quantity and an intensive quantity or its reciprocal, such as pV or V/T, is extensive; (3) the ratio of two extensive quantities, such as W/V or N_i/W_s, is intensive; (4) the derivative of an extensive quantity with respect to an intensive quantity, such as $(dV/dT)_{p,N}$ or $(dV/dp)_{T,N}$, is extensive; (5) the derivative of an extensive quantity with respect to another extensive quantity, such as $(dV/dW_i)_{T,p,w}$ or $(dV/dN_i)_{T,p,N}$, is intensive.

1-7.   Appendices

The organization of the main body of this work is developed logically, and may be followed from the table of contents. The reader's attention is called to the appendices, however, which should be consulted as necessary. Appendix A (pp. 167–173) gives definitions of symbols used throughout. Appendix B discusses the determination of analytic expressions from experimental measurements. It also contains a table of numerical values of various constants, and a short table of series expansions. Appendix C contains the thermodynamic relations important in the theory of solutions. Appendix D contains a summary of the classification of molecules and of their interactions.

2. Ideal and Species-Ideal Gases

2-1. Ideal Gases

The essential characteristic of an ideal gas is that each molecule behaves as though it were alone in the container. Statistical mechanics enables us to derive the macroscopic properties of a gas from this fact alone, but we shall take this derivation for granted. The importance of the concept of ideal gases comes from the facts, verified by a large number of experimental observations, that the behavior of practically all one-component gases and of most gaseous mixtures approaches that of an ideal gas in some important respects as the concentration is decreased, and that for many gases the differences from ideal gases are small even at atmospheric pressure.

We shall define an ideal gas by the relation for the work content A:

$$A = \Sigma_i N_i[G_i - RT + RT \ln RT\, N_i/V]$$
$$= RTV \Sigma_i C_i[G_i/RT - 1 + \ln RTC_i], \qquad (2\text{-}1)$$

in which N_i is the number of moles of component i, G_i is a function of the temperature only, characteristic of the component i, R is the gas constant, T the absolute temperature, and V the volume, and C_i [$= N_i/V$] is the concentration of species i. This is a fundamental equation in the sense of Gibbs, for all the other thermodynamic functions can be calculated from it:

The pressure is

$$p = (-\partial A/\partial V)_{T,N} = RT\Sigma_i N_i/V = RT\Sigma_i C_i; \qquad (2\text{-}2)$$

the entropy is

$$S = (-\partial A/\partial T)_{V,N} = -\Sigma_i N_i(dG_i/dT + R \ln RT\, N_i/V); \quad (2\text{-}3)$$

and the energy is

$$E = A + TS = \Sigma_i N_i(G_i - T\, dG_i/dT - RT). \qquad (2\text{--}4)$$

Obviously, the energy is a function of the temperature only.

The chemical potential of a typical component k is

$$\bar{G}_k = (\partial A/\partial N_k)_{T,V,N} = G_k + RT \ln RT\, N_k/V = G_k + RT \ln p_k, \qquad (2\text{--}5)$$

in which p_k $[= pN_k/\Sigma_i N_i]$ is the partial pressure of component k. Eq. (2–5) is also a fundamental equation and might have been used as the definition of an ideal gas. Eq. (2–1) was chosen instead because it is more easily developed to include nonideal gases. Eqs. (2–2) and (2–4), which are often given as definitions of an ideal gas, are not fundamental equations and do not together constitute a fundamental equation.

The heat capacity at constant volume is

$$C_V = (\partial E/\partial T)_{V,N} = -\Sigma_i N_i(Td^2G_i/dT^2 + R). \qquad (2\text{--}6)$$

The free energy is

$$G = A + PV = \Sigma_i N_i(G_i + RT \ln p_i); \qquad (2\text{--}7)$$

the enthalpy is

$$H = E + PV = \Sigma_i N_i(G_i - T\, dG_i/dT); \qquad (2\text{--}8)$$

and the heat capacity at constant pressure is

$$C_p = (\partial H/\partial T)_{p,N} = -\Sigma_i N_i T\, d^2G_i/dT^2. \qquad (2\text{--}9)$$

The coefficient of thermal expansion is

$$\alpha = (\partial \ln V/\partial T)_{p,N} = 1/T, \qquad (2\text{--}10)$$

and the coefficient of compressibility is

$$\beta = -(\partial \ln V/\partial p)_{T,N} = 1/p. \qquad (2\text{--}11)$$

2–2. Standard States

We select a standard temperature T^0, and at T^0 we let $G_i = G_i^0$, $dG_i/dT = G'_i$, $d^2G_i/dT^2 = G''_i$, ... Then we develop G_i in a Taylor's series about T^0, to give

$$G_i = G_i^0 + G'_i(T - T^0) + G''_i(T - T^0)^2/2 + G'''_i(T - T^0)^3/6 + \cdots, \qquad (2\text{--}12)$$

$$dG_i/dT = G'_i + G''_i(T - T^0) + G'''_i(T - T^0)^2/2 + \cdots, \qquad (2\text{--}13)$$

$$\begin{aligned} G_i - TdG_i/dT = {} & G_i^0 - G'_iT^0 - G''_i(T^2 - T^{0^2})/2 \\ & - G'''_i(2T + T^0)(T - T^0)^2/6 + \cdots, \end{aligned} \qquad (2\text{--}14)$$

$$d^2G_i/dT^2 = G_i'' + G_i'''(T-T^0) + \cdots. \tag{2-15}$$

From these relations it is a simple matter to express each of the quantities A, S, E, G, H, $\bar{G}_k$, C_V, C_p in terms of the G_i^0, G_i', G_i'', ... If we further define the standard state as unit partial pressure of the component i and temperature T^0, we have

$$\bar{G}_i^0 = G_i^0, \tag{2-16}$$

$$\bar{S}_i^0 = -G_i', \tag{2-17}$$

$$\bar{H}_i^0 = G_i^0 - G_i'T^0. \tag{2-18}$$

This standard-state treatment is arbitrary in many ways. The first is the choice of the chemical potential and its temperature derivatives as the basic quantities. The energy, the entropy, and the temperature derivatives of the former might seem more fundamental. However, the chemical potential is the most important quantity in chemical thermodynamics, and it is useful to keep it simple. The choice of unit pressure rather than unit concentration, or some other value of either, is completely arbitrary, but it is the almost universal practice to take the standard pressure as 1 atmosphere. It is also the usual practice, following G. N. Lewis, to take 25° C as the standard temperature. Finally, the choice of the state of zero free energy, zero entropy, and zero enthalpy is also arbitrary. It is again usual to follow Lewis and to assume that the free energy and the enthalpy of each element are each zero at 25° C and 1 atm pressure in the state that is stable under these conditions. The tables of standard free energies and standard enthalpies of gases therefore give $\bar{G}_i^0$ and $(G_i^0 - G_i'T^0)$, with these quantities zero for those elements that are gaseous in the standard state, except for small corrections for the differences between real gases and perfect gases, which will be discussed in the next chapter. The tables of standard entropies, on the other hand, utilize the third law of thermodynamics and list $G' = 0$ at $T = 0$. It is sometimes useful to expand Eq. (2–12) and to collect the coefficients of each power of T to give

$$\begin{aligned} G_i &= [G_i^0 - G_i'T^0 + G_i''T^{0^2}/2 - G_i'''T^{0^3}/6 + \cdots] \\ &\quad + [G_i' - G_i''T^0 + G_i'''T^{0^2}/2 + \cdots]T \\ &\quad + [G_i''/2 - G_i'''T^0/2 + \cdots]T^2 - [G_i'''/6]T^3 + \cdots \\ &= \Gamma_i^0 + \Gamma_i'T + \Gamma_i''T^2/2 + \Gamma_i'''T^3/6 + \cdots. \end{aligned} \tag{2-19}$$

The second form of Eq. (2–19) retains no evidence of the standard temperature T^0. The first form, however, makes it obvious that Γ_i^0 is not the value of G_i when $T = 0$, but is a parameter for the expression of G in the neighborhood of T^0. The range of validity of Eq. (2–19) is exactly the same as that of Eq. (2–12).

This choice of standard states of $G_i^0 = 0$ and $G_i' = 0$ for an element in its stable state at T^0 corresponds to the determination of the difference in height between two points by determining first the height of each above sea level. This is useful for determining the difference in the heights of two mountains, much less useful for determining the thickness of ice on a mountain lake, and ridiculous for determining the thickness of a sheet of gold foil or of a mono-molecular surface layer. In solution chemistry we have many problems that do not involve chemical changes other than association, for which it may be convenient to assume G_i^0 and G_i' to be zero for each component, or to make other equally arbitrary assumptions for each particular problem. This corresponds to measuring the height of each surface above the table top, or above some surface even nearer those to be measured.

2–3.   Chemical Equilibrium and Species-Ideal Gases

In the treatment of chemical equilibrium we assume that it is possible to freeze every chemical reaction, to change the temperature and pressure without changing the number of moles of any species, and to add or subtract one species without changing the number of moles of any other species. Then the system is treated as though each species were an independent component.

We define ν_i as the number of moles of species i produced by the reaction as written, and consider that 1 mole of a species used up is -1 mole of that species produced. If the extent of the reaction is α, $\nu_i \alpha$ moles of species i are produced. For example, in the reaction

$$aA + bB + \cdots = eE + fF + \cdots, \qquad (2\text{–}20)$$

$\nu_A = -a$, $\nu_E = e$, and so on. If α is $\frac{1}{2}$, $-a/2$ moles of A and $e/2$ moles of E are produced, and so on. The third condition of equilibrium, Eq. (1–1), is

$$\Sigma_i \nu_i \bar{G}_i = 0.$$

We shall call N_i' the number of moles of species i to distinguish it from N_i, the stoichiometric number of moles of component i. If there are chemical reactions occurring of which r is the typical reaction, with ν_{ir} the number of moles of component i produced in the reaction r and α_r the extent of the reaction,

$$N_i' = N_i + \Sigma_r \nu_{ir} \alpha_r, \quad r = 1, \ldots, z. \qquad (2\text{–}21)$$

The chemical potential of any species is the same whether the species is regarded as a component or not. Suppose that dN_k moles of component k are added to the system at constant temperature and pressure, with all chemical equilibria frozen. The change in free energy is $\bar{G}'_k dN_k$. If now the restraints on the chemical actions are removed, the additional change in free energy is

$$(\bar{G}_k - \bar{G}'_k)\, dN_k = [\Sigma_r(\partial\alpha_r/\partial N_k)\,\Sigma_i\nu_{ir}\bar{G}_i]\, dN_k = 0, \qquad (2\text{–}22)$$

since the system is in equilibrium with respect to each reaction r, so that $\Sigma_i\nu_{ir}\bar{G}_i = 0$.

We define a species-ideal gas as one for which Eq. (2–1) for the work content is applicable in terms of the species actually present. If the chemical equilibria are considered frozen, the ideal-gas equations for the pressure, (2–2), the chemical potential, (2–5), and the free energy, (2–7), are also applicable in terms of the species actually present.

The value of the concept of species-ideal gases is very different from that of ideal gases. The behavior of a real gas does not generally approach that of a species-ideal gas at low pressures unless it also approaches that of an ideal gas. However, there are a large number of gas mixtures for which the deviations from the behavior of an ideal gas due to chemical actions are so much greater than those from other causes that the latter may be neglected to a good approximation. Moreover, if there is a chemical reaction in which the total number of moles is unchanged, such as $H_2 + I_2 = 2HI$, the behavior of the mixture will approach that of a species-ideal gas at low pressures, but not that of an ideal gas.

If we replace each value of $\bar{G}_i$ in Eq. (1–1) by the last expression of Eq. (2–5), we obtain

$$\Sigma_i\nu_i\bar{G}_i = \Sigma_i\nu_i(G_i + RT \ln p'_i) = 0,$$

or

$$\Sigma_i\nu_i \ln p'_i = \Sigma_i\nu_iG_i/RT = \ln K_p, \qquad (2\text{–}23)$$

where K_p is the equilibrium constant in terms of partial pressures. Similarly, for the next to the last expression of Eq. (2–5),

$$\Sigma_i\nu_i\bar{G}_i = \Sigma_i\nu_i(G_i + RT \ln RTC'_i) = 0,$$

or

$$\Sigma_i\nu_i \ln C'_i = -(\Sigma_i\nu_iG_i/RT) - \Sigma_i\nu_i \ln RT = \ln K_c, \qquad (2\text{–}24)$$

where K_c is the equilibrium constant in terms of concentrations.

If we let Π_i represent a continued product such that

$$\Sigma_i \nu_i \ln p'_i = \ln \Pi_i p'^{\nu_i}_i, \qquad (2\text{--}25)$$

then

$$K_p = \Pi_i p'^{\nu_i}_i \qquad (2\text{--}26)$$

and

$$K_c = \Pi_i C'^{\nu_i}_i. \qquad (2\text{--}27)$$

These are the more familiar forms of the law of mass action.

The early history of the law of mass action is interesting. The discovery of this law is usually credited to Guldberg and Waage, "Etudes sur les affinités chimique" (1864 and 1867); *J. Prakt. Chem.* (2) *19*, 69 (1879). In their first paper (1864) the ν's were nonintegral and determined empirically; in their second paper (1867) ν was -1 for each reactant and $+1$ for each product, which is correct for the esterification reaction that they were then studying. It was not until the third paper (1879) that they related the ν's properly to the chemical equation. In the meantime Gibbs, *Trans. Connecticut Acad. 3*, 168 (1875–76), derived the equation for "ideal gas mixtures," as well as for real mixtures. He also derived at the same time the relation between the temperature coefficient and the heat of reaction that is generally credited to van't Hoff, who obtained it much later. Indeed, Gibbs's expression goes further than that of van't Hoff in that it takes into account the variation of the heat of reaction with temperature.

The variations of the equilibrium constants with temperature are derived by differentiating Eqs. (2–23) and (2–24) with respect to $1/T$:

$$d \ln K_p/d(1/T) = - (\Sigma_i \nu_i (G_i - T dG_i/dT))/R, \qquad (2\text{--}28)$$

$$d \ln K_c/d(1/T) = - (\Sigma_i \nu_i (G_i - T dG_i/dT - RT \, \Sigma_i \nu_i))/R. \qquad (2\text{--}29)$$

The right-hand sides of Eqs. (2–28) and (2–29) are, respectively, $-1/R$ times the change in enthalpy and in energy when α is unity, that is, for the change of state described by the chemical equation.

It is a simple matter to express the equilibrium constants and the changes in enthalpy, or in energy, in terms of the G_i^0 and their temperature derivatives. Usually, the heat capacity of the products is not very different from that of the reactants, so that the variation of enthalpy, or of energy, with the temperature is small and the graph of $\ln K$ as a function of $1/T$ is nearly a straight line. For precise work, however, the change in heat capacity must be taken into account.

2-4. An Associating Species-Ideal Gas. The Virial Expansion

If a gas exists as a monomer A_1, a dimer A_2, a trimer A_3, and so forth, the concentration $C = N/V$ is

$$C_A = C'_{A_1} + 2C'_{A_2} + 3C'_{A_3} + \cdots$$
$$= C'_A + 2K_{A_2}C'^2_A + 3K_{A_3}C'^3_A + \cdots \qquad (2\text{--}30)$$

and the pressure is given by

$$p/RT = C'_{A_1} + C'_{A_2} + C'_{A_3} + \cdots$$
$$= C'_A + K_{A_2}C'^2_A + K_{A_3}C'^3_A + \cdots$$
$$= C_A - K_{A_2}C'^2_A - 2K_{A_3}C'^3_A - \cdots. \qquad (2\text{--}31)$$

These equations illustrate the general truth that the simplest variable is one closely related to the chemical potential or the activity:

$$C'_A = a_A = \exp[(\bar{G}_A - G_A - RT \ln RT)/RT]. \qquad (2\text{--}32)$$

If there are only monomers and dimers, C'_A can be expressed as a function of either C'_A or p/RT by the general quadratic equation. However, we shall discuss only the method of repeated substitution, which can be used for any degree of association:

$$C'_A = C_A - 2K_{A_2}C'^2_A - 3K_{A_3}C'^3_A - 4K_{A_4}C'^4_A + \cdots$$
$$= C_A - 2K_{A_2}C^2_A - (3K_{A_3} - 8K^2_{A_2})C^3_A$$
$$- (4K_{A_4} - 30K_{A_2}K_{A_3} + 40K^3_{A_2})C^4_A + \cdots, \qquad (2\text{--}33)$$

$$p_A/RTC_A = 1 - K_{A_2}C_A - (2K_{A_3} - 4K^2_{A_2})C^2_A$$
$$- (3K_{A_4} - 18K_{A_2}K_{A_3} + 20K^3_{A_2})C^3_A + \cdots. \qquad (2\text{--}34)$$

This expression of p_A/RTC_A as a power series in C_A is known as a virial expansion and the coefficients of the various powers of C_A are called the virial coefficients. The first virial coefficient is always unity by this convention. The second virial coefficient depends upon the interactions of the molecules in pairs, expressed for species-ideal gases as $-K_{A_2}$. The third virial coefficient depends upon the interactions of three molecules at a time, but it depends not only upon the formation of trimers but also upon the fact that the number of monomers is decreased by the formation of dimers.

2-5. Species-Ideal Gas Mixtures

If a mixture of gases A and B exists as the monomers A_1 and B_1, the dimers A_2 and B_2, and the mixed dimer AB, we have

$$C_A = C'_{A_1} + 2C'_{A_2} + C'_{AB}$$
$$= C'_{A_1} + 2K_{A_2}C'^2_A + K_{AB}C'_A C'_B,$$
$$C_B = C'_{B_1} + 2K_{B_2}C'^2_B + K_{AB}C'_A C'_B; \qquad (2\text{--}35)$$

$$C'_A = C_{A_1} - 2K_{A_2}C_A^2 - K_{AB}C_AC_B + 4K_{A_2}^2C_A^3$$
$$+ (4K_{A_2}K_{AB} + K_{AB}^2)C_A^2C_B + (2K_{B_2}K_{AB} + K_{AB}^2)C_AC_B^2 + \cdots,$$
$$C'_B = C_{B_1} - 2K_{B_2}C_B^2 - K_{AB}C_AC_B + 4K_{B_2}C_B^3 \qquad (2\text{-}36)$$
$$+ (4K_{B_2}K_{AB} + K_{AB}^2)C_AC_B^2 + (2K_{A_2}K_{AB} + K_{AB}^2)C_A^2C_B + \cdots;$$

$$p/RT = C_{A_1} + C_{B_1} - K_{A_2}C_A^2 - K_{B_2}C_B^2 - K_{AB}C_AC_B$$
$$+ 4K_{A_2}C_A^3 + 4K_{B_2}C_B^3 + (4K_{A_2} + K_{AB})K_{AB}C_A^2C_B$$
$$+ (4K_{B_2} + K_{AB})K_{AB}C_AC_B^2 + \cdots. \qquad (2\text{-}37)$$

If the complex splits into new molecules, the picture is more complicated even for a one-component gas. For example, ozone in equilibrium with oxygen would give $\frac{3}{2} O_2 = O_3$:

$$C'_{O_3} = KC'^{3/2}_{O_2}, \qquad (2\text{-}38)$$

$$C_{O_2} = C'_{O_2} + \tfrac{3}{2} C'_{O_3} = C'_{O_2} + \tfrac{3}{2} KC'^{3/2}_{O_2}, \qquad (2\text{-}39)$$

$$p/RT = C'_{O_2} + KC'^{3/2}_{O_2} = C_{O_2} - \tfrac{1}{2} KC'^{3/2}_{O_2} + \cdots. \qquad (2\text{-}40)$$

This does not give an integral power-series virial expansion, but the second term is proportional to the square root of the concentration. Ozone is not normally a complication in measuring the density of oxygen, but similar cases might arise.

A mixture of hydrogen and oxygen in equilibrium with water gives $H_2 + \tfrac{1}{2} O_2 = H_2O$:

$$C'_{H_2O} = K_{H_2O}C'_{H_2}C'^{1/2}_{O_2}, \qquad (2\text{-}41)$$

$$C_{H_2} = C'_{H_2} + C'_{H_2O} = C'_{H_2} + K_{H_2O}C'_{H_2}C'^{1/2}_{O_2}, \qquad (2\text{-}42)$$

$$C_{O_2} = C'_{O_2} + \tfrac{1}{2} C'_{H_2O} = C'_{O_2} + \tfrac{1}{2} K_{H_2O} C'_{H_2}C'^{1/2}_{O_2}, \qquad (2\text{-}43)$$

$$p/RT = C'_{H_2} + C'_{O_2} + C'_{H_2O} = C_{H_2} + C_{O_2} - \tfrac{1}{2} K_{H_2O}C'_{H_2}C'^{1/2}_{O_2} + \cdots. \qquad (2\text{-}44)$$

Again the virial has the second term proportional to the square root of the concentration. Similarly, a mixture of hydrogen and nitrogen in equilibrium with ammonia gives

$$p/RT = C_{H_2} + C_{N_2} - K_{NH_3}C_{H_2}^{3/2}C_{N_2}^{1/2}. \qquad (2\text{-}45)$$

The second term in the virial is proportional to the first power of the total concentration, but only to the square root of the nitrogen mole fraction. Equilibrium in the formation of water or ammonia is attained only in the presence of catalysts. Hydrobromic and hydriodic acids are formed without catalysts, however, and yield a more serious complication. For the reaction

$$\tfrac{1}{2}H_2 + \tfrac{1}{2}X_2 = HX, \quad C'_{HX} = K_{HX}(C'_{H_2}C'_{X_2})^{1/2}:$$
$$C_{H_2} = C'_{H_2} + \tfrac{1}{2} K_{HX}(C'_{H_2}C'_{X_2})^{1/2},$$
$$C_{X_2} = C'_{X_2} + \tfrac{1}{2} K_{HX}(C'_{H_2}C'_{X_2})^{1/2}; \qquad (2\text{-}46)$$

$$C'_{H_2} = C_{H_2} - \frac{K_{HX}}{2} (C_{H_2} C_{X_2})^{1/2},$$

$$C'_{X_2} = C_{X_2} - \frac{K_{XH}}{2} (C_{H_2} C_{X_2})^{1/2}; \tag{2-47}$$

$$p/RT = C'_{H_2} + C'_{X_2} + K_{HX}(C'_{H_2} C'_{X_2})^{1/2} = C_{H_2} + C_{X_2}. \tag{2-48}$$

The pressure is the same as for an ideal gas, and the extent of the reaction is independent of the total concentration insofar as the mixture is species-ideal. There are a color change and a heat effect on mixing hydrogen with the halogen, and the activities of the components are smaller than the concentrations, as shown in Eqs. (2–47). The case of a chemical reaction without change in the total number of moles is the only one that approaches a species-ideal gas, but not an ideal gas, as the concentration decreases. If there is a change in the total number of moles, the components can always be chosen as the species that give the larger number of moles. Then the behavior of the mixture will approach that of an ideal gas as the concentration approaches zero.

Although the greatest usefulness of the concept of species-ideal gases is in the treatment of interactions that are definitely chemical, it may be used also for the total behavior of gases below their critical temperatures, which were once called vapors. The chief limitations of the concept are that, except at zero concentration, it makes the virial, p/RTC, always less than unity and always decreasing as p increases. For real gases the virial passes through a minimum and then increases almost linearly with the pressure, for the concentration is almost independent of the pressure at very high pressures. Moreover, at temperatures more than about 5/2 the critical temperature, even the second virial coefficient is positive. The limitations obviously arise from the fact that a real molecule has an intrinsic volume which the concepts of an ideal gas and a species-ideal gas ignore.

Further disadvantages of the species-ideal gas concept, or of any chemical explanation of gas behavior, are that it gives no indication of the relation of the third virial coefficient to the second, or of the second virial coefficient of mixtures to those of the pure components. In the next chapters we consider the physical theories, which are much better in these respects for simple molecules.

3. Real Gases. Physical Models

3-1. Virial Expansion

We shall express the properties of a real gas or gas mixture by adding to Eq. (2–1) an integral power series in the concentrations of the components:

$$
\begin{aligned}
A &= \Sigma_i N_i[G_i - RT + RT \ln RT\, N_i/V] + RT\, \Sigma_{ij}\beta_{ij}N_iN_j/V \\
&\quad + RT\, \Sigma_{ijk}\gamma_{ijk}N_iN_jN_k/2V^2 + \cdots \\
&= RTV\, \Sigma_i C_i(G_i/RT - 1 + \ln RTC_i) \\
&\quad + RTV\Sigma_{ij}\beta_{ij}C_iC_j + RTV\Sigma_{ijk}\gamma_{ijk}C_iC_jC_k/2 + \cdots, \qquad (3\text{–}1)
\end{aligned}
$$

$$
\begin{aligned}
p &= -(\partial A/\partial V)_{T,N_i} = RT(\Sigma_i C_i + \Sigma_{ij}\beta_{ij}C_iC_j + \Sigma_{ijk}\gamma_{ijk}C_iC_jC_k + \cdots) \\
&= RTC\,(1 + C\Sigma_{ij}\beta_{ij}\bar{N}_i\bar{N}_j + C^2\Sigma_{ijk}\gamma_{ijk}\bar{N}_i\bar{N}_j\bar{N}_k + \cdots). \qquad (3\text{–}2)
\end{aligned}
$$

This method of expression is known as a virial expansion, and in our expansion the first virial coefficient is unity, the second is $\Sigma_{ij}\beta_{ij}\bar{N}_i\bar{N}_j$, the third is $\Sigma_{ijk}\gamma_{ijk}\bar{N}_i\bar{N}_j\bar{N}_k$, and so on. To make Eq. (3–1) a fundamental equation in the meaning of Gibbs, each β_{ij}, γ_{ijk}, ..., as well as each G_i, must be known as a function of temperature.

The most important characteristic of Eqs. (3–1) and (3–2) is that the first term added to Eqs. (2–1) and (2–2) is proportional to the square of the concentration. The virial expansion is obviously convenient for integration and differentiation at constant temperature, but sometimes other relations that meet this specification will express the behavior at high concentrations with a smaller number of parameters.

The virial expansion does not express the behavior of all gases. If the gas contains ions there will be a term proportional to the square root of the ion concentrations. The coefficient of this term can be calculated by the Debye theory. This deviation is seldom met in practice. There will also be a term with exponent less than 2 if there is a chemical reaction in which more than half a mole is formed from

each mole of component. Examples of such reactions have been given
in Sec. 2–5. For oxygen in equilibrium with ozone, the exponent is 3/2.
For a mixture of oxygen and hydrogen in equilibrium with water, the
exponent is 1 for the hydrogen and 1/2 for the oxygen, whereas for
hydrogen and bromine or iodine in equilibrium with the corresponding
halogen acid, the exponent is 1/2 for each, or the exponent of the
total concentration is 1. In the first two cases the chemical reactions
are not likely to disturb an ordinary p–V–T measurement, but other
reactions of the same type might occur, as the hydrogen-halogen
reaction does. If there are such reactions, it is necessary to consider
the product as a component and to take the equilibrium constant
into account. For almost all gases and gas mixtures, however, Eq. (3–1)
expresses the behavior satisfactorily up to fairly high concentrations.

3–2. The Van der Waals Equation and Corresponding States

Although there were many earlier attempts to improve the ideal gas
laws, the equation of van der Waals[1] is so superior that it is not worth
while to consider its predecessors.

With two specific parameters, a to account for the attraction between molecules at small distances that is necessary for liquefaction
and b to account for the repulsion at very small distances that is
necessary to prevent total collapse, van der Waals obtained an
equation that approximates the behavior of gases and liquids and of
the equilibrium between them. It may be written

$$p = \frac{RT}{\bar{V} - b} - \frac{a}{\bar{V}^2} = \frac{RTC}{1 - bC} - aC^2,$$

in which $\bar{V} = 1/C$, or

$$\frac{p}{RTC} = 1 + (b - \frac{a}{RT})C + \frac{b^2C^2}{1 - bC}. \tag{3–3}$$

The second virial coefficient is a linear function of the reciprocal
temperature, all the other coefficients are independent of the temperature, and expansion of $1/(1-bC)$ shows that the nth virial
coefficient, $n \neq 2$, is b^{n-1}.

Figures 3–1 and 3–2 illustrate the principle of continuity. The
reduced pressure p_r is plotted against the reduced concentration C_r
or the reduced volume V_r for three values of the reduced temperature
T_r. These "reduced" quantities are the absolute values p, C, $\bar{V}$, T

[1] J. D. van der Waals, dissertation, Leyden (1873); *Die Kontinuitat des gas-
formigen und flüssigen Zustands* (Leipzig, 1881–1899).

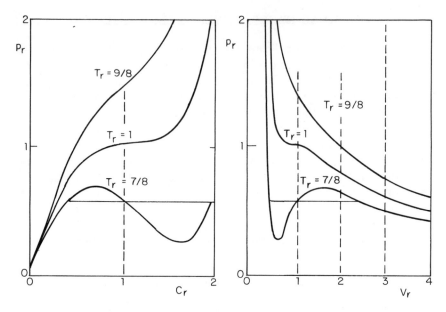

Figure 3–1. (left) Van der Waals gas. Reduced pressure vs. C_r

Figure 3–2. (right) Van der Waals gas. Reduced pressure vs. V_r

each divided by the corresponding critical value p_c, C_c, $\bar{V}_c$, T_c. For values of T_r less than 1, the curves show a maximum and a minimum, between which is the region of absolute instability where the calculated pressure and free energy decrease as the concentration increases. The horizontal line marks the two-phase system liquid-vapor. The portions of the curve between this line and the extrema are the metastable regions in which the liquid is stable in the absence of vapor, or vice versa. The principle of continuity says that the curves for the liquid and the vapor are segments of the same curve, although this curve does pass through an intermediate region of absolute instability.

The position of the horizontal line is determined so that the two areas between that line and the curve in Fig. 3–2 are equal. This may be visualized by making the left-hand edge of the figure the bottom and integrating Vdp from right to left, with increasing p. The areas below the curve cancel out as the path is reversed, and the enclosed area at smaller volume is negative. Obviously, the pressure and the temperature are the same for the two phases in equilibrium.

Figure 3–3 illustrates the different regions of the preceding curves as well as the method of determining precisely the equilibrium pressure.[2] It is not convenient to express $\Delta G = \int dp/C$ for a van der

[2] J. C. Slater, *Introduction to Chemical Physics* (McGraw-Hill, New York, 1939), p. 188.

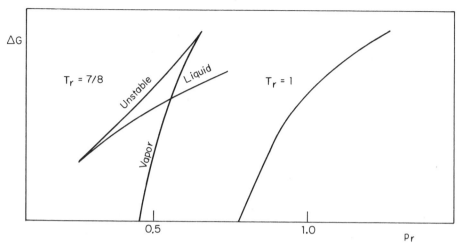

Figure 3–3. Gibbs free energy of Van der Waals gas.

Waals fluid as an analytic function of p, so both ΔG and p are calculated as functions of C from Eq. (3–3), and ΔG is plotted as a
function of p. The upper line, labeled "Unstable," marks the region
of absolute instability. The parts of the other curves between their
intersections with each other and with the unstable curve mark the
metastable regions, and the parts outside mark the stable vapor and
liquid regions. At the intersection of the vapor and liquid curves, the
pressure and the free energy are the same in the two phases, which
are therefore in equilibrium. Except for purposes of illustration, the
important parts of these curves are those near the equilibrium point.
The calculation of two points on each curve suffices to determine the
intersection approximately, and the calculation of more points closer
to the intersection will determine it with any desired precision.

As the reduced temperature approaches unity, the cusps in Fig. 3–3
come nearer to the intersection, and the extremes in Figs. 3–1 and
3–2 approach the inflection point, until at $T_r = 1$ the latter curves
show a horizontal inflection, or $dp/dc = d^2p/dc^2 = 0$. These two
relations with Eq. (3–3) give three relations among the quantities R,
a, b, p_c, V_c, and T_c. We shall use them to eliminate a and b and to
determine the value of RT_cC_c/p_c. From Eq. (3–3),

$$dp/dC = RT/(1 - bC)^2 - 2aC$$

and

$$d^2p/dC^2 = 2\,RTb/(1 - bC)^3 - 2a.$$

It follows that, for a van der Waals fluid,

$$RT_cC_c/p_c = 8/3,$$
$$b = 1/3C_c = RT_c/8p_c, \tag{3-4}$$
$$a = 9RT_c/8C_c = 27(RT_c)^2/64 \ p_c = 3p_c/C_c^2.$$

If we substitute Eqs. (3–4) and the definitions of the reduced quantities in Eq. (3–3), we obtain

$$p_r = \frac{8 \ T_r C_r/3}{1 - C_r/3} - 3 \ C_r^2, \tag{3-5}$$

which contains no specific parameters. Two fluids in states such that they have the same values for two of the corresponding properties also have the same value for the third, and are therefore said to be in corresponding states.

By partial substitution we obtain

$$\frac{p}{RTC} = \frac{8p_c}{3RT_cC_c} \ [1 + \frac{C_r/3}{1 - C_r/3} - \frac{27}{8T_r} \ C_r/3]$$
$$= 1 + bC\left(1 - \frac{27}{8T_r}\right) + \frac{b^2C^2}{1 - bC}, \tag{3-6}$$

with $8 \ p_c = 3 \ RT_cC_c$ to make the first virial coefficient unity, and $b = 1/3 \ C_c$. This expression leads to

$$\beta = b(1 - 27/8T_r)$$
$$= (1/3C_c)(1 - 27/8T_r)$$
$$= (RT_c/8 \ p_c)(1 - 27/8T_r). \tag{3-7}$$

The third form is more often used than the second, because the critical temperature and pressure are often better known than the critical volume. The two must not be used together because the third form gives values about a quarter larger. One of van der Waals's triumphs was the calculation from p–V–T properties at relatively hig is temperatures of the critical temperature and boiling point of hydrog before it had been liquefied. We see something of the limitations of the method, however, from the fact that for most substances RT_cC_c/p_c is about 3.4 instead of the 2.67 of van der Waals.

We shall also examine the behavior of two typical gases. Figure 3–4 shows $(p - RTC)/C^2$ as a function of T and C for xenon[3] and Fig. 3–5 shows the same quantity for water.[4] These figures represent surfaces in three dimensions and probably cannot be visualized without some practice. The perpendicular lines represent the coordinates. The curves rising from left to right are the isotherms, each starting

[3] J. A. Beattie, R. J. Barriault, and J. S. Brierley, *J. Chem. Phys. 19,* 1219, 1222 (1951); H. P. Julien, Ph.D. thesis, M.I.T. (1955).
[4] F. G. Keyes, L. B. Smith, and H. T. Gerry, *Proc. Am. Acad. 70,* 319 (1936).

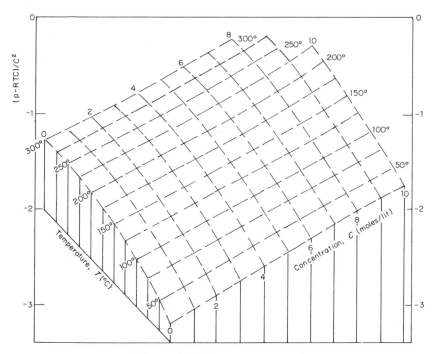

Figure 3–4. Pressure of xenon.

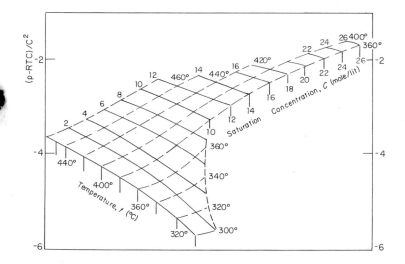

Figure 3–5. Pressure of steam.

at zero concentration. The curves rising from right to left are the isochors. At low concentrations the isochors have positive curvature, at high concentrations they appear to have slight negative curvature, and at about the critical concentration they are almost straight. This appears to be a quite general phenomenon. The curvature of the isotherms of xenon also changes sign at about 100°C, but the isotherms of water are much more complicated. The values of p/RTC for xenon are listed in Table 3–1.

Since van der Waals's time there has been an enormous amount of work on gases, which may be divided into three types. The first is the attempt to improve upon Eq. (3–3) as an analytic expression for $p/RTC = f(T,C)$; the second is the attempt to improve upon Eq. (3–6) as an expression for $p/RTC = f(T_r,b,C)$, though not necessarily as an analytic function; and the third is the statistical-mechanical study of simple models to obtain the lower virial coefficients. The third is particularly important for the consideration of mixed gases. We shall use it as the framework for surveying the first two.

For molecules with spherical symmetry, the second virial coefficient is

$$\beta = 2\pi N_A \int_0^\infty (1 - e^{-u/kT}) r^2 dr, \tag{3–8}$$

in which N_A is Avogadro's number, and u is the mutual potential energy of two molecules at a distance r. For nonspherical molecules u is also a function of one or more of the five angles that determine the relative orientation of the molecules.[5]

3–3. Rigid Spheres and the Square-Well Potential

For a rigid sphere, $u = 0$ when $r > \sigma$ and $u = \infty$ when $r < \sigma$, if σ the sum of the radii of the two spheres, and

$$\beta = b_0 = 2\pi N \sigma^3/3. \tag{3–9}$$

If the two molecules are the same size, the volume of either is $\pi\sigma^3/6$, so b_0 is four times the actual volume of the molecules in 1 mole. For ternary and higher encounters, there is the complication that a molecule is excluded from some parts of space by more than one other molecule. The next coefficients for two molecules of the same size are:

$$\gamma = 0.625\ b_0^2, \quad \delta = 0.2869\ b_0^3, \quad \epsilon \cong 0.110\ b_0^4. \tag{3–10}$$

---

[5]An excellent and much fuller discussion of the material in the following sections is to be found in J. O. Hirschfelder, C. F. Curtiss, and R. B. Bird, *Molecular Theory of Gases and Liquids* (Wiley, New York, 1954).

Table 3-1. Values of p/RTC as a function of C for xenon.[a]

t (C)	16.60	25	50	75	100	125	150	175	200	225	250	275	300
T (K)	289.754	298.153	323.151	348.153	373.160	398.170	423.183	448.197	473.213	498.229	523.245	548.260	573.274
$1000/T$ (K^{-1})	3.45120	3.35398	3.09452	2.87230	2.67982	2.51149	2.36304	2.23116	2.11321	2.00711	1.91115	1.82395	1.74437
RT (lit/mole K)	23.7756	24.4648	26.5160	28.5675	30.6194	32.6716	34.7241	36.7766	38.8292	40.8819	42.9346	44.9872	47.0397
C (mole/lit)													
1.0	0.8685	0.8756	0.8941	0.9090	0.9221	0.9328	0.9418	0.9495	0.9564	0.9622	0.9672	0.9715	0.9751
1.5	.8063	.8171	.8448	.8671	.8857	.9015	.9142	.9263	.9364	.9451	.9526	.9592	.9644
2.0	.7481	.7621	.7985	.8278	.8520	.8725	.8904	.9053	.9183	.9300	.9396	.9481	.9554
2.5	.6928	.7102	.7547	.7903	.8208	.8463	.8677	.8854	.9025	.9163	.9275	.9386	.9476
3.0	.6412	.6616	.7135	.7557	.7911	.8211	.8468	.8685	.8885	.9043	.9185	.9310	.9407
3.5	.5925	.6159	.6756	.7239	.7641	.7983	.8278	.8534	.8750	.8934	.9105	.9244	.9356
4.0	.5474	.5734	.6401	.6943	.7392	.7775	.8109	.8382	.8639	.8849	.9033	.9190	.9325
4.5	.5056	.5343	.6072	.6676	.7167	.7587	.7953	.8265	.8546	.8774	.8986	.9160	.9308
5.0	.4672	.4985	.5775	.6418	.6960	.7428	.7829	.8166	.8469	.8723	.8959	.9127	.9304
6.0	.4003	.4354	.5251	.5991	.6616	.7149	.7624	.8026	.8375	.8682	.8930	.9164	.9334
7.0	.3460	.3840	.4823	.5651	.6352	.6963	.7498	.7949	.8354	.8702	.8998	.9229	.9455
8.0	.3031	.3427	.4486	.5397	.6179	.6865	.7464	.7986	.8423	.8807	.9133	.9379	.9626
9.0	.2694	.3103	.4245	.5242	.6106	.6860	.7519	.8094	.8595	.9027	.9347	.9604	—
10.0	.2428	.2862	.4097	.5189	.6144	.6968	.7689	.8332	.8855	.9331	—	—	—
$-\beta$ (lit/mole)[b]	0.1384	0.1309	0.1114	0.0959	0.0828	0.0716	0.0621	0.0544	0.0476	0.0411	0.0359	0.0316	0.0275

[a] H. P. Julien, Ph.D. thesis, M.I.T. (1955).
[b] In determining β the results at $C = 1$ were ignored.

Each of the virial coefficients is independent of temperature. The importance of the higher virial coefficients is much less than for van der Waals's equation, for which each numerical coefficient is unity.

The square-well potential is defined as being infinite when $r < \sigma$, $-\epsilon$ when $\sigma < r < R\sigma$, and zero when $r > R\sigma$, so each rigid sphere is surrounded by a shell to which other molecules are attracted. The second virial coefficient is

$$\beta = b_0 \left[1 - (R^3 - 1)(e^{\epsilon/kT} - 1)\right]$$
$$= b_0 R^3 - b_0(R^3 - 1)e^{\epsilon/kT}. \tag{3-11}$$

The higher virial coefficients are much more complicated because, in addition to the exclusion from parts of space by more than one molecule, there will be exclusion by one molecule from the potential well of another, and also some regions of space in which the wells of two molecules overlap, so the potential energy of the third molecule is -2ϵ. This is illustrated in Fig. 3–6, in which two molecules are

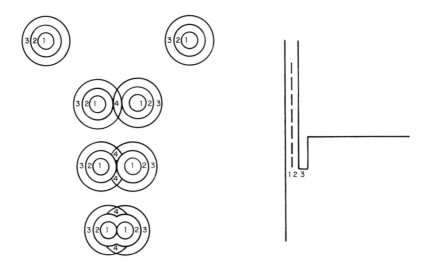

Figure 3–6. Square-well potential.

each represented by a circle (numbered 1) for the hard sphere itself, a ring (numbered 2) for the spherical shell from which the centers of other molecules are excluded, and a ring (numbered 3) for the shell of negative potential $-\epsilon$. The distance between the two molecules varies from practically infinity at the top to zero (contact) at the bottom. The regions with potential -2ϵ for a third molecule are numbered 4.

If we write Δ for $e^{\epsilon/kT} - 1$, the third virial coefficient is

$$\gamma = \tfrac{1}{8}b_0^2[5 - (R^6 - 18R^4 + 32R^3 - 15)\Delta$$
$$- (2R^6 - 36R^4 + 32R^3 + 18R^2 - 16)\Delta^2$$
$$- (6R^6 - 18R^4 + 18R^2 - 6)\Delta^3], \qquad R \leq 2$$

and

$$\gamma = \tfrac{1}{8}b_0^2[5 - 17\Delta - (-32R^3 + 18R^2 + 48)\Delta^2$$
$$- (5R^6 - 32R^3 + 18R^2 + 26)\Delta^3], \qquad R \geq 2. \qquad (3\text{--}12)$$

If we expand the exponential in a power series,

$$\Delta = \epsilon/kT + \tfrac{1}{2}(\epsilon/kT)^2 + \tfrac{1}{6}(\epsilon/kT)^3 + \cdots$$

and

$$\beta = b_0 - b_0(R^3 - 1)(\epsilon/kT)[1 + \tfrac{1}{2}(\epsilon/kT) + \cdots]. \qquad (3\text{--}13)$$

If we stop with the first term of the expansion, the form is identical with that of van der Waals, with $b_0 = b$ and $b_0(R^3 - 1)\epsilon N = a$. Therefore, to this approximation the principle of corresponding states applies. For the higher terms in the expansion, however, a third parameter ϵ/k is introduced, and the principle of corresponding states does not apply unless $(R^3 - 1)$ is determined by b_0 and ϵ.

The square-well potential describes the second virial coefficients of gases so much better than other equally simple expressions that we shall use it as the norm with which to compare other expressions. The three parameters may be determined from the second virial coefficient and its first two derivatives with respect to $1/T$:

$$\beta = b_0 - b_0(R^3 - 1)(e^{\epsilon/kT} - 1), \qquad (3\text{--}11)$$

$$\beta' = d\beta/d(1/T) = - b_0(R^3 - 1)\epsilon^{\epsilon/kT}\epsilon/k, \qquad (3\text{--}14)$$

$$\beta'' = d^2\beta/d(1/T)^2 = - b_0(R^3 - 1)\epsilon^{\epsilon/kT}(\epsilon/k)^2 = \beta'\epsilon/k. \qquad (3\text{--}15)$$

If we consider β/b_0 as a function of $(R^3 - 1)\epsilon/kT$ with $(R^3 - 1)$ constant we have

$$\beta/b_0 = 1 - (R^3 - 1)(e^{\epsilon/kT} - 1), \qquad (3\text{--}16)$$

$$d(\beta/b_0)/d[(R^3 - 1)\epsilon/kT] = - e^{\epsilon/kT}, \qquad (3\text{--}17)$$

$$d^2(\beta/b_0)/d[(R^3 - 1)\epsilon/kT]^2 = - e^{\epsilon/kT}/(R^3 - 1) \qquad (3\text{--}18)$$
$$= d(\beta/b_0)/d[(R^3 - 1)\epsilon/kT]/(R^3 - 1).$$

The function is unity and its first derivative is -1 when $1/T$ is zero, and the ratio of the second derivative to the first is $1/(R^3 - 1)$. A van der Waals gas is the limiting case with $1/(R^3 - 1) = 0$ since the second derivative is zero.

Table 3-2. Parameters for equations of state.[a]

Gas	Square-Well $b_0 \dfrac{cm^3}{mole}$	$1/(R^3 - 1)$	ϵ/k (K)	Lennard-Jones $b_0 \dfrac{cc}{mole}$	ϵ/k
Ne	17.05	0.1805	19.5	27.10	34.9
				26.21	35.60
A	39.87	.1875	69.4	49.80	119.8
				49.58	122
Kr	47.93	.1875	98.3	58.86	171
				58.7	158
X_e				86.94	221
				78.5	217
N_2	45.29	.1805	53.7	63.78	95.05
				64.42	95.9
CO_2	75.79	.1950	119	113.9	189
				85.05	205
methane				70.16	148.2
ethane	55.72	.2850	244	78	243
propane	108.8	.4678	347	226	242
n-butane	140.6	.4513	387	155	297
n-heptane	330.6	.7882	629	884	282
ethene	47.28	.2691	222	116.7	199.2
propene	101.4	.4735	339		
propadiene	115.8	.6296	382		
1-butene	220.6	1.0544	492		
2-methylpropane	218.0	1.0289	490		
trans-2-butene	185.3	0.7571	465		
cis-2-butene	239.5	1.2601	537		
CCl_3F	117.6	0.3720	399		
$CHCl_2F$	27.60	.0869	306		
CCl_2F_2	140.6	.5852	345		
$CCl_2F.CClF_2$	63.73	.1260	335		
CH_3Cl	99.90	.7194	469		
NH_3	30.83	.9627	692		
H_2O	22.33	1.3818	1260		

[a] Hirschfelder, Curtiss, and Bird, ref. 5.

Table 3–2 gives the values of the parameters b_0, $1/(R^3 - 1)$, and ϵ/k for several gases from the parameters of Hirschfelder, Curtiss, and Bird.[5] For the noble gases, $1/(R^3 - 1)$ lies between 0.18 and 0.19. The two values less than 0.18 for polyatomic gases probably arise from experimental errors, but it is dangerous to exclude values that are not in agreement with our generalizations. The usual trend is that $1/(R^3 - 1)$ increases with increasing complexity, and particularly with increasing polarity. In the latter case, at least, this does not mean that $(R^3 - 1)$ is small, but rather that the energy of attraction is large only for a small fraction of the possible orientations.

The Boyle point, the temperature at which β is zero, is unity for a van der Waals gas in the β/b_0 vs. $(R^3 - 1)\epsilon/kT$ diagram. For any other gas it is $(R^3 - 1) \ln [1 + 1/(R^3 - 1)]$, which is 0.92 for $1/(R^3 - 1)$ = 0.1805 (neon) and 0.63 for $1/(R^3 - 1) = 1.3818$ (water).

There is no practical advantage in choosing the intercept and the negative of the slope to be unity when $1/T$ is zero, for the square-well potential does not agree well with the few measured values of the second virial coefficients for temperatures higher than $3T_B$. The parameters must be determined from experimental measurements at lower temperatures. The use of Eqs. (3–11), (3–14), and (3–15) is equivalent to determining the three parameters from measurements of β at three different temperatures, very close together, with the reciprocal of the intermediate temperature equal to the average of the reciprocals of the extreme temperatures. If the expression is to be used for interpolation over a large temperature range, it is necessary to have the extremes near the end of the range. The application of these methods to an analytical expression for β is illustrated in Sec. 3–7.

3–4. Chemical Association

A reexamination of Secs. 2–3 and 3–3 indicates that for rigid spheres that can react chemically when in contact, but have no other deviations from the perfect gas laws, the virial coefficients should be the sum of those for rigid spheres without chemical action and those for species-perfect gases. For the second virial coefficient,

$$\beta = b - K_2. \tag{3-19}$$

Comparison of this expression with the second virial coefficient for the square-well potential, Eq. (3–11), allows the identifications

$$b = b_o R^3 \; ; \; K_2 = b_o(R^3 - 1)e^{\epsilon/kT} \tag{3-20}$$

and from Eq. (2–29) we get $\Delta E = -N_A\epsilon$ for the association reaction. The value of K_3 that corresponds to the square-well potential can be calculated from the third virial coefficient, but the expression is complicated.

3–5. Inverse Sixth-Power Attraction and Lennard-Jones Potential

Real molecules are not rigid spheres, and their softness shows up in the second virial coefficient's passing through a maximum at high temperatures. This softness may be accounted for by an energy of repulsion that varies as a high power of the reciprocal distance of separation or as an exponential function of this distance (see Appendix D-4). The energy of attraction does not change suddenly to zero at a distance $R\sigma$. For two neutral molecules it varies as the inverse sixth power of the distance at large distances (Appendix D-5). The sum of attractive and repulsive energies may be used to compute the virial coefficients (Appendix D-6).

We shall consider only the Lennard-Jones 6-12 potential, Eq. (D-22),

$$u = 4\epsilon[(\sigma/r)^{12} - (\sigma/r)^6],$$

in which σ is the distance at which u becomes zero. Values of the second and third virial coefficients are tabulated by Hirschfelder, Curtiss, and Bird.[5] The second virial coefficient may be expressed as a series of Γ functions:

$$\beta = 2^{1/2}b_0\left(\frac{\theta}{T}\right)^{1/4}[\Gamma\left(\frac{3}{4}\right) - \sum_{n=1}^{\infty} \frac{2^{n-2}}{n!} \Gamma\left(\frac{2n-1}{4}\right)\left(\frac{\theta}{T}\right)^{n/2}],$$

$$(3\text{-}21)$$

in which $b_0 = 2\pi N_A\sigma^3/3$ and $\theta = \epsilon/k$.

Stockmayer and Beattie[6] give an approximate solution, valid for values of θ/T from 0.2 to 1 and of T_r from approximately 0.75 to 4:

$$\beta = \left(\frac{\theta}{T}\right)^{1/4}b_0(1.064 - 3.602\,\theta/T) \qquad (3\text{-}22)$$

or

$$\beta T^{5/4} = -3.602\,\theta^{1/4}\,b_0\theta + 1.064\,\theta^{1/4}\,b_0T. \qquad (3\text{-}23)$$

Eq. 3-21 gives a very convenient method of calculating θ and b_0 from measurements in this range.

To compare the Lennard-Jones expressions with those for the square well, we reduce the parameters of the square-well equations to two by making $R = 1.8$, and then choose the ratios σ_{SW}/σ_{LJ} and $\epsilon_{SW}/\epsilon_{LJ}$ to give the best overall agreement for the second and third virial coefficients. Fig. 3-7 shows β/b_0 and γ/b_0^2 vs. log T/T_B, in which T_B is the Boyle temperature. The solid curves are for a Lennard-Jones gas and the broken curves for a square-well-potential gas with $R = 1.8$, $\sigma_{SW}/\sigma_{LJ} = 0.913$, and $\epsilon_{SW}/\epsilon_{LJ} = 0.643$. The inset shows the mutual energy of two molecules as a function of r/σ_{LJ} for these two cases. The agreement is good to about twice the Boyle temperature. At

---

[6]W. H. Stockmayer and J. A. Beattie, *J. Chem. Phys.* **10**, 476 (1942).

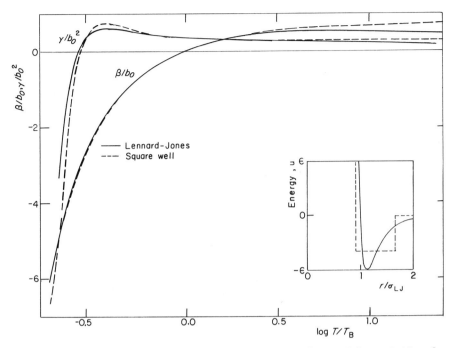

Figure 3-7. Second and third virial coefficients from Lennard-Jones 6–12 and square-well $R = 1.8$ potentials.

about seven times the Boyle temperature the Lennard-Jones β passes through a flat maximum whereas the square-well potential continues to increase slowly to b_0.

The second virial coefficient of a gas consisting of polar molecules was calculated by Stockmayer[7] by combining the Lennard-Jones energy with that of two point dipoles. The third virial coefficient has been computed by Rowlinson.[8] The values of both are tabulated in Hirschfelder, Curtiss, and Bird.[5]

3-6. Equations of State

Very many equations of state have been proposed. We shall list enough to illustrate the historical developments since van der Waals gave his equation,

$$p = RTC/(1 - bC) - aC^2.$$

[7] W. H. Stockmayer, *J. Chem. Phys. 9*, 398 (1941).

[8] J. S. Rowlinson, *J. Chem. Phys. 19*, 827 (1951).

Clausius[9] changed the form by dividing a by T and adding a third constant, to give

$$p = RTC/(1 - bC) - aC^2/T \ (1 + \delta C)^2. \tag{3-24}$$

Dieterici[10] used two constants but expressed the attractive term as an exponential:

$$p = RTCe^{-aC/RT}/(1 - bC). \tag{3-25}$$

Berthelot[11] eliminated the third constant of the Clausius equation:

$$p = RTC/(1 - bC) - aC^2/T. \tag{3-26}$$

Keyes[12] extended the van der Waals equation by adding two constants to give

$$p = RTC/(1 - bCe^{-\beta C}) - aC^2/(1 + \alpha C)^2. \tag{3-27}$$

Beattie and Bridgeman[13] used the virial form with five constants:

$$p = RT(C + B_0C^2 - B_0bC^3) - (A_0C^2 - A_0aC^3) \tag{3-28}$$
$$- (R/T^2)(c_0C^2 + B_0c_0C^3 - B_0c_0bC^4).$$

Their equation is valid to about the critical concentration.

The Beattie-Bridgeman parameters for several gases are given in Table 3-3. For consistency they are all given with T (°K) = T(°C) + 273.13, which was the accepted value when most of them were determined, and the corresponding value of R.

Benedict, Webb, and Rubin[14] have added three more constants to the Beattie-Bridgeman equation:

$$p = RT \ (C + B_0C^2 - B_0bC^3) - (A_0C^2 - A_0aC^3 - A_0a\alpha C^6) \tag{3-29}$$
$$- (R/T^2)[c_0C^2 - c_0cC^2(1 + \gamma C^2)e^{-\gamma C^2}].$$

This equation is valid to about twice the critical density and gives satisfactorily the vapor pressures and the saturated liquid and vapor densities.

The second virial coefficient of the van der Waals, the Dieterici, and the Keyes equations is

$$\beta = b - a/RT, \tag{3-30}$$

that of the Clausius and the Berthelot equations is

$$\beta = b - a/RT^2, \tag{3-31}$$

[9]R. Clausius, *Ann. Phys.* (3) *9*, 377 (1880).
[10]C. Dieterici, *Ann. Phys.* *69*, 685 (1899).
[11]D. Berthelot, *Trav. Bur. Int. Poids Més.* *13* (1907).
[12]F. G. Keyes, *Proc. Nat. Acad. Sci.* *3*, 323 (1917).
[13]J. A. Beattie and O. C. Bridgeman, *J. Am. Chem. Soc.* *49*, 1665 (1927).
[14]M. Benedict, G. B. Webb, and L. C. Rubin, *J. Chem. Phys.* *8*, 334 (1940).

Table 3-3. Values of the constants of the Beattie-Bridgeman equation of state for several gases:[a] $p = [RT(1 - \epsilon)/V^2] [V + B] - A/V^2$; $A = A_0(1 - a/V)$; $B = B_0(1 - b/V)$; $\epsilon = c_0/VT^3$. Units: normal atmospheres; liters per mole; K (T K $= t$ C $+ 273.13$); $R = 0.08206$.

Gas	A_0	a	B_0	b	$10^{-4} c_0$	Molecular weight
He	0.0216	0.05984	0.01400	0.0	0.0040	4.00
Ne	.2125	.02196	.02060	.0	.101	20.2
Ar	1.2907	.02328	.03931	.0	5.99	39.91
Kr[b]	2.4281	.02866	.05267	.0	14.92	83.8
Xe[b]	4.4600	.03390	.06800	.0	30.00	131.3
H_2	0.1975	−0.00506	.02096	−0.04359	0.0504	2.0154
N_2	1.3445	0.02617	.05046	.00691	4.20	28.016
O_2	1.4911	.02562	.04624	+0.004208	4.80	32
Air	1.3012	.01931	.04611	−0.01101	4.34	28.964
I_2	17.0		.325		4000	253.864
CO_2	5.0065	.07132	.10476	+0.07235	66.00	44.000
NH_3	2.3930	.17031	.03415	.19112	476.87	17.0311
CH_4	2.2769	.01855	.05587	−0.01587	12.83	16.0308
C_2H_4	6.1520	.04964	.12156	+0.03597	22.68	28.0308
C_2H_6	5.8800	.05861	.09400	.01915	90.00	30.0462
C_3H_8	11.9200	.07321	.18100	.04293	120.00	44.0616
l-C_4H_8	16.6979	.11988	.24046	.10690	300.00	56.0616
iso-C_4H_{10}	16.9600	.10860	.24200	.08750	250.00	56.0616
n-C_4H_{12}	17.7940	.12161	.24620	.09423	350.00	58.077
iso-C_4H_{18}	16.6037	.11171	.23540	.07697	300.00	58.077
n-C_5H_{10}	28.2600	.15099	.39400	.13960	400.00	72.0924
neo-C_5H_{12}	23.3300	.15174	.33560	.13358	400.00	72.0924
n-C_7H_{16}	54.520	.20066	.70816	.19179	400.00	100.1232
CH_3OH	33.309	.09246	.60362	.09929	32.03	32.0308
$(C_2H_5)_2O$	31.278	.12426	.45446	.11954	33.33	74.077

[a] J. A. Beattie, *Chem. Rev.* **44**, 141 (1949).
[b] Reference 3.

and that of the Beattie-Bridgeman and the Benedict-Webb-Rubin equations is

$$\beta = B_0 - A_0/RT - c_0/T^3. \qquad (3\text{--}32)$$

The van der Waals and Keyes equations give linear isometrics, or p is a linear function of T at constant C. The Dieterici equation at finite concentrations gives positive curvature, so the curves are concave upward. The Clausius, Berthelot, and Beattie-Bridgeman equations give negative curvature. The Benedict-Webb-Rubin equation agrees with the behavior of real gases in giving negative curvature for concentrations less than the critical, positive curvature between one and two times the critical concentration, and negative curvature again at higher concentrations.

The second virial coefficient for the two-constant equations may be expressed in terms of the critical temperature and pressure and the reduced temperature. The van der Waals equation gives

$$\beta = (T_c/p_c)(10.26 - 34.62/T_r) \ cm^3/mole. \tag{3-33}$$

The Berthelot equation gives

$$\beta = (T_c/p_c)(10.26 - 34.62/T_r^2). \tag{3-34}$$

Berthelot also proposed an empirical equation for β, which is the one usually given his name and often used to determine β at low temperatures:

$$\beta = (T_c/p_c)(5.77 - 34.62/T_r^2). \tag{3-35}$$

Keyes[15] obtained the equations

$$\beta = (T_c/p_c)(13.29 - 34.9/T_r - 5.47/T_r^3) \tag{3-36}$$

for nonpolar gases and

$$\beta = (T_c/p_c)(11.36 - 25.77/T_r - 11.8/T_r^3) \tag{3-37}$$

for the polar gases water and ammonia.

By a combination of dimensional analysis with an empirical study of the behavior of a large number of gases, Su and Chang[16] obtain the following expressions for the Beattie-Bridgeman constants in terms of the critical constants:

$$A_0 = 0.4758 \ (RT_c/p_c)^2 \ p_c, \qquad a = 0.1127 \ (RT_c/p_c),$$
$$B_0 = 0.18764 \ (RT_c/p_c), \qquad b = 0.0383 \ (RT_c/p_c),$$
$$c_0 = 0.05 \ (RT_c/p_c) \ T_c^3.$$

These give the second virial coefficient,

$$\beta = (T_c/p_c)(15.40 - 39.04/T_r - 4.76/T_r^3) \ cm^3/mole. \tag{3-38}$$

At higher pressures it is often convenient to have a graphic representation. If C is taken as one of the independent variables, it is customary to use the pseudo-reduced quantity $C'_r = C/(p_c/RT_c)$.[17] Then $p/RTC = p_r/T_rC'_r$, and should be independent of the substance. Figure 3–4 corresponds to a plot of $(R^2T_c^2/p_c)(p_r - T_rC'_r)/C_r'^2$. For

[15] F. G. Keyes, *J. Am. Chem. Soc. 60*, 1761 (1938).

[16] G. J. Su and C. H. Chang, *J. Am. Chem. Soc. 68*, 1080 (1946).

[17] J. A. Beattie and W. H. Stockmayer, *Reports on Progress in Physics*, 7 (Physical Society, London, 1940), p. 195.

chemical-engineering purposes[18] it is convenient to have a chart of p/RTC against p_r or against the logarithm of p_r. The area between the latter curve, the zero abscissa, and two perpendiculars is the logarithm of the fugacity ratio for the corresponding values of p_r.

3-7.   Other Thermodynamic Functions

For constant composition we may rewrite Eqs. (3–1) and (3–2) as

$$A = \Sigma_i N_i(G_i - RT + RT \ln RT\, N_i/V) + RT\, \beta\, N^2/V$$
$$+ RT\gamma N^3/2V^2 + RT\delta N^4/3V^3 + \cdots,$$
$$(3\text{–}39)$$

$$pV/NRT = 1 + \beta N/V + \gamma(N/V)^2 + \delta(N/V)^3 + \cdots, \quad (3\text{–}40)$$

in which $N = \Sigma_i N_i$, $\beta = \Sigma_{ij}\beta_{ij}\bar{N}_i\bar{N}_j$, and so forth.

Then, as in Eqs. (2–3) to (2–12),

$$S = - \Sigma_i N_i\left(\frac{dG_i}{dT} + R \ln RT\, \frac{N_i}{V}\right) - R\left(\beta + T\,\frac{d\beta}{dT}\right)\frac{N^2}{V} - \cdots.$$
$$(3\text{–}41)$$

The temperature differentiation of the higher terms is analogous to that of the β term:

$$E = \Sigma_i N_i(G_i - T\, dG_i/dT - RT) - RT^2(d\beta/dT)N^2/V - \cdots,$$
$$(3\text{–}42)$$

$$C_V = - \Sigma_i N_i(T\, d^2G_i/dT^2 + R) - R[2T(d\beta/dT)$$
$$+ T^2\, d^2\beta/dT^2]\, N^2/V - \cdots, \quad (3\text{–}43)$$

$$G = \Sigma_i N_i(G_i + RT \ln RT\, N_i/V) + 2\, RT\beta N^2/V + 3RT\gamma N^3/2V^2$$
$$+ 4RT\delta N^4/3V^3 + \cdots, \quad (3\text{–}44)$$

$$H = \Sigma_i N_i(G_i - T\, dG_i/dT) + RT(\beta - T\, d\beta/dT)N^2/V + \cdots,$$
$$(3\text{–}45)$$

$$C_p = - \Sigma_i N_i Td^2G_i/dT^2 - RT^2(d^2\beta/dT^2)N^2/V - \cdots. \quad (3\text{–}46)$$

To use p as independent variable we solve for C in terms of p by repeated substitution to obtain

$$C = p/RT - \beta(p/RT)^2 - (\gamma - 2\beta^2)(p/RT)^3$$
$$- (\delta - 5\beta\gamma + 5\beta^3)(p/RT)^4 + \cdots, \quad (3\text{–}47)$$

---

[18]The recent revival of interest in corresponding states was initiated by W. K. Lewis and his collaborators for engineering computations. See H. P. Meissner and P. W. Keyser, M.I.T. thesis for B.S. in Chem. Eng. (1929); J. Q. Cope, W. K. Lewis, and H. C. Weber, *J. Ind. Eng. Chem.* **23**, 8 (1931); W. K. Lewis and C. D. Luke, *Trans. Am. Soc. Mech. Eng.* **54**, 55 (1933), *Ind. Eng. Chem.* **25**, 725 (1933).

$$pV/NRT = 1 + \beta(p/RT) + (\gamma - \beta^2)(p/RT)^2$$
$$+ (\delta - 3\beta\gamma + 2\beta^3(p/RT)^3 + \cdots$$
$$= 1 + \beta'p + \gamma'p^2 + \delta'p^3 + \cdots, \qquad (3\text{--}48)$$

in which $\beta' = \beta/RT$, $\gamma' = (\gamma - \beta^2)/(RT)^2$, $\delta' = (\delta - 3\beta\gamma + 2\beta^3)/(RT)^3$, and so on. Then

$$\bar{V} = RT/p + RT\beta' + RT\gamma'p + RT\delta'p^2 + \cdots. \qquad (3\text{--}49)$$

Equation (3–48) does not converge nearly as well as Eq. (3–2). From Eq. (3–49) we see that $\bar{V} - RT/p$ approaches $RT\beta' = \beta$, not zero, at zero concentration or pressure.

In terms of pressure,

$$G = \Sigma_i N_i(G_i + RT \ln p\bar{N}_i) + RTN\beta'p + RTN\gamma'p^2/2$$
$$+ RTN\delta'p^3/3 + \cdots, \qquad (3\text{--}50)$$

$$S = -\Sigma_i N_i(dG_i/dT + R \ln p\bar{N}_i) - RN(\beta' + T\,d\beta'/dT)p - \cdots, \qquad (3\text{--}51)$$

$$H = \Sigma_i N_i(G_i - T\,dG_i/dT) - RT^2N(d\beta'/dT)p - \cdots, \qquad (3\text{--}52)$$

$$C_p = -\Sigma_i N_i T d^2 G_i/dT^2 - RN(2T\,d\beta'/dT + T^2\,d^2\beta'/dT^2)p - \cdots, \qquad (3\text{--}53)$$

$$\alpha = \left(\frac{\partial \ln V}{\partial T}\right)_p = \frac{1}{T} + \frac{p\,d\beta'/dT + p^2\,d\gamma'/dT + p^3\,d\delta'/dT + \cdots}{1 + \beta'p + \gamma'p^2 + \delta'p^3 + \cdots}, \qquad (3\text{--}54)$$

$$\beta = -\left(\frac{\partial \ln V}{\partial p}\right)_T = \frac{1}{p} - \frac{\gamma'p + 2\delta'p^2 + \cdots}{1 + \beta'p + \gamma'p^2 + \delta'p^3 + \cdots}. \qquad (3\text{--}55)$$

We shall consider the chemical potential in the next chapter.

3–8. Determination of Parameters

The results expressed in Figs. 3–4 and 3–5 and in Table 3–1 are representative of the information available in the more thoroughly studied cases, and the equations of the preceding sections are typical of those for which parameters are to be calculated.

The first steps in the determination of the parameters are the choice of the form of the equation and the number of parameters, and the decision as to the range of measurements to be represented. Then the decision must be made whether to smooth first with respect to temperature, to concentration, or to both together.

For the van der Waals and the Keyes equations the isochors are linear in the temperature, so it is simple to smooth first with respect to temperature. If p or p/C is chosen so as to give the proper relative weights to the different isochors, it is not hard to determine the range of fit for any desired precision. It is also possible to divide by T, or RT, to obtain linear functions of $1/T$. For the Clausius and the Berthelot equations, p/T is a linear function of $1/T^2$, or pT is a linear function of T^2.

Beattie and Bridgeman use the isochors and choose the value of c_0 so as to make all the isochors as straight as possible. The extent to which this can be done to the desired precision may limit the range to which the equation is applicable, or it may be less restrictive than making A_0 and B_0 quadratic functions of the concentration.

More often, however, the virial coefficients are determined for each isotherm, and each coefficient is then smoothed with temperature, perhaps in accordance with some theory. If the concentration range to be covered is determined first, the number of parameters may be determined by the method of the first paragraph of Appendix B–2. If the range and parameters are to be selected at the same time, the method of the second paragraph is useful.

The concentrations of Table 3–1 start too high for the best adaptation to the determination of the second virial coefficient, especially since the author places less reliance on the measurements at 1 mole/lit than on the others. However, they are satisfactory for purposes of illustration. For any of the equations we have discussed in this section, the virial coefficients may be treated in the same way as the individual isometrics.

For the Lennard-Jones 6–12 second virial coefficient in the range in which the Beattie-Stockmayer approximation, Eq. (3–22), is valid, $\beta T^{5/4}$ is a linear function of T, and the parameters are easily determined from this straight line. The parameters for the square-well potential may be determined by successive approximations to the values at three representative temperatures, or at two temperatures if the value of R is predetermined. The computation is simplified if the Boyle temperature is chosen as one of the temperatures. In any case, the results obtained from representative points must be confirmed with all the measurements.

4. Gas Mixtures

4-1. Pressure–Volume–Temperature Relations

For a binary mixture,

$$\beta = \Sigma_{ij}\beta_{ij}\bar{N}_i\bar{N}_j = \beta_{11}\bar{N}_1^2 + \beta_{12}\bar{N}_1\bar{N}_2 + \beta_{21}\bar{N}_2\bar{N}_1 + \beta_{22}\bar{N}_2^2. \quad (4\text{--}1)$$

Since we cannot distinguish between $\bar{N}_1\bar{N}_2$ and $\bar{N}_2\bar{N}_1$, however, we cannot distinguish between β_{12} and β_{21}. Therefore we write

$$\beta = \beta_{11}\bar{N}_1^2 + 2\beta_{12}\bar{N}_1\bar{N}_2 + \beta_{22}\bar{N}_2^2. \quad (4\text{--}2)$$

Similarly, $\quad\quad\quad\quad\quad\quad\quad\quad\quad\quad\quad\quad\quad\quad\quad\quad\quad\quad\quad (4\text{--}3)$

$$\gamma = \Sigma_{ijk}\gamma_{ijk}\bar{N}_i\bar{N}_j\bar{N}_k = \gamma_{111}\bar{N}_1^3 + 3\gamma_{112}\bar{N}_1^2\bar{N}_2 + 3\gamma_{122}\bar{N}_1\bar{N}_2^2 + \gamma_{222}\bar{N}_2^3,$$

in which $3\gamma_{112}$ represents $\gamma_{112} + \gamma_{121} + \gamma_{211}$, and so on. In general, the nth virial coefficient of a binary system is obtained by multiplying each term in the expansion of $(\bar{N}_1 + \bar{N}_2)^n$ by a parameter, which is designated by subscripts that indicate its coefficient. This generalization may be extended to any number of components. The terms in the nth virial coefficient are the corresponding terms in $(\Sigma_i\bar{N}_i)^n$, each multiplied by an appropriate parameter. For a three-component system

$$\beta = \beta_{11}\bar{N}_1^2 + 2\beta_{12}\bar{N}_1\bar{N}_2 + \beta_{22}\bar{N}_2^2 + 2\beta_{23}\bar{N}_2\bar{N}_3 + \beta_{33}\bar{N}_3^2 + 2\beta_{13}\bar{N}_1\bar{N}_3 \quad (4\text{--}4)$$

and

$$\gamma = \gamma_{111}\bar{N}_1^3 + 3\gamma_{112}\bar{N}_1^2\bar{N}_2 + 3\gamma_{122}\bar{N}_1\bar{N}_2^2 + \gamma_{222}\bar{N}_2^3 + 3\gamma_{223}\bar{N}_2^2\bar{N}_3 + 3\gamma_{233}\bar{N}_2\bar{N}_3^2$$
$$+ \gamma_{333}\bar{N}_3^3 + 3\gamma_{133}\bar{N}_1\bar{N}_3^2 + 3\gamma_{113}\bar{N}_1^2\bar{N}_3 + 6\gamma_{123}\bar{N}_1\bar{N}_2\bar{N}_3, \quad (4\text{--}5)$$

in which $6\gamma_{123} = \gamma_{123} + \gamma_{132} + \gamma_{213} + \gamma_{231} + \gamma_{312} + \gamma_{321}$.

To determine β for a mixture we need the β's for the components and one additional parameter for each binary system, but nothing more for ternary or higher systems. For γ we need the γ's for the components, two additional parameters for each binary system, and one additional for each ternary system. For δ we need the δ's for the components, three additional parameters for each binary system,

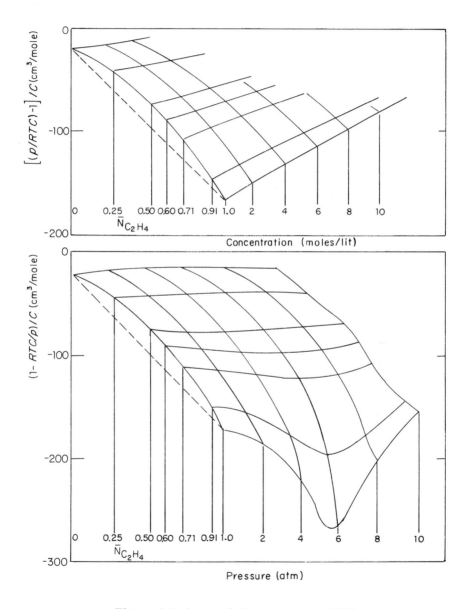

Figure 4–1. Argon-ethylene mixtures at 25 °C.

three additional for each ternary system, and one additional for each quaternary system.

Figure 4–1 shows the properties of argon-ethylene mixtures at 25°C,[1] first with $(p/RTC\text{-}1)/C$ as a function of total concentration and composition, and next with $(1\text{-}RTC/p)/C$ as a function of pressure and composition. The absolute temperature is about twice the critical temperature of argon and about 3 percent above the critical temperature of ethylene. The lines all stop at 125 atm, which is about 2.5 times the critical pressure of either gas. These two figures show the great advantage of concentration over pressure as independent variable. The constant composition lines are remarkably straight in the concentration diagram. The line at zero concentration is β, and its deviation from the straight line between the end points is

$$\beta - \beta_{11}\bar{N}_1 - \beta_{22}\bar{N}_2 = (2\beta_{12} - \beta_{11} - \beta_{22})\ \bar{N}_1\bar{N}_2 = B^0_{12}\bar{N}_1\bar{N}_2. \quad (4\text{--}6)$$

Measurement at low pressures of the change of volume on mixing at constant pressure and extrapolation to zero pressure is an excellent method of determining the second virial coefficient of a mixture if the coefficients of the components are known.

4–2. Combination of Constants

Van der Waals calculated the constants for gaseous mixtures by the relations

$$b\ \ = \Sigma_i b_i \bar{N}_i, \quad (4\text{--}7)$$

$$a^{1/2} = \Sigma_i a_i^{1/2}\bar{N}_i. \quad (4\text{--}8)$$

For a binary mixture,

$$b = b_1\bar{N}_1 + b_2\bar{N}_2,\ b_{12} = (b_{11} + b_{22})/2,$$
$$a = a_{11}\bar{N}_1^2 + 2(a_1a_2)^{1/2}\ \bar{N}_1\bar{N}_2 + a_{22}\bar{N}_2,\ a_{12} = (a_1a_2)^{1/2}.$$

For rigid spheres Lorentz[2] logically took additive radii:

$$\sigma_{ij} = (\sigma_{ii} + \sigma_{jj})/2,\ \text{or}$$
$$b_{ij} = \Sigma_{ij}\bar{N}_i\bar{N}_j(b_{ii}^{1/3} + b_{ij}^{1/3})^3/8. \quad (4\text{--}9)$$

It is customary to refer to Eq. (4–7) as linear combination of b, Eq. (4–9) as Lorentz combination of b, and Eq. (4–8) as quadratic combination of a. Other treatments of the energy of attraction of nonpolar molecules vary the function to be combined quadratically from ab to a/b^2. Quadratic combination of a/b is the current favorite. In other

---

[1]I. Masson and L. G. F. Dolley, *Proc. Roy. Soc. (London)* **103**, 524 (1923).
[2]H. A. Lorentz, *Ann. Phys.* **12**, 127, 660 (1881).

equations of state the parameters that appear in the second virial coefficient are treated as the a or b of van der Waals, or as some combination of them, and those that do not appear in the second virial coefficient are usually combined linearly because that is the simplest way, we do not know how to do better, and these terms are usually not important. For instance, in the Beattie-Bridgeman equation $B_0^{1/3}$ is combined linearly, A_0 and c_0 (or A_0/B_0 and c_0/B_0) are combined quadratically, and a and b are combined linearly.

Beattie and Stockmayer have applied the Beattie-Bridgeman equation to mixtures of methane and n-butane. Their results for the second virial coefficient with $B_0^{1/3}$ combined linearly and with various treatments of the other parameters are given in Table 4–1, in which a represents both A_0 and c_0. The temperatures 150°C and 300°C represent the temperature range of their measurements. The measured values fall just halfway between those for a/B_0 and those for a/B_0^2 combined quadratically. Guggenheim and McGlashan[3] claim that the agreement of these measurements with the equivalent of a/B_0 combined quadratically is excellent if the parameters for the pure components are obtained at the same reduced temperatures instead of at about the same absolute temperatures. They also claim that no analytic equation of state gives sufficiently meaningful parameters for combination.

Table 4-1. Values of the second virial coefficient for mixtures of methane (1) and n-butane (2).[a]

			$-\beta_{12}/RT$	
			150°C	300°C
β_{11}		$0.05587 - 27.75/T - 12.83 \times 10^4/T^3$		
β_{22}		$0.16020 - 152.09/T - 975 \times 10^4/T^3$		
β_{12} (measured)		$0.1102 - 78/T - 54 \times 10^4/T^3$	0.081	0.029
β_{12} (aB_0 quadratic)	$0.09906 -$	$62.04/T - 107 \times 10^4/T^3$	.062	.015
β_{12} (a quadratic)	$0.09906 -$	$64.97/T - 112 \times 10^4/T^3$	.068	.020
β_{12} (a/B_0 quadratic)	$0.09906 -$	$68.02/T - 117 \times 10^4/T^3$	.077	.026
β_{12} (a/B_0^2 quadratic)	$0.09906 -$	$71.23/T - 123 \times 10^4/T^3$	.085	.032

[a]J. A. Beattie and W. H. Stockmayer, *J. Chem. Phys. 10*, 473 (1942).

For Lennard-Jones 6–12 gases we should combine δ linearly and ϵ quadratically, and for square-well-potential gases we should combine δ and $R\delta$ linearly and ϵ quadratically. These combination rules should be expected to be less satisfactory for the higher virial coefficients.

The situation is much more complicated if one or both of the

[3]E. A. Guggenheim and M. L. McGlashan, *Proc. Roy. Soc. (London)* [A] *206*, 448 (1951).

components are polar, so that dipole-dipole and dipole-polarization energies must be taken into account. Both vary as the inverse sixth power of the distance at large distances, but both give larger deviations at smaller distances than do the London energies (Appendix D). If we combine Eqs. (D–19), (D–20), and (D–21), we obtain

$$-r^6(\bar{u}_L + \bar{u}_{DD} + \bar{u}_{DP}) = K[\alpha^{2/3}\alpha' + \alpha_1\alpha'^{2/3}] + 2\mu^2\mu'^2/3kT + (\mu^2\alpha' + \mu'^2\alpha). \tag{4-10}$$

The second term, like the first, might be expected to lead to quadratic combination, but the third does not. However, it is small and is therefore usually ignored. Even so, the sum of the first two terms will not combine quadratically.

Stockmayer[4] treats mixtures of polar gases by using an equation of the type

$$\beta = B - Ae^{D/T^2}/RT, \tag{4-11}$$

which yields the Beattie-Bridgeman equation if the exponential is expanded and all terms after the second are discarded. Then $B_0 = B$, $A_0 = A$, and $c_0 = AD/R$. Stockmayer then splits A and D into non-polar and polar parts: $A = A_n + A_p$, $D = D_n + D_p$, with

$$A_n = \tfrac{1}{2}[A + (A^2 - 0.756\,\mu^4)^{1/2}], \tag{4-12}$$

$$A_p = \tfrac{1}{2}[A - (A^2 - 0.756\,\mu^4)^{1/2}], \tag{4-13}$$

$$D_n = 0.048\,(A_n/RB)^2, \tag{4-14}$$

$$D_p = 13.6\,(\mu^2/B)^2. \tag{4-15}$$

For mixtures, $B^{1/3}$ is combined linearly and A_n, A_p, D_n, D_p are each combined quadratically. The results are much more successful than those for quadratic combination of A and D.

By the use of Eqs. (4–7) and (4–8), or their analog for other parameters, the practical chemist may determine the properties of a gas mixture of any number of components from the properties of the individual components.

4-3.   Chemical Potentials

For convenience we repeat Eq. (3–1):

$$A = \Sigma_i N_i[G_i - RT + RT \ln RT\,N_i/V] + RT\,\Sigma_{ij}\beta_{ij}N_iN_j/V + RT\,\Sigma_{ijk}\gamma_{ijk}N_iN_jN_k/2V^2 + \cdots; \tag{3-1}$$

---

[4] W. H. Stockmayer, *J. Chem. Phys.* **9**, 863 (1941).

also,

$$\bar{G}_k = \left(\frac{\partial A}{\partial N_k}\right)_{T,V,N} = G_k + RT \ln RT\, C_k + 2RT\, \Sigma_i \beta_{ik} C_i$$
$$+ 3RT\, \Sigma_{ij}\gamma_{ijk}C_iC_j/2 + \cdots. \qquad (4\text{-}16)$$

For a three-component system this reduces to

$$\bar{G}_1 = G_1 + RT \ln RTC_1 + 2RT(\beta_{11}C_1 + 3\gamma_{111}C_1^2/2 + \cdots$$
$$+ 2\beta_{12}C_2 + 3\gamma_{112}C_1C_2 + 3\gamma_{122}C_2^2/2 + \cdots$$
$$+ 2\beta_{13}C_3 + 3\gamma_{113}C_1C_3 + 3\gamma_{133}C_3^2/2 + \cdots$$
$$+ 3\gamma_{123}C_2C_3 + \cdots). \qquad (4\text{-}17)$$

We may replace C_1 by $C\bar{N}_1$ and subtract $\bar{G}_{kk}$, the value of $\bar{G}_k$ when $\bar{N}_k = 1$ at the same temperature and total concentration, to give

$$\bar{G}_k - \bar{G}_{kk} = RT \ln \bar{N}_k + 2RTC\Sigma_i(\beta_{ik} - \beta_{kk})\,\bar{N}_i$$
$$+ (3C^2/2)RT\Sigma_{ij}(\gamma_{ijk} - \gamma_{kkk})\,\bar{N}_i\bar{N}_j + \cdots. \qquad (4\text{-}18)$$

It is obvious that a term in Eq. (4-18) that contains only k in the subscript of the virial coefficient must vanish, and one that does not contain k in the subscript does not appear in $\bar{G}_k$.

If we replace C_k in Eq. (4-16) by repeated substitution, we obtain

$$\bar{G}_k = G_k + RT \ln p\bar{N}_k + p\Sigma_{ij}(2\beta_{ik} - \beta_{ij})\bar{N}_i\bar{N}_j$$
$$+ (p^2/RT)[\Sigma_{ij}(\tfrac{3}{2}\gamma_{ijk} - \gamma_{ijj})\bar{N}_i\bar{N}_j\bar{N}_j$$
$$- \Sigma_{ij}(\beta_{ij}\bar{N}_i\bar{N}_j)(\Sigma_{ij}(2\beta_{ik} - \tfrac{3}{2}\beta_{ij})\bar{N}_i\bar{N}_j] + \cdots. \qquad (4\text{-}19)$$

Subtracting the value of $\bar{G}_k$ when $\bar{N}_k = 1$ at the same temperature and pressure gives

$$\bar{G}_k^M = RT \ln \bar{N}_k + p\,\Sigma_{ij}(2\beta_{ik} - \beta_{ij} - \beta_{kk})\,\bar{N}_i\bar{N}_j + \cdots. \qquad (4\text{-}20)$$

It is well worth noting the differences between Eq. (4-20) and Eq. (4-18) chopped off at the β term:

$$\bar{G}_k - \bar{G}_{kk} = RT \ln \bar{N}_k + 2\,RTC\,\Sigma_i(\beta_{ik} - \beta_{kk})\,\bar{N}_i + \cdots. \qquad (4\text{-}21)$$

As Eq. (4-20) is derived above, the differences appear to arise from the complexity of the substitution. They are, however, much more fundamental. The corresponding equations for the free energy and for the work content are much more alike:

$$G = \Sigma_i N_i[G_i + RT \ln p\bar{N}_i + p\Sigma_j\beta_{ij}\bar{N}_j] + \cdots, \qquad (4\text{-}22)$$

$$A = \Sigma_i N_i[G_i - RT + RT \ln RTC\bar{N}_i + RTC\Sigma_j\beta_{ij}\bar{N}_j] + \cdots$$
$$(\ = \Sigma_i N_i[G_i - RT + RT \ln RT\, N_i/V + RT\,\Sigma_j\beta_{ij}N_j/V] + \cdots). \qquad (4\text{-}23)$$

In the differentiation with respect to N_k to obtain $\bar{G}_k$, p is held constant in Eq. (4-22), but it is V in the second form, not C in the first form, that is kept constant in Eq. (4–23). It is possible to add component k at constant volume without changing the concentration of any other component, but it is not possible to add component k at constant pressure without changing the mole fraction, and therefore the partial pressure, of every other component.

For a three-component system,

$$\bar{G}_1 - \bar{G}_{11} = RT \ln \bar{N}_1 + 2RTC[(\beta_{12} - \beta_{11})\bar{N}_2 + (\beta_{13} - \beta_{11})\bar{N}_3].$$
$$(4\text{–}24)$$

For the change on mixing at total constant pressure we make use of B^0_{ij} defined in Eq. (4–6) to obtain

$$\bar{G}^M_1 = RT \ln \bar{N}_1 + p[B^0_{12}\bar{N}^2_2 + B^0_{13}\bar{N}^2_3 - (B^0_{23} - B^0_{12} - B^0_{13})\bar{N}_2\,\bar{N}_3]$$
$$= RT \ln \bar{N}_1 + p[B^0_{12}\bar{N}_2(1 - \bar{N}_1) + B^0_{13}\bar{N}_3(1 - \bar{N}_1) - B^0_{23}\bar{N}_2\bar{N}_3].$$
$$(4\text{–}25)$$

The term in Eq. (4–20) in which both i and j are 1 obviously vanishes; that in which i is 1 and j is 2 and that in which i is 2 and j is 1 combine to give

$$2\beta_{11} - \beta_{12} - \beta_{11} + 2\beta_{12} - \beta_{12} - \beta_{11} = 0.$$

4–4. Physical Equilibria

The usual statements of Dalton's law have little relation to the problem that interested Dalton particularly, which was the effect of foreign gases upon the equilibrium between liquid water and water vapor. One of his statements of his law is that a gas behaves as a vacuum to another gas. Gibbs[5] interprets this as equivalent to saying that the potential of any component depends only upon the temperature and its own concentration, but not upon the concentrations of other gases, and notes that this does not depend upon the ideal gas laws, which Dalton also assumed, but that it may be consistent with any equation of state. Gillespie[6] called this the Gibbs-Dalton rule.

The assumption that gases should behave as ideal solutions, or that $\bar{G}^M_k = RT \ln \bar{N}_k$ is called the Lewis and Randall rule, since they were the first to use it for gases that were not assumed ideal.

We may readily compare these two rules with the results of combining constants of the van der Waals equation by the van der Waals method as far as the β term, which we may take as sufficient indication of the behavior of real gases. The Gibbs-Dalton rule assumes that β_{12} is zero, whereas the Lewis and Randall rule assumes that it is

[5]J.W. Gibbs, *Collected Works*, Vol. 1, p. 157.
[6]L. J. Gillespie, *Phys. Rev. 36*, 121 (1930).

$\frac{1}{2}(\beta_{11} + \beta_{22})$. The difference of the van der Waals β_{12} from the Gibbs-Dalton β_{12} is $\frac{1}{2}(b_{11} + b_{22}) - (a_{11}a_{22})^{1/2}/RT$, which is negative except at high temperatures. The difference of the van der Waals β_{12} from the Lewis and Randall β_{12} is $\frac{1}{2}(a_{11}^{1/2} - a_{22}^{1/2})^2/RT$, which is always positive. At low temperatures, where the correction is important, it is much smaller than the Gibbs-Dalton difference.

The effect of another gas on the vapor pressure of slightly volatile solids, such as iodine, or of mixtures of a pair of hydrates or amines may be determined experimentally,[7,8,9] and the change in $\bar{G}_k$ may be calculated from the volume of the solid phase or phases. For slightly volatile liquids a further correction, which is usually negligible for solids, must be made for the effect of the gas dissolved in the liquid on the vapor pressure of the liquid. Except for these corrections, the solid, the pair of solids, or the liquids act essentially as a membrane permeable only to one gas.

The Gibbs-Dalton rule says that the concentration C_1 will be independent of C_2. The Lewis and Randall rule says that the partial pressure will usually increase, but that the effect will be independent of the second component. The usually accepted rules for combining constants say that the concentration and the pressure will both usually increase and that the increases will be greater the higher is the critical temperature of the second component. For iodine at room temperature, the partial pressure decreases with added hydrogen, increases with air, and increases much more with carbon dioxide. The results have been used to calculate the Beattie-Bridgeman constants for iodine that are given in Table 3–3.[10]

The vapor pressure of a component of a liquid or solid solution is often measured to determine the potential of the component in the condensed phase. We shall discuss this use in Chapter 6.

4–5. Chemical Equilibria

The treatment of chemical equilibria is based on Gibbs's equation,

$$\Sigma_i \nu_i \bar{G}_i = 0, \tag{1–1}$$

which involves treating each species as a component, with an equation of condition for each chemical reaction (see Sec. 2–3). Substitution of $\bar{G}_i$ from Eq. (4–16) gives

---

[7] H. Braune and F. Strassmann, *Z. phys. Chem. 143A*, 225 (1929).

[8] G. P. Baxter and M. R. Grose, *J. Am. Chem. Soc. 37*, 1061 (1915).

[9] L. J. Gillespie and E. Lurie, *J. Am. Chem. Soc. 53*, 2978 (1931).

[10] H. T. Gerry and L. J. Gillespie, *Phys. Rev. 40*, 269 (1932).

$$\ln K_C = \Sigma_i \nu_i \ln C'_i$$
$$= \ln K_{C_0} - \Sigma_i \nu_i (2\Sigma_j \beta_{ij} C'_j + 3\Sigma_{jk}\gamma_{ijk} C'_j C'_k / 2$$
$$+ 4\Sigma_{jkl}\delta_{ijkl} C'_j C'_k C'_l / 3 + \cdots)$$
$$= \ln K_{C_0} - \Sigma_i \nu_i (2C'\Sigma_j \beta_{ij} \bar{N}'_j + 3C'^2 \Sigma_{jk}\gamma_{ijk} \bar{N}'_j \bar{N}'_k / 2 + \cdots),$$

in which $\bar{N}'_i$, like C'_i, refers to the species i, and   (4–26)

$$\ln K_{C_0} = - \Sigma_i \nu_i (G_i / RT + \ln RT). \tag{4–27}$$

Substitution of $\bar{G}_i$ from Eq. (4–19) gives

$$\ln K_p = \Sigma_i \nu_i \ln p \bar{N}'_i$$
$$= \ln K_{p_0} - \Sigma_i \nu_i \{\Sigma_{jk}(\beta_{ij} - \beta_{jk}) \bar{N}'_j \bar{N}'_k p / RT$$
$$+ [\Sigma_{jkl}(3\gamma_{ijk}/2 - \gamma_{jkl}) \bar{N}'_j \bar{N}'_k \bar{N}'_l$$
$$- (\Sigma_{jk}\beta_{jk} \bar{N}'_j \bar{N}'_k)(\Sigma_{jk}(2\beta_{ij} - 3\beta_{jk}/2) \bar{N}'_j \bar{N}'_k](p/RT)^2\}, \tag{4–28}$$

$$\ln K_{p_0} = - \Sigma_i \nu_i G_i / RT. \tag{4–29}$$

The quantity desired by the practical chemist is the yield, or extent of the reaction, which is more closely related to

$$\Sigma_i \nu_i \ln \bar{N}'_i = \ln K_C - \Sigma_i \nu_i \ln C'$$
$$= \ln K_p - \Sigma_i \nu_i \ln P. \tag{4–30}$$

Since the composition depends upon the extent of the reaction, it must be determined by successive approximation. It seems, therefore, that the most convenient method would be to work with Eqs. (4–26) and (4–30) and then to determine the pressure from Eq. (3–2), even though the practical variables are temperature, composition, and pressure. This is not the customary procedure, however.

The interaction coefficients must usually be determined from the properties of the pure components by some combination rules. Unless the reaction proceeds only in the presence of a catalyst, it may be impossible to isolate some of the species to determine their coefficients. We can see, however, that there will be compensation between the effects on the products and those on the reactants. The numbers of atoms and of electrons are unchanged by the reaction, so that the changes in interaction should come almost entirely from changes in polarity, particularly from changes in hydrogen bonding. So the equilibrium constants obey the laws of species-ideal gases much better than do the fugacities of the individual components. Much of the success of the law of mass action depends upon this compensation.

There are very few measurements of chemical equilibria in gases under high pressure. The ammonia synthesis,

$$\tfrac{3}{2}H_2 + \tfrac{1}{2}N_2 = NH_3,$$

has been studied at high pressures because it is economically very important and because the yield is much increased with increasing pressure. Most of this increase is due to the decrease in the total number of moles by 1 for each mole of ammonia formed, but there is an additional gain due to the increase in K_p with increasing pressure caused by the high polarity of the ammonia.

Many of the experiments were carried out with an original mixture of hydrogen, nitrogen, and argon in the ratios 762:235:3. If x moles of ammonia are formed, the mole fractions are

$$\bar{N}'_{H_2} = (762 - 1.5x)/(1000 - x), \quad \bar{N}'_{N_2} = (235 - 0.5x)/(1000 - x),$$
$$\bar{N}'_{NH_3} = x/(1000 - x), \quad \bar{N}'_A = 3/(1000 - x).$$

From the constants of the Beattie-Bridgeman equation for the individual gases, Gillespie and Beattie calculate

$$(d \log K_p/dp)_{T, N_0} = \frac{0.119849}{T} + \frac{91.87212}{T^2} + \frac{25122730}{T^4}, \quad (4-31)$$

in which N_0 is the initial composition, given above, and p is in atmospheres. Although this expression corresponds only to the second virial coefficient, it agrees excellently with the experimental measurements throughout the range of the latter, which varied from 325 to 950°C for 30 atm. pressure, and from 10 to 1000 atm. at about 500°C.

5. Liquid Mixtures. Mathematical Relations

5-1. Extension of Equations for Gases to Liquid Mixtures

It is possible, at least in principle, to add enough terms to Eq. (3–1) that it may include both gaseous and liquid states. We are usually less ambitious and are satisfied with a comparison of a liquid mixture with its unmixed components as liquids at the same temperature and pressure, or with infinitely dilute liquid solutions in one of them. Therefore, we use Eq. (3–44) for the Gibbs free energy, but with $\beta, \gamma, \ldots$ expanded:

$$G = \Sigma_i N_i[G_i + RT \ln RTN_i/V + 2\,RT\,\Sigma_j\beta_{ij}N_j/V \\ + 3\,RT\Sigma_{jk}\gamma_{ijk}N_jN_k/2V^2 + 4\,RT\,\Sigma_{jkl}\delta_{ijkl}N_jN_kN_l/3V^3 + \cdots].$$

(5–1)

For the unmixed components (see p. 188),

$$G^0 = \Sigma_i N_i[G_i + RT \ln RT/\bar{V}_i^i + 2\,RT\beta_{ii}/\bar{V}_i^i \\ + 3RT\,\gamma_{iii}/2\bar{V}_i^{i2} + 4RT\delta_{iii}/3\bar{V}_i^{i3} + \cdots],$$

(5–2)

$$G^M = G - G^0 = RT\,\Sigma_i N_i[\ln N_i\bar{V}_i^i/V + 2(\Sigma_j\beta_{ij}N_j/V - \beta_{ii}/\bar{V}_i^i) \\ + 3(\Sigma_{jk}\gamma_{ijk}N_jN_k/V^2 - \gamma_{iii}/\bar{V}_i^{i2})/2 \\ + 4(\Sigma_{jkl}\delta_{ijkl}N_jN_kN_l/V^3 - \delta_{iiii}/\bar{V}_i^{i3})/3].$$

(5–3)

The volumes of liquid mixtures are nearly additive, and there is a great practical advantage in replacing the volume V by the volume of the components, $V_0 = \Sigma_j N_j\bar{V}_j^j$. If the temperature or the pressure is changed, the $\bar{V}_j^j$ may be measured at a standard temperature and pressure. This use of V_0 involves the expansions of $\ln V_0/V$, $(1/V_0 - 1/V)$, and so on, and will change slightly the coefficients of the β's and higher terms. We shall then use the volume fraction $\phi_i = N_i\bar{V}_i^i/V_0$ and the definitions $\beta'_{ij} = \beta_{ij}/\bar{V}_i^i\bar{V}_j^j$, $\gamma_{ijk} = \gamma'_{ijk}/\bar{V}_i^i\bar{V}_j^j\bar{V}_k^k$, corrected for the expansions discussed above. This gives

$$G^M = RT \, \Sigma_i N_i \ln \phi_i + RT \, V_0[2\Sigma_{ij}(\beta'_{ij} - \beta'_{ii}) \, \phi_i\phi_j$$
$$+ 3\Sigma_{ijk}(\gamma'_{ijk} - \gamma'_{iii}) \, \phi_i\phi_j\phi_k/2$$
$$+ 4\Sigma_{ijkl}(\delta'_{ijkl} - \delta'_{iiii}) \, \phi_i\phi_j\phi_k\phi_l/3 + \cdots]. \tag{5-4}$$

At constant pressure the coefficients β, γ, ... are degenerate and the sum may be replaced by the highest term with new coefficients obtained by multiplying each lower term by the appropriate power of $\Sigma_i\phi_i = 1$, and adding all the terms. Thus

$$\beta_{ij}\phi_i\phi_j = \beta_{ij}\phi_i^2\phi_j + \beta_{ij}\phi_i\phi_j^2 + \beta_{ij}\phi_i\phi_j\phi_k + \cdots.$$

For binary mixtures covering a large range of composition it is more convenient to expand around the midpoint to give

$$G^M = RT(N_i \ln \phi_i + N_j \ln \phi_j) + V_0\phi_i\phi_j \sum_{\nu=0}^{n} G^\nu_{ij}(\phi_i - \phi_j)^\nu. \tag{5-5}$$

For more complex mixtures we add $RT \, N_i \ln \phi_i$ for each component,

$V_0\phi_i\phi_j \sum_{\nu=0}^{n} G^\nu_{ij}(\phi_i - \phi_j)^\nu$ for each binary mixture $(i \neq j)$, and terms

like those in Eq. (5–4) involving γ_{ijk}, with all three indices different, for each ternary mixture, and so on.

5-2. Flory-Huggins and Ideal Entropy

If the $\bar{V}^j_i$ are measured at a standard temperature and pressure, the first term in Eq. (5–4) is proportional to the absolute temperature, and therefore arises from the entropy of mixing. This entropy term was independently derived by Huggins and by Flory and is known as the Flory-Huggins entropy:[1,2]

$$-S^{FH}/R = G^{FH}/RT = \Sigma_i N_i \ln \phi_i = \Sigma_i N_i(\ln \bar{N}_i + \ln \Sigma_j\phi_j\bar{V}^i_i/\bar{V}^i_j)$$
$$= \Sigma_i N_i(\ln \bar{N}_i + \ln (\Sigma_j\bar{N}_j\bar{V}^i_i/\Sigma_j\bar{N}_jV^i_j). \tag{5-6}$$

If the components have equal volumes, this reduces to the ideal free energy G^I, and

$$G^I/RT = -S^I/R = \Sigma_i N_i \ln \bar{N}_i. \tag{5-7}$$

The corresponding potentials are

[1]M. L. Huggins, *J. Chem. Phys.* **9**, 440 (1941); *J. Phys. Chem.* **46**, 151 (1942); *Ann. New York Acad. Sci.* **41**, 11 (1942); *J. Am. Chem. Soc.* **64**, 1712 (1942);
[2]P. J. Flory, *J. Chem. Phys.* **9**, 660 (1941); *10*, 51 (1942); *13*, 453 (1945).

$$\bar{G}_k^{\mathrm{I}}/RT = \ln \bar{N}_k,$$

$$\begin{aligned}
\bar{G}_k^{\mathrm{FH}}/RT &= \ln \phi_k + \Sigma_i \phi_i (\bar{V}_i^i - \bar{V}_k^k)/\bar{V}_i^i \\
&= \ln \phi_k + (1 - \bar{V}_k^k/\bar{V}_0) \\
&= \ln \bar{N}_k + (1 - \bar{V}_k^k/\bar{V}_0) + \ln \bar{V}_k^k/\bar{V}_0 \\
&= \ln \bar{N}_k + (1 - \bar{V}_k^k/\bar{V}_0) + \ln [1 - (1 - \bar{V}_k^k/\bar{V}_0)] \\
&= \ln \bar{N}_k - (1 - \bar{V}_k^k/\bar{V}_0)^2/2 + \cdots.
\end{aligned} \qquad (5\text{--}8)$$

As ϕ_k, or $\bar{N}_k$, approaches unity, $(1 - \bar{V}_k^k/\bar{V}_0)$ approaches zero, so $\bar{G}_k^{\mathrm{FH}}/RT$ approaches $\ln \bar{N}_k$ much more rapidly than it does $\ln \phi_k$. In other words, a Flory-Huggins solution obeys Raoult's law that $d\bar{G}_k/d\bar{G}_k^{\mathrm{I}}$ approaches unity as $\bar{N}_k$ approaches unity.

If one solvent is in excess, we shall usually call it component S, but sometimes component 1. If we take for the standard state of reference an infinitely dilute solution in S, we shall use the partial molal volumes in the standard state, $\bar{V}_i^S$. For the solvent, $\bar{G}_S^{\mathrm{FHS}}$ is still given by Eq. (5–8), but the potential of a solute becomes

$$\bar{G}_k^{\mathrm{FHS}}/RT = \ln \phi_k + \bar{V}_k^S (\bar{V}_0 - \bar{V}_S^S)/\bar{V}_0 \bar{V}_S^S \qquad (5\text{--}9)$$

$$= \ln \bar{N}_k + \bar{V}_k^S (\bar{V}_0 - \bar{V}_S^S)/\bar{V}_0 \bar{V}_S + \ln \bar{V}_k^S/\bar{V}_0. \qquad (5\text{--}10)$$

Both the volume-fraction and the mole-fraction methods have the same disadvantage as the partial-pressure treatment of gases. The addition of any component changes the fraction of every other component. This may be avoided, as it is avoided in the concentration treatment of gases, by expressing compositions in moles per unit mass of solvent $N_i/N_S\bar{W}_S$ or per mole of solvent. If $\bar{W}_S$ is in kilograms per mole, $N_i/N_S\bar{W}_S = m_i$. The convenience of the method decreases as the fraction of the solvent decreases, but not much until that fraction becomes very small.

We take for a solute

$$\bar{G}_k^{is}/RT = \ln (N_k/N_S\bar{W}_S) = \ln m_k. \qquad (5\text{--}11)$$

By the Gibbs-Duhem relation, Eq. (C–25),

$$\bar{G}_S^{is}/RT = 1 - \Sigma_i N_i/N_S = - \bar{W}_S \Sigma_{i \neq S} m_i = - \bar{W}_S M \qquad (5\text{--}12)$$

and

$$\begin{aligned}
G^{is}/RT &= \sum_{i \neq S} N_i (\ln N_i/N_S\bar{W}_S - 1) \\
&= \sum_{i \neq S} N_i (\ln m_i - 1).
\end{aligned} \qquad (5\text{--}13)$$

The chemical potential of a solute is the same as that of a perfect gas in a volume $N_S\bar{W}_S$, and the potential of unit mass of solvent $\bar{G}_S^{is}/\bar{W}_S$ is the same as the total gas pressure. From Eqs. (5–8), (5–11), and (5–12), if $M = \Sigma_{i \neq S} m_i$,

$$\bar{G}_k^{\mathrm{I}}/RT = \bar{G}_k^{is}/RT - \ln (1 + \bar{W}_S M) + \ln \bar{W}_S,$$

or

$$\bar{G}^{Is}_k/RT = \bar{G}^{is}_k/RT - \ln (1 + \bar{W}_sM), \qquad (5\text{--}14)$$

since $\ln \bar{W}_S$ persists in the limit. Then

$$\bar{G}^{1}_S/RT = -\ln (1 + \bar{W}_sM) = \bar{G}^{is}_S/RT + \bar{W}_sM - \ln (1 + \bar{W}_sM). \qquad (5\text{--}15)$$

Similarly, from Eqs. (5–10), (5–11), and (5–12),

$$\bar{G}^{\mathrm{FHs}}_k/RT = \bar{G}^{is}_k/RT + \frac{\bar{V}^S_k}{\bar{V}^S_S}\frac{\bar{W}_s\Sigma_j m_j(1 - \bar{V}^S_j/\bar{V}^S_S)}{1 + \bar{W}_s\Sigma_j m_j\bar{V}^S_j/\bar{V}^S_S}$$
$$- \ln (1 + \bar{W}_s\Sigma_j m_j\bar{V}^S_j/\bar{V}^S_S) \qquad (5\text{--}16)$$

and

$$\bar{G}^{\mathrm{FH}}_S/RT = \bar{G}^{is}_S/RT + \frac{(1 + \bar{W}_sM)\bar{W}_s\Sigma_j m_j\bar{V}^S_j/\bar{V}^S_S}{1 + \bar{W}_s\Sigma_j m_j\bar{V}^S_j/\bar{V}^S_S}$$
$$- \ln (1 + \bar{W}_s\Sigma_j m_j\bar{V}^S_j/\bar{V}^S_S). \qquad (5\text{--}17)$$

The corresponding osmotic coefficients are

$$\phi^I = 1 - \bar{W}_sM/2 + \cdots, \qquad (5\text{--}18)$$
$$\phi^{\mathrm{FH}} = 1 - (\bar{W}_s\Sigma_j m_j\bar{V}^S_j/\bar{V}^S_S)[\Sigma_j m_j(1 - \bar{V}^S_j/2\bar{V}^S_S)]/M + \cdots. \qquad (5\text{--}19)$$

For a single solute the Flory-Huggins solution behaves the same as an ideal solution if $\bar{V}^S_k/\bar{V}^S_S = 1$. For all other ratios the second virial coefficient of a Flory-Huggins solution is less negative than that of an ideal solution, and for $\bar{V}^S_k/\bar{V}^S_S > 2$ it becomes positive, so that the molality treatment represents a Flory-Huggins solution more accurately than does the ideal-solution treatment if $\bar{V}^S_k/\bar{V}^S_S \geq 2$. Although we believe that the Flory-Huggins relation gives better expressions for the entropy and free energy of solutions that have no enthalpy or volume change on mixing at constant pressure, we shall often use the molality treatment for dilute solutions and the mole-fraction treatment for concentrated solutions, throwing the deviations into the other terms. We call these others the enthalpy terms because the enthalpy must come entirely from them, although they need not come entirely from the enthalpy.

5-3.   The Enthalpy Terms

If we subtract the entropy term in Eq. (5–5) and divide by V_0, we obtain

$$[G^M - RT(N_i \ln \phi_i + N_j \ln \phi_j)]/V_0 = \phi_i\phi_j \sum_{\nu=0}^{n} G^\nu_{ij}(\phi_i - \phi_j)^\nu. \qquad (5\text{--}20)$$

Similarly for the enthalpy change of mixing per unit volume,

$$H^M/V_0 = \phi_i\phi_j \sum_{\nu=0}^{n} H^{\nu}_{ij}(\phi_i - \phi_j)^{\nu}. \qquad (5\text{-}21)$$

We so often have similar expressions for quantities per mole in terms of mole fractions with the ideal entropy terms that we define excess free energy and so on by relations such as

$$\bar{G}^E = \bar{G}^M - RT(\bar{N}_i \ln \bar{N}_i + \bar{N}_j \ln \bar{N}_j) = \bar{N}_i\bar{N}_j \sum_{\nu=0}^{n} G^{\prime\nu}_{ij}(\bar{N}_i - \bar{N}_j)^{\nu},$$
$$(5\text{-}22)$$

$$\bar{H}^E = \bar{H}^M = \bar{N}_i\bar{N}_j \sum_{\nu=0}^{n} H^{\prime\nu}_{ij}(\bar{N}_i - \bar{N}_j)^{\nu}.$$
$$(5\text{-}23)$$

Other forms have also been used. Van Laar,[3] following van der Waals, treated a liquid as a van der Waals fluid condensed to a volume b, and obtained the equivalent of

$$G^E/\Sigma_i N_i b_{ii} = \Sigma_{ij}\beta_{ij}N_i b_{ii}N_j b_{jj}/(\Sigma_{\iota}N_{\iota}b_{\iota\iota})^2, \qquad (5\text{-}24)$$

with no higher terms. Langmuir[4] treated a liquid as a surface and assumed that the surface area of a molecule is proportional to the two-thirds power of the volume, to obtain the equivalent of

$$G^E/\Sigma_i N_i \bar{V}^{i2/3} = \Sigma_{ij}\beta_{ij}N_i\bar{V}_i^{i2/3}N_j\bar{V}_j^{j2/3}/(\Sigma_{\iota}N_{\iota}\bar{V}_{\iota}^{i2/3})^2, \qquad (5\text{-}25)$$

and Heitler[5] treated a liquid as a cubic crystal lattice and obtained the equivalent of

$$G^E/\Sigma_i N_i = \Sigma_{ij}\beta_{ij}N_iN_j/(\Sigma_{\iota}N_{\iota})^2 = \Sigma_{ij}\beta_{ij}\bar{N}_i\bar{N}_j. \qquad (5\text{-}26)$$

I reviewed their results from a more general point of view,[6] and concluded that volumes were the best units and that a single term might be sufficient for nonpolar mixtures

$$\bar{G}^E/\Sigma_i N_i\bar{V}_i^i = \Sigma_{ij}\beta_{ij}N_i\bar{V}_i^iN_j\bar{V}_j^j/(\Sigma_{\iota}N_{\iota}\bar{V}_{\iota}^i)^2. \qquad (5\text{-}27)$$

The β's have different values in each of Eqs. (5–24) to (5–27). Obviously, each of us assumed the ideal-solution form for the first term.

In Heitler's treatment, each molecule occupies one site in the crystal lattice. Huggins and Flory also used a crystal model in which each molecule of one component occupies one lattice site, but each molecule of the other component occupies a series of sites, as a string of beads

---

[3] J. J. van Laar, *Z. phys. Chem.* **72**, 723 (1910); **83**, 599 (1913).

[4] *Third Colloid Symposium Monograph* (Chemical Catalogue Company. New York, 1925, p. 3.

[5] W. Heitler, *Ann. Phys.*[4] *80*, 630 (1926).

[6] G. Scatchard, *Chem. Rev. 8*, 321 (1931).

would if each bead occupied a site and there were no unoccupied sites between adjacent beads in the chain. For high polymers each molecule occupies a very large number of sites. Eq. (5–6) is an approximation of their answers, which depend somewhat upon the number of contacts between adjacent sites. The same approximation gives the form of Eq. (5–27) for the enthalpy term, although this term is not $\bar{G}^E/\Sigma_i N_i \bar{V}_{ii}$. This treatment is discussed elegantly by Guggenheim[7].

A difficulty with any lattice model of liquids is that a pair of real liquids does not usually have commensurable lattices. A generalization of the treatment would still make the number of contacts the important measure. For nonpolar liquids these should be approximately proportional to the surface area, but not to the two-thirds power of the volume. For chains of beads the surface area is nearly proportional to the volume itself. This applies to chain polymers and to aliphatic carbon chains, and more closely than the two-thirds power to most molecules. We may expect deviations with mixtures of molecules with incommensurable lattices because the number of contacts any molecule makes with other molecules will depend upon the nature of the molecules around it. We shall expect more difficulty with polar molecules, which have very different energies for different points of contact. We shall require good evidence to believe that any other expression fits better than the volume fraction, but we shall continue to use sometimes the mole-fraction or molality forms for convenience. Unless Eq. (5–5) can be expressed with only one or two G^ν_{ij}, it makes little difference what units are used.

The form of this expression is so useful for mixtures that we shall consider it in detail for Eq. (5–21). The form will be the same for any extensive property if the fraction units correspond to the denominator on the left:

$$H^M/V_0 = \phi_1 \phi_2 \sum_{\nu=0}^{n} H^\nu_{12}(\phi_1 - \phi_2)^\nu = \phi_1(1 - \phi_1) \sum_{\nu=0}^{n} H^\nu_{12}(2\phi_1 - 1)^\nu,$$

$$(5\text{–}21)$$

$$d(H^M/V_0)/d\phi_1 = \bar{H}^M_1/\bar{V}^1_1 - \bar{H}^M_2/\bar{V}^2_2 = \sum_{\nu=0}^{n} H^\nu_{12}[- (2\phi_1 - 1)^{\nu+1}$$

$$+ 2\nu\phi_1(1 - \phi_1)(2\phi_1 - 1)^{\nu-1}],\qquad (5\text{–}28)$$

$$H^M/V_0 + (1 - \phi_1)\, d(H^M/V_0)/d\phi_1 = \bar{H}^M_1/\bar{V}^1_1$$

$$= (1 - \phi_1)^2 \sum_{\nu=0}^{n} H^\nu_{12}[(2\phi_1 - 1)^\nu + 2\nu\phi_1(2\phi_1 - 1)^{\nu-1}],$$

$$(5\text{–}29)$$

---

[7] E.A. Guggenheim, *Proc. Roy. Soc. (London)* A183, 203, 213 (1944); *Trans. Faraday Soc. 41*, 107 (1945).

$$H^M/V_0 - \phi_1 d(H^M/V_0)/d\phi_1 = \bar{H}_2^M/\bar{V}_2^2$$

$$= \phi_1^2 \sum_{\nu=0}^{n} H_{12}^\nu [(2\phi_1 - 1)^\nu - 2\nu(1 - \phi_1)(2\phi_1 - 1)^{\nu-1}].$$

$$(5\text{–}30)$$

The coefficients of H_{12}^ν for the first five terms are given in Fig. 5–1 for H^M/V_0, in Fig. 5–2 for $\bar{H}_1^M/\bar{V}_1^1$ and $\bar{H}_2^M/\bar{V}_2^2$, and in Fig. 5–3 for $\bar{H}_1^M/\bar{V}_1^1 - \bar{H}_2^M/\bar{V}_2^2$.

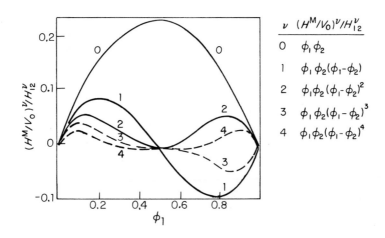

Figure 5–1. Expansion of H^M/V_0.

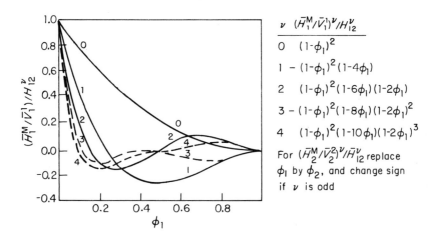

Figure 5–2. Expansion of $\bar{H}_1^M/\bar{V}_1^1$.

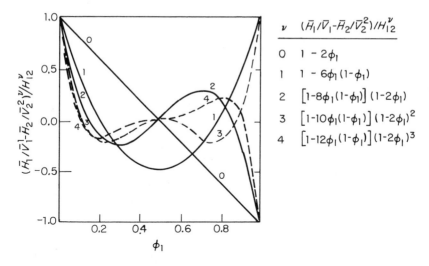

Figure 5–3. Expansion of $\bar{H}_1/\bar{V}_1^1 - \bar{H}_2/\bar{V}_2^2$.

The justification for the first equalities of Eqs. (5–28) to (5–30) is

$$\frac{d(H^M/V_0)}{d\phi_1} = \frac{d[H^M/(N_1\bar{V}_1^1 + N_2\bar{V}_2^2)]/d(N_1\bar{V}_1^1)}{d[N_1\bar{V}_1^1/(N_1\bar{V}_1^1 + N_2\bar{V}_2^2)]/d(N_1\bar{V}_1^1)}$$

$$= (\bar{H}_1^M/\bar{V}_1^1 - H^M/V_0)/(1 - \phi_1), \qquad (5\text{–}31)$$

$$\bar{H}_1^M/\bar{V}_1^1 = H^M/V_0 + (1 - \phi_1)\, d(H^M/V_0)/d\phi_1, \qquad (5\text{–}32)$$

$$\bar{H}_2^M/\bar{V}_2^2 = H^M/V_0 - \phi_1\, d(H^M/V_0)d\phi_1, \qquad (5\text{–}33)$$

$$\bar{H}_1^M/\bar{V}_1^1 - \bar{H}_2^M/\bar{V}_2^2 = d(H^M/V_0)/d\phi_1. \qquad (5\text{–}34)$$

The geometric representation of these equations is shown in Fig. 5–4.

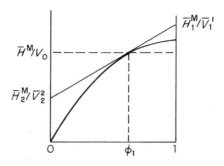

Figure 5–4. Derivation of partial quantities.

When we have an analytic expression we may determine any H_{12}^{ν} from the value at $\phi_1 = 0.5$ of the corresponding derivative $d^{\nu}(H^M/V_0\phi_1\phi_2)/d\phi_1^{\nu}$ by the relations

$$H_{12}^0 = 4(H^M/V_0)_{0.5},$$

$$H_{12}^{\nu} = \frac{1}{\nu!}\ [d^{\nu}(H^M/V_0\phi_1\phi_2)/d\phi_1^{\nu}]_{0.5}. \tag{5-35}$$

From the slopes of the curve at $\phi_1 = 0$ and $\phi_1 = 1$ we may determine the sum of the terms with ν even and the sum with ν odd. The terminal slopes are

$$\frac{d(H^M/V_0)}{d\phi_1} = \begin{cases} -\ \Sigma H_{12}^{\nu} & \text{when } \phi_1 = 1, \\ \Sigma H_{12}^{\nu}(-1)^{\nu} & \text{when } \phi_1 = 0. \end{cases} \tag{5-36}$$

The sum of these two terminal slopes is seen to give

$$\Sigma H_{12}^{\nu}[(-1)^{\nu} - 1] = -\ 2\Sigma_{odd} H_{12}^{\nu} \tag{5-37}$$

and the difference between them is

$$\Sigma H_{12}^{\nu}[(-1)^{\nu} + 1] = 2\Sigma_{even} H_{12}^{\nu}. \tag{5-38}$$

We may also use a number of values of $(H^M/V_0) - 4\phi_1(1 - \phi_1)$ $(H^M/V_0)_{1/2}$ equal to the number of parameters desired, just as we would with a smooth curve through measured points. For a finite number of terms, no two of these three methods are exactly consistent.

The application of this expansion to volumes is shown in Fig. 5-5, where the molal volume is plotted against the mole fraction. For benzene-cyclohexane, the molecular weights differ only by 8 percent,

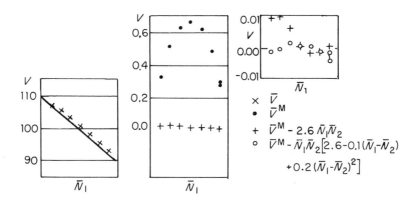

Figure 5-5. Volumes of benzene (1)–cyclohexane (2) mixtures at 30°C.

so the curves of specific volume against weight fraction would be but little different.

For dilute solutions we would probably prefer a virial-type expression, such as

$$\tilde{V} = \tilde{V}^0 + \tilde{V}'\tilde{W}_2 + \tilde{V}''\tilde{W}_2^2 + \tilde{V}'''\tilde{W}_2^3 + \cdots, \qquad (5\text{-}39)$$

$$\frac{d\tilde{V}}{d\tilde{W}_2} = \tilde{V}' + 2\tilde{V}''\tilde{W}_2 + 3\tilde{V}'''\tilde{W}_2^2 + \cdots, \qquad (5\text{-}40)$$

and, analogously to Eqs. (5–32) and (5–33),

$$\tilde{V}_1 = \tilde{V} - \tilde{W}_2 \frac{d\tilde{V}}{d\tilde{W}_2} = \tilde{V}^0 - \tilde{V}''\tilde{W}_2^2 - 2\tilde{V}'''\tilde{W}_2^3 - \cdots, \qquad (5\text{-}41)$$

$$\tilde{V}_2 = \tilde{V} + (1 - \tilde{W}_2)\frac{d\tilde{V}}{d\tilde{W}_2} = \tilde{V}^0 + \tilde{V}' + 2\tilde{V}''\tilde{W}_2$$
$$- (\tilde{V}'' - 3\tilde{V}''')\tilde{W}_2^2 - \cdots. \qquad (5\text{-}42)$$

5-4. Electrolyte Solutions

We shall treat electrolyte solutions only by virial-type expansions for dilute solutions. We cannot calculate the second virial coefficient of two ions by Eq. (3–8) because the total energy of the central ion with all the ions of type i in a spherical shell of thickness dr is proportional to r, and therefore diverges. Classical statistical mechanics is unable to take advantage of the fact that the sum for ions of all types does converge in an electrically neutral solution. Debye and Hückel[8] solved this problem by unorthodox methods. Their model is a structureless solvent medium of uniform dielectric constant D_0, in which is dissolved a set of ions that are rigid nonpolarizable spheres with spherically symmetric charge and dielectric constant.

The Debye-Hückel limiting laws for activity coefficient and osmotic coefficient are

$$(\ln \gamma_k)_L = - z_k^2 A(I/2)^{1/2}, \qquad (5\text{-}43)$$

$$(\phi - 1)_L = - (I/M)(A/3)(I/2)^{1/2}, \qquad (5\text{-}44)$$

in which z_k is the valence of the kth ion $I = \Sigma_i m_i z_i^2$, so $I/2$ is the ionic strength as defined by Lewis and Randall, M is the total molality, and

$$A = N_A^2 \epsilon^3 (\rho_0 \pi/500)^{1/2}/(RTD_0)^{3/2}, \qquad (5\text{-}45)$$

in which N_A is Avogadro's number, ϵ is the protonic charge, ρ_0 is the

---

[8] P. J. W. Debye and W. Hückel, *Phys. Z.* **24**, 185 (1923).

density of the solvent, D_0 is its dielectric constant, and R, T, and π have their usual significance. For water solutions,[9]

$$A = 1.1244 \, [1 + 0.15471(t/100) + 0.03569(t/100)^2 + 0.02389(t/100)^3], \quad (5\text{--}46)$$

which gives $A = 1.1708$ at 25°C and 1.3653 at 100°C.

Since it is not possible to measure exactly the work of adding a single ionic species, the potential or activity of a single ion species does not have exact meaning. Therefore Eqs. (5–43) and (5–44) have only conventional significance. We adopt the convention that they are to be used only in combinations that lead to a zero net change in charge. These equations do not give the best possible values of single-ion activities, but any difference vanishes for the combinations permitted by this convention.

The Debye-Hückel limiting laws are now generally accepted, and the same answers have been obtained by more orthodox methods. There is more diversity of opinion about the extension to higher concentrations.

It is possible to express the activity and osmotic coefficients of more concentrated solutions by adding to the limiting laws power series in $(I/2)^{1/2}$. The coefficients are inconveniently large, however, and it is preferable to add power series in $I/2$ to the more extended Debye-Hückel expressions; then the coefficients are about the same as for nonelectrolytes. The extended equations can be expressed in terms of

$$x = \kappa a = a(4\pi N_A^2 \epsilon^2 \Sigma_i C_i z_i^2/1000 \; RTD)^{1/2}, \quad (5\text{--}47)$$

in which $1/\kappa$ is the "thickness of the ion atmosphere." Then

$$\bar{G}_k^{es}/RT = \ln \gamma_k = -z_k^2 \, A(I/2)^{1/2}/(1 + x) + 2\Sigma_i \beta_{ik} m_i + \cdots$$
$$= -z_k^2 \, AY/\alpha + 2\Sigma_i \beta_{ik} m_i + \cdots, \quad (5\text{--}48)$$

$$\alpha = x/(I/2)^{1/2}, \quad (5\text{--}49)$$

$$Y = x/(1 + x), \quad (5\text{--}50)$$

$$\phi - 1 = -AZI/\alpha M + \Sigma_{ij} \beta_{ij} m_i m_j/M + \cdots, \quad (5\text{--}51)$$

$$Z = [1 + x - 1/(1 + x) - 2 \ln (1 + x)]/x^2$$
$$= x/3 - 2x^2/4 + 3x^3/5 - 4x^4/6 + \cdots. \quad (5\text{--}52)$$

Debye and Hückel assumed that $D = D_0$, independent of the concentration, and therefore that κ is proportional to $(\Sigma_i C_i z_i^2)^{1/2}$. If we make the more reasonable assumption that the ions behave like holes of very small dielectric constant, so that D is proportional to the

[9] G. Scatchard, *J. Am. Chem. Soc.* 65, 1249 (1943).

volume concentration of solvent, we obtain the more convenient
result[10] that κ is proportional to $(\Sigma_i m_i z_i^2)^{1/2}$, or

$$x = \kappa a = a(4\pi N_A^2 \epsilon^2 \rho_0 \Sigma_i m_i z_i^2 / 1000 \, RT \, D_0)^{1/2}. \tag{5-53}$$

Then for water solutions,[9]

$$\alpha = 0.3240 \times 10^8 \, a[1 + 0.05217(t/100) - 0.00916(t/100)^2$$
$$+ 0.00888(t/100)^3]. \tag{5-54}$$

The numerical coefficient of $10^8 a$ is then 0.3281 at 25°C and 0.3408
at 100°C.

We shall assume that x is independent of the pressure at constant
temperature and ionic strength, which is equivalent to assuming that
D_0 is proportional to the concentration of water as the concentration
is altered by changing pressure. We also assume that at constant
pressure and ionic strength α is independent of the temperature. This
would imply that the distance of closest approach a of the ions
decreases by about 6 percent from 0°C to 100°C. This gives the very
practical advantage that the Debye-Hückel part of every ϕ_k^{es} is proportional to $z_k^2 Y$, that of ϕ_s^e is proportional to Z, and that of ϕ^E is
proportional to $X = Y - Z$, and any constant of proportionality may
be determined by the appropriate differentiation of A, just as the
corresponding analogue of β_{ij} is determined by differentiation of β_{ij}.

Figure 5.6 shows x, Y, X, and Z as functions of x, and Fig. 5.7

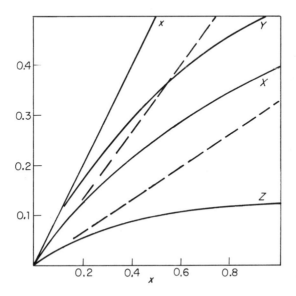

Figure 5-6. x, Y, X, and Z vs. x.

[10]G. Scatchard and L. F. Epstein, *Chem. Rev. 30*, 211 (1942).

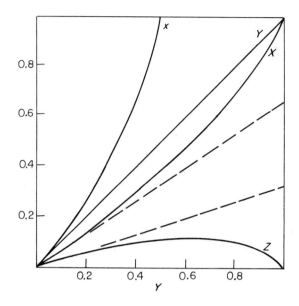

Figure 5–7. x, Y, X, and Z vs. Y.

shows them as functions of Y, which is unity when $x = \infty$, so that Fig. 5.7 covers the whole concentration range. The broken lines are the limiting laws, with slopes 2/3 for X and 1/3 for Z.

Eq. (5–52) is not convenient for computation at any value of x, and when x is small it requires the expansion of the logarithm and carrying many terms in the series. It is therefore very useful to tabulate Z as a function of Y. Table 5–1 gives Z for every step of 0.01 in Y from 0 to 1, which covers the whole range of possible concentrations, but in small enough intervals that linear interpretation is accurate enough for most purposes.

For all small ions in water we shall assume that $\alpha = 1.5$. This corresponds to $a \cong 4.5$ Å and to a fair average of the values computed for simple electrolytes. The use of the same value of all electrolytes makes computations for solutions of single salts somewhat easier, and those for mixtures very much simpler. This method was first proposed by Guggenheim,[11] who used $\alpha = 1$. We followed him at first,[12] but then found a larger value more desirable.[10,13] Figure 5–8 shows $-0.7805\ Z$ and $-0.7805\ Y$ as functions of ionic strength, $I/2$. These are the Debye-Hückel $(1 - \phi)/k$ and $\ln \gamma_{\pm}/k$, for $A = 1.1708$ and $\alpha = 1.5$,

[11]E. A. Guggenheim, *Phil. Mag. 19*, 588 (1935).
[12]G. Scatchard, *Chem. Rev. 19*, 309 (1936).
[13]G. Scatchard and R. E. Breckenridge, *J. Phys. Chem. 58*, 596 (1954).

Table 5-1. Values of the Debye function Z.[a]

Y	0	0.10	0.20	0.30	0.40	0.50	0.60	0.70	0.80	0.90
0.00	0.00000	0.03160	0.05941	0.08287	0.10129	0.11371	0.11885	0.11487	0.09882	0.06537
	332	296	255	209	152	86	5	−100	−244	−467
.01	.00332	.03456	.06196	.08496	.10281	.11457	.11890	.11387	.09638	.06070
	328	292	251	203	147	80	−4	−112	−261	−498
.02	.00660	.03748	.06447	.08699	.10428	.11537	.11886	.11275	.09377	.05572
	325	288	246	197	141	72	−14	−125	−280	−531
.03	.00985	.04036	.06693	.08896	.10569	.11609	.11872	.11150	.09097	.05041
	321	284	242	193	134	64	−23	−138	−299	−567
.04	.01306	.04320	.06935	.09089	.10703	.11673	.11849	.11012	.08798	.04474
	318	281	237	187	128	56	−34	−152	−320	−607
.05	.01624	.04601	.07172	.09276	.10831	.11729	.11815	.10860	.08478	.03867
	314	276	233	182	122	48	−44	−165	−341	−651
.06	.01938	.04877	.07405	.09458	.10953	.11777	.11771	.10695	.08137	.03216
	311	272	228	176	115	40	−54	−180	−363	−701
.07	.02249	.05149	.07633	.09634	.11068	.11817	.11717	.10515	.07774	.02515
	307	268	223	171	108	31	−66	−195	−387	−759
.08	.02556	.05417	.07856	.09805	.11176	.11848	.11651	.10320	.07387	.01756
	304	264	218	165	101	23	−77	−211	−412	−830
.09	.02860	.05681	.08074	.09970	.11277	.11871	.11575	.10109	.06975	.00926
	300	260	213	159	94	14	−88	−227	−438	−926
.10	.03160	.05941	.08287	.10129	.11371	.11885	.11487	.09882	.06537	.00000

[a]The value of Y is the sum of the numbers at the left of the row and the top of the column. The lower entry in each cell is the change in Z for a change of 0.01 in Y.

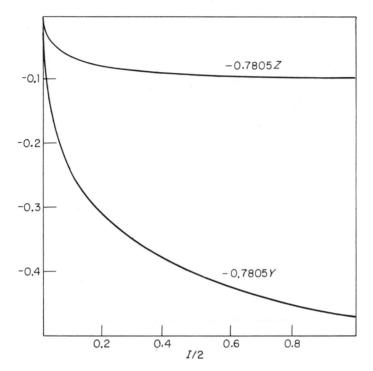

Figure 5–8. Debye-Hückel and ideal functions.

if k is $-z_+z_-$. Table 5–2 gives these values and that of $0.3390\ Y$ [$= \log \gamma_\pm/k$] for certain round values of $I/2$. For other values of A or α, $I/2$ should be multiplied by $\alpha^2/2.25$, and $0.7805\ Z$, $0.7805\ Y$, and $0.3390\ Y$ should be multiplied by $1.5A/1.1708\alpha$. The values of the ionic strength are chosen to give values of m_s for 1-1, 2-1, 2-2, and 3-1 electrolytes in steps of 0.1 from 0 to 1, of 0.2 from 1 to 2, of 0.5 from 2 to 5, and of 1 at higher concentrations.

5–5. Properties of Nonpolar Mixtures from Those of Their Components

In order to determine the properties of nonpolar liquid mixtures from those of their components, we shall assume that the energy E and entropy S of a liquid mixture are given by the relations

$$E = \Sigma_i N_i E_i + V\Sigma_{ij} b_{ij}\phi_i\phi_j, \tag{5–55}$$

$$S = \Sigma_i N_i [S_i - R \ln \phi_i]. \tag{5–56}$$

The E_i are the energies in the ideal-gas state and are functions only

Table 5-2. Debye functions for $A = 1.1708$ and $\alpha = 1.5$.

$I/2$	$0.7805Z$	$0.7805Y$	$0.3390Y$	$I/2$	$0.7805Z$	$0.7805Y$	$0.3390Y$
0.1	0.06816	0.25112	0.10907	6.0	0.07955	0.61354	0.26647
.2	.07924	.31338	.13611	6.4	.07865	.61774	.26830
.3	.08464	.35204	.15290	7.0	.07736	.62344	.27077
.4	.08777	.37999	.16504	7.2	.07695	.62520	.27154
.5	.08972	.40175	.17449	7.5	.07633	.62773	.27263
.6	.09099	.41949	.18219	8.0	.07537	.63165	.27434
.7	.09179	.43440	.18867	8.4	.07462	.63456	.27560
.8	.09231	.44721	.19423	9.0	.07357	.63861	.27736
.9	.09261	.45840	.19909	9.6	.07255	.64232	.27897
1.0	.09277	.46832	.20340	10.0	.07190	.64463	.27998
1.2	.09276	.48523	.21075	10.5	.07115	.64735	.28116
1.4	.09249	.49924	.21683	10.8	.07069	.64889	.28183
1.5	.09228	.50541	.21951	11	.07039	.64989	.28226
1.6	.09206	.51114	.22200	12	.06900	.65456	.28429
1.8	.09154	.52143	.22647	13	.06769	.65873	.28610
2.0	.09096	.53046	.23039	13.5	.06707	.66066	.28694
2.1	.09064	.53459	.23219	14	.06648	.66249	.28773
2.4	.08972	.54570	.23701	15	.06535	.66591	.28922
2.5	.08939	.54903	.23846	16	.06428	.66902	.29057
2.7	.08876	.55525	.24116	17	.06329	.67189	.29182
2.8	.08843	.55816	.24242	18	.06231	.67500	.29317
3.0	.08780	.56360	.24478	19	.06144	.67699	.29403
3.2	.08718	.56862	.24696	20	.06060	.67927	.29502
3.5	.08625	.57546	.24994	21	.05977	.68140	.29595
3.6	.08595	.57758	.25086	22	.05900	.68339	.29681
4.0	.08477	.58540	.25425	23	.05827	.68527	.29763
4.2	.08418	.58895	.25579	24	.05759	.68704	.29840
4.5	.08335	.59389	.25794	25	.05690	.68870	.29912
4.8	.08253	.59843	.25991	26	.05625	.69028	.22980
5.0	.08203	.60127	.26114	27	.05564	.69178	.30045
5.4	.08100	.60653	.26343	28	.05505	.69320	.30107
5.6	.08052	.60897	.26449	29	.05450	.69455	.30166
				30	.05394	.69583	.30222

$A = 1.1328$ $[1 + 0.1369$ $(t/100) + 0.0669$ $(t/100)^2 + 0.0073$ $(t/100)^3]$; $10^{-8} \kappa/(I/2)^{1/2} = 0.3249$ $[1 + 0.04625$ $(t/100) + 0.00157$ $(t/100)^2 + 0.00321$ $(t/100)^3]$. At 25°C., $A = 1.1764$; $10^{-8} \kappa/(I/2)^{1/2} = 0.3287$

of the temperature. The b_{ij} are functions of both temperature and a pressure. If the two subscripts are the same, b_{ii} is the difference in energy of unit volume of component i as a liquid and as an ideal gas at the same temperature, or the negative of the energy of evaporation to the ideal-gas state of unit volume of liquid component i at the given pressure. The S_i are also functions of temperature and pressure, characteristic of the pure liquids. As in (5-4) ϕ_i and ϕ_j are volume fractions.

We need to consider two changes, each of which has the same initial state: pure liquid components, each at the temperature T_0 and pressure p_0, with a total volume V_0. For change I, the final state is the mixture at T_0, p, V_0; and for change II, the final state is the mixture at T_0, p_0, V [$= V_0 + V^M$]. We assume that the pressure that determines the values of the b_{ij} and the S_i is p_0 for the first change of state.

Hildebrand[14] assumed that for nonpolar mixtures the entropy of mixing might be the same as for ideal solutions, that is, with $\bar{N}_i\bar{N}_j$ replacing $\phi_i\phi_j$ in Eq. (5-56), and gave two definitions of "regular solutions." For the first the ideal entropy held for mixing at constant total volume (change I), and the second for mixing at constant pressure (change II). We now know that the first is better than the second, that volume fractions are better than mole fractions, but that Eqs. (5-55) and (5-56) are incompatible. The seriousness of this error is difficult to determine in general, but Rushbrooke,[15] Kirkwood,[16] and Guggenheim[7] have studied the special case of binary solutions of spheres of equal volume, and have obtained substantially the same results. We shall use the expression of Kirkwood, with $(2b_{12} - b_{11} - b_{22}) = B_{12}$, and σ the number of contacts a molecule makes with its neighbors. Then

$$\bar{E}_V^M = B_{12}\bar{V}\ \bar{N}_1\bar{N}_2(1 - 2B_{12}\bar{V}\ \bar{N}_1\bar{N}_2/\sigma RT), \qquad (5\text{-}57)$$

$$\bar{S}_V^M - \bar{S}^{\text{FH}} = -\ (B_{12}\bar{V}\ \bar{N}_1\bar{N}_2)^2/\sigma R, \qquad (5\text{-}58)$$

$$\bar{A}_V^M - \bar{A}^{\text{FH}} = B_{12}\bar{V}\ \bar{N}_1\bar{N}_2(1 - B_{12}\bar{V}\ \bar{N}_1\bar{N}_2/\sigma RT). \qquad (5\text{-}59)$$

These expressions are symmetric. When the expression in $\bar{G}_p^M$ corresponding to $B_{12}\bar{V}\ \bar{N}_1\bar{N}_2/\sigma RT$ exceeds $[1 + (1 - 4\sigma)^{1/2}]/4$, the solution separates into two immiscible phases. The critical value is 0.046 if $\sigma = 12$, 0.073 if $\sigma = 8$, and 0.106 if $\sigma = 6$. The probable value of σ is larger than 8, so the maximum error is less than 7 percent in $\bar{A}_V^M$. The form of Eq. (5-55) appears to be satisfactory.

We next assume that

$$b_{ij} = (b_{ii}b_{jj})^{1/2}. \qquad (5\text{-}60)$$

This is largely by analogy with the treatment of gas mixtures. We know that the energy of mixing approaches zero as $(b_{ii} - b_{jj})$, not $(b_{ii}\bar{V}_i - b_{jj}\bar{V}_j)$ or any other such function, approaches zero. For a van der Waals fluid with volume reduced to b, $b_{ii} = a_i/b_i^2$, so our

[14]J. H. Hildebrand, *J. Am. Chem. Soc.* **51**, 66 (1929).
[15]G. S. Rushbrooke, *Proc. Roy. Soc. (London)* **A166**, 296 (1938).
[16]J. G. Kirkwood, *J. Phys. Chem.* **43**, 97 (1939).

combination corresponds to the last one in Sec. 4–2. With this assumption, the calculation of b_{ij} is moderately accurate, but usually b_{jj} is not very different from b_{ii}, so the relative error in B_{ij} is large. Eqs. (5–55), (5–56), and (5–60) permit the calculation of the energy, the entropy, and the work content of mixing at constant total volume from the molal volumes and the energies of evaporation of the components.

To study change II, at constant pressure, we define the compressibility of the mixture as β and that of the unmixed components as β_0. Then

$$P - P_0 = (1/\beta)\ln(V/V_0) = (1/\beta)\ln(1 + V^M/V_0), \quad (5\text{–}61)$$

$$G_V^M - A_V^M = (P - P_0)V_0 = (V_0/\beta)\ln(1 + V^M/V_0), \quad (5\text{–}62)$$

$$G_p^M - G_V^M = \int_{V_0}^{V} - (1/\beta)dV = -V^M/\beta, \quad (5\text{–}63)$$

$$\begin{aligned}
G_p^M - A_V^M &= (V_0/\beta)[\ln(1 + V^M/V_0) - V^M/V_0] \\
&= -(V_0/2\beta)(V^M/V_0)^2 + \cdots;
\end{aligned} \quad (5\text{–}64)$$

V^M/V_0 is usually so small that the higher terms in the expansion of the logarithm may be neglected. Often all of Eq. (5–64) may be neglected, and G_p^M may be calculated directly from Eqs. (5–55), (5–56), and (5–60). For example, for the three mixtures benzene–cyclohexane, cyclohexane–carbon tetratrachloride, and benzene–carbon tetrachloride,[17] V^M/V_0 is 0.0065, 0.0016, and 0.00003; and the ratio $(G_p^M - A_V^M)/E_V^M$ is 0.037, 0.000, and 0.000, respectively.

Carrying out a process for the entropy similar to that of Eqs. (5–62) to (5–64) yields

$$\begin{aligned}
S_p^M - S_V^M &= V^M\left(\frac{dP_0}{dT}\right)_V + \frac{V_0 d\ln\beta}{\beta Td\ln T}\left[\ln\left(1 + \frac{V^M}{V_0}\right) - \frac{V^M}{V_0}\right] \\
&= \frac{V^M\alpha_0}{\beta_0} - \frac{V_0 d\ln\beta}{2\beta T\,d\ln T}\left[\left(\frac{V^M}{V_0}\right)^2 + \cdots\right],
\end{aligned} \quad (5\text{–}65)$$

in which α_0 is the coefficient of thermal expansion. Multiplying Eq. (5–65) by T and adding it to Eq. (5–64) gives $(H_p^M - E_V^M)$, which is very much larger than $(G_p^M - A_V^M)$. For the three systems mentioned, $(H_p^M - E_V^M)/E_V^M$ is 0.465, 0.652, and 0.041, respectively. Although the quadratic combination is sufficient to give b_{ij} to 0.025, 0.005, and 0.005, respectively, the calculated value of E_V^M is only a fifth to a third of the measured value. Furthermore, for these systems, Eq. (5–56) is badly in error.

---

[17] G. Scatchard, S. E. Wood, and J. M. Mochel, *J. Phys. Chem.* **43**, 119 (1939); *J. Am. Chem. Soc.* **61**, 3206 (1939); **62**, 712 (1940).

We may also calculate approximately the volume change on mixing,

$$V^M = \frac{dG_p^M}{dP_0} \simeq \frac{dE_V^M}{dP_0} = -\beta_0\left(\frac{VdE}{dV} - \frac{V_0dE}{dV_0}\right) \simeq \beta_0 E_V^M.$$
(5-66)

The last equality depends upon the approximate relation that the internal pressure is equal to the cohesive energy density; see Eq. (D-9). For spheres of the same volume, Prigogine and Mathot[18] and Salsburg and Kirkwood[19] calculate from statistical mechanics that V^M is negative whereas E^M is positive. For small differences in size, however, V^M becomes approximately equal to $\beta_0 E_V^M$, in accordance with Eq. (5-66). Experimental results confirm their conclusions. The only part of Eq. (5-66) that can be enough in error to account for a change in sign, or even for a change in the order of magnitude, is the last approximate equality. There is still much to learn about even the simplest solutions.

---

[18]I. Prigogine and V. Mathot, *J. Chem. Phys. 20*, 49 (1952).
[19]Z. W. Salsburg and J. G. Kirkwood, *J. Chem. Phys. 21*, 2189 (1953).

6. Measurements of Chemical Potentials in Liquid Mixtures[1]

6-1. Comparison of Vapor Pressures at Constant Temperature

One method of determining the difference of potential of a substance in two liquid solutions at the same temperature is to determine its vapor pressure in each, that is, the partial pressure of the substance in a vapor that is in equilibrium with the solution, utilizing the fact that at equilibrium the potential is the same in the liquid as in the gas phase, and to determine the difference in potential of that substance in the two gas phases by the methods of Chapters 3 and 4. The pure liquid may be considered as one of the solutions. These relations hold for either solvent or solute in solutions of any concentration.

If the total gas pressure is small enough that the vapor may be treated as an ideal gas and the volume of the liquid may be neglected relative to that of the vapor, the difference in potential may be written

$$(\bar{G}_i - \bar{G}_i^0)/RT = \ln p\,\bar{N}_i' - \ln p^0\bar{N}_i'^0 \qquad (6\text{-}1)$$
$$= \ln p_i/p_i^0.$$

In this chapter the prime is used to indicate the gas phase, and the superscript ^{0} to indicate a standard reference state in general. Eq. (6-1) is sufficiently accurate for all but the more precise measurements.

When Eq. (6-1) is insufficient, we try the next approximation, in which only the β terms are used for the vapor, as in Eqs. (4-19) and (4-20), and the partial volumes in the liquid are considered as con-

---

[1]Good discussions of liquid solutions are: J. H. Hildebrand and R. L. Scott, *The Solubility of Non-Electrolytes* (Reinhold, New York, 3d ed., 1950), and E. A. Guggenheim, *Mixtures* (Clarendon Press, Oxford, 1952); also D. A. Mac Innes, *The Principles of Electrochemistry* (Reinhold, New York, 1939); H. S. Harned and B. B. Owen, *The Physical Chemistry of Electrolytic Solutions* (Reinhold, New York, 1950); R. A. Robinson and R. H. Stokes, *Electrolyte Solutions* (Academic Press, New York, 1955).

stants, independent of both pressure and composition. We wish to determine the potential difference when both solutions are at the same pressure P, and we do this by a five-stage process:

(1) in the change of the standard solution from the pressure P to its equilibrium pressure p_ϕ^0,

$$\Delta_1 \bar{G}_i / RT = \bar{V}_i (p_\phi^0 - P)/RT;$$

(2) in the change from the liquid to the vapor at the equilibrium pressure,

$$\Delta_2 \bar{G}_i / RT = 0;$$

(3) in the change from the vapor in equilibrium with the standard solution to that in equilibrium with the other solution,

$$\Delta_3 \bar{G}_i / RT = \ln p \, \bar{N}_i' / p^0 \bar{N}_i'^0 + \beta_{ii}'(p - p^0)/RT \\ + \Sigma_{jk}(2\beta_{ij}' - \beta_{ii}' - \beta_{jk}')(p\bar{N}_j'\bar{N}_k' - p^0\bar{N}_j'^0\bar{N}_k'^0/RT; \quad (6\text{-}2)$$

(4) in the change from vapor to liquid at the equilibrium pressure,

$$\Delta_4 \bar{G}_i / RT = 0;$$

(5) in the change of the solution from the equilibrium pressure p to the pressure P,

$$\Delta_5 \bar{G}_i / RT = \bar{V}_i (P - p)/RT.$$

In the sum the pressure P drops out, or the change is independent of the pressure P, because of our assumption that $\bar{V}_i$ is constant. The sum is

$$(\bar{G}_i - \bar{G}_i^0)/RT = \ln p \, \bar{N}_i' / p^0 \bar{N}_i'^0 + (\beta_{ii}' - \bar{V}_i)(p - p^0)/RT \\ + \Sigma_{jk}(2\beta_{ij}' - \beta_{ii}' - \beta_{jk}')(p\bar{N}_j'\bar{N}_k' - p^0\bar{N}_j'^0\bar{N}_k'^0)/RT.$$

If two or more components are volatile, the full Eq. (6-2) is used, but if only component i is volatile, the last term disappears. In the gas-current method the total pressure is constant, so the second term vanishes. If the vapor contains only component i and the carrier gas g, the last term becomes

$$(2\beta_{ig}' - \beta_{ii}' - \beta_{gg}')(\bar{N}_g'^2 - \bar{N}_g'^{02})p/RT$$

or

$$- (2\beta_{ig}' - \beta_{ii}' - \beta_{gg}')(\bar{N}_i' - \bar{N}_i'^0)(2 - \bar{N}_i' - \bar{N}_i'^0)p/RT.$$

If the mole fraction of i is small, this becomes

$$- 2(2\beta_{ig}' - \beta_{ii}' - \beta_{gg}')(p_i - p_i^0)/RT,$$

so the form becomes the same as when i is the only volatile component.

6-2. Vapor Pressures of Dilute Solutions

We may write the potential of a nonelectrolyte solute as

$$(\bar{G}_i - \bar{G}_i^0)/RT = \ln (m_i/m_i^0) + 2\Sigma_j \beta_{ij}(m_j - m_j^0)$$
$$+ 3\Sigma_{jk}\gamma_{ijk}(m_j m_k - m_j^0 m_k^0)/2 + \cdots$$
$$\cong \ln (p_i/p_i^0) + \beta'(p_i - p_i^0)/RT$$
$$\cong \ln (p_i/p_i^0) + \beta' p_i^0/m_i^0 RT(m_i - m_i^0) + \cdots,$$

$$(6-3)$$

in which $\beta' = (\beta'_{ii} - \bar{V}_i)$ if i is the only volatile component, and $\beta' = -(2\beta'_{ig} - \beta'_{ii} - \beta'_{gg})$ if g is a carrier gas in large excess. We shall use the superscript s when every m_i^0 becomes so small that $m_j^0 \beta$, $m_j^0 \beta'$ are negligible. Then

$$\ln(p_i/m_i) = \ln (p_i^s/m_i^s) + 2\Sigma_j \beta_{ij} m_j - (\beta' p_i^s/RT m_i^s) m_i$$
$$+ 3\Sigma_{jk}\gamma_{ijk} m_j m_k/2 + \cdots.$$

$$(6-4)$$

The form is the same as for the potential, but the interpretation is somewhat different.

The corresponding equation for the solvent is

$$-(\bar{G}_S - \bar{G}_S^s)/RT = \bar{W}_S(\Sigma_i m_i + \Sigma_{ij}\beta_{ij} m_i m_j + \Sigma_{ijk}\gamma_{ijk} m_i m_j m_k)$$
$$\cong -\ln(p_S/p_S^s) - \beta'(p_S - p_S^s)/RT$$
$$\cong -\ln(p_S/p_S^s) + \beta' p_S^s \bar{W}_S \Sigma_i m_i/RT,$$

$$(6-5)$$

and for the osmotic coefficient for vapor pressure, ϕ_{vp}:

$$-\phi_{vp} = \frac{-\ln(p_S/p_S^s)}{\bar{W}_S \Sigma_i m_i} = \left(1 - \frac{\beta' p_S^s}{RT}\right) + \frac{\Sigma_{ij}\beta_{ij} m_i m_j}{\Sigma_i m_i}$$
$$+ \frac{\Sigma_{ijk}\gamma_{ijk} m_i m_j m_k}{\Sigma_i m_i} + \cdots.$$

$$(6-6)$$

The osmotic coefficient for vapor pressure does not approach exact unity as the concentrations approach zero unless the vapor obeys the ideal-gas law. If S is the only volatile component, the difference from unity usually increases with increasing temperature because the decrease in β' with increasing temperature does not compensate for the increase in p_S^s. For water the value is 1.0013 at 20°C and 1.0148 at 100°C. If the partial pressure is measured by the gas-current method, the limit depends upon the nature of the carrier gas, and it may be less than unity.

If a solute has a high vapor pressure, the pressure may be measured from very dilute solutions. Except for the solubilities of gases at high pressures, the deviations from Eq. (6-1) are usually less than the experimental error, as in the measurements of the vapor pressure of hydrogen chloride in various solutions.[2] Figure 6-1 shows log p/m vs.

---

[2] M. Randall and L. E. Young, *J. Am. Chem. Soc.* 50, 1001 (1928).

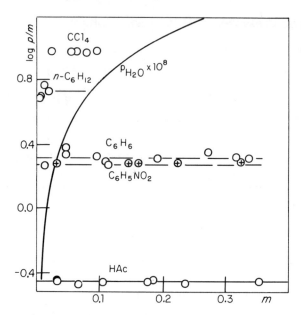

Figure 6–1. Vapor pressure of hydrogen chloride in various solutions.

m for various solvents, with p in atmospheres. Except for water the values show no trend with concentration. The solutions would also appear ideal in this concentration range in terms of mole fractions or volume fractions. The values are different for the different solutes, however. They are the logarithms of the Henry's-law constants. No one attaches theoretical significance to moles per kilogram. Table 6–1

Table 6–1. Vapor pressure of hydrogen chloride.

Solvent	$\log (p/m)$	$\log (p/N)$	$\log (p/C)$
Ideal	−0.08	1.48	(0.00)
Carbon tetrachloride[a]	− .97	1.78	1.17
Chloroform[a]	− .65	1.65	0.83
n-Hexane[b]	− .72	1.80	.54
Benzene[c]	− .32	1.43	.26
Nitrobenzene[c]	− .28	1.17	.36
Acetic Acid[d]	− .45	0.87	−0.43

[a]J. J. Howland, Jr., D. R. Miller, and J. E. Willard, *J. Am. Chem. Soc. 63*, 2807 (1941).
[b]S. J. O'Brien and C. L. Kenny, *ibid. 62*, 1189 (1940).
[c]S. J. O'Brien, C. L. Kenny, and R. A. Ziercher, *ibid. 61*, 2524 (1939).
[d]W. H. Rodebush and R. H. Ewart, *ibid. 54*, 419 (1932).

shows not only log p_2/m_2, but also log $p_2/\bar{N}_2$ and log p_2/C_2. The ideal values are calculated from the logarithm of the fugacity for liquid hydrogen chloride, $(\bar{G}_i^i - G_i^i)/2.3\ RT$. The concentration value is in parentheses because the density was extrapolated linearly with an equation determined between -104 and $-83°C$.

The values for water are very much smaller than those for the other solvents and vary rapidly with the concentration. They are not measured directly but are calculated from vapor-pressure measurements in more concentrated solutions and electromotive-force measurements, as discussed in Secs. 6–3 and 6–8. In dilute aqueous solutions, hydrogen chloride exists as ions, probably hydrated. Consideration of this fact and of the Debye-Hückel interaction explains the measurements. Fig. 6–2 shows the deviations of electromotive-

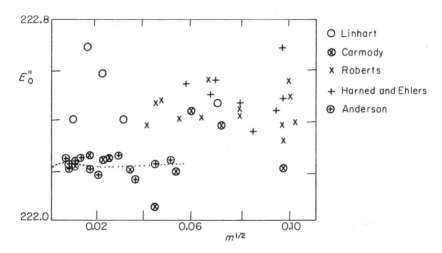

Figure 6-2. Electromotive force of hydrochloric acid concentration cells: $E_0'' = E + 118.3 \log m - 59.62\ m^{1/2} + 90\ m$ (mV).

force measurements in very dilute solutions from an equation that in this concentration range corresponds closely to the equation on which Table 5–2 is based. The deviations are plotted against $m^{1/2}$ merely to spread out the results in dilute solutions.

Sinclair, Robinson, and Stokes[3] have developed a method—which they name "isopiestic" and we[4] called "isotonic," after the earlier

[3]D. A. Sinclair, *J. Phys. Chem. 37*, 495 (1933). See also R. A. Robinson and D. A. Sinclair, *J. Am. Chem. Soc. 56*, 1830 (1934), whose results on a very large number of solutes are summarized by R. H. Stokes, *Trans. Faraday Soc. 44*, 295 (1948), and by R. A. Robinson and R. H. Stokes, *ibid. 45*, 612 (1949).
[4]G. Scatchard, W. J. Hamer, and S. E. Wood, *J. Am. Chem. Soc. 60*, 3061 (1938).

name of de Vries—of determining the potential or the vapor pressure of the solvent in a solution of a nonvolatile solute by comparison with a solution of another nonvolatile solute with which it is in equilibrium through the vapor phase. If the vapor pressure or the potential of the second solution is known as a function of the concentration, that of the first solution is determined. The method is more precise in dilute solutions than the direct measurements of the vapor pressure, and is calibrated from other types of measurement.

Fig. 6–3 shows the osmotic coefficients at 25°C of some aqueous nonelectrolytes and electrolytes of different valence types, and for the electrolytes the osmotic coefficients with the electrostatic term deducted are also shown. With this correction, the electrolytes behave essentially like the nonelectrolytes.

The vapor pressure of a solute may also be used to determine the effect of another solute on the potential of the first. The effects of adding salts on the vapor pressures of many solute gases have been determined, usually as the change in concentration of the non-electrolyte at constant potential, or salting out. If the concentration of the gas is always small, Eq. (6–3) yields at constant potential

$$\ln m_i/m_{i0} = -2\,\beta_{ij}m_j - 3\,\gamma_{ijj}m_j^2/2 + \cdots. \qquad (6\text{–}7)$$

Usually the β term is sufficient for salt concentrations up to 1 or 2 molal. Salting out will be discussed in Sec. 8–4, when the theory for β_{ij} is considered.

6–3.   Vapor Pressures of Concentrated Solutions

In a mixture of two or more volatile components we measure the equilibrium temperature, pressure, and compositions of the liquid and vapor phases. We keep the temperature constant, consider the liquid composition as independent variable, and calculate the partial pressures $p_i[= p\,\bar{N}_i']$ from the pressure and vapor composition. Then we calculate the potentials of the components by Eq. (6–1) or (6–2). The system is overspecified because we can calculate the potential of one component by the Gibbs-Duhem relation if the potentials of all the other components are known. The potentials are determined less accurately for components present only in small quantities. The free energy of mixing, however, is determined with about the same precision over the whole range if the pure components have about the same vapor pressures.

We consider first a two-component system, and compare it with Fig. 5–4, in which the ordinate is $\bar{G}^M$ and the abscissa is $\bar{N}_1$. We determine at a given value of $\bar{N}_1$ the values of $\bar{G}_1^M$ and $\bar{G}_2^M$, which are

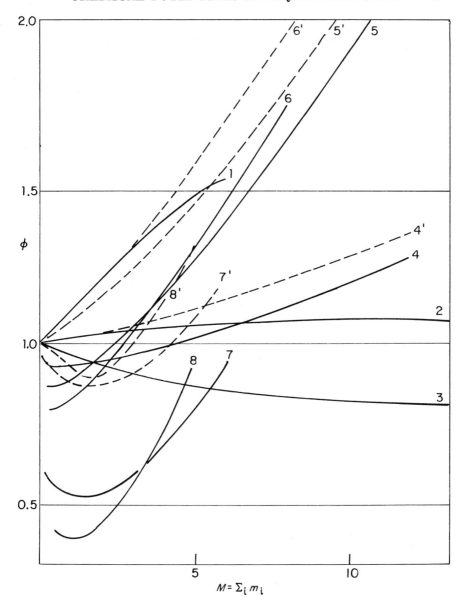

Figure 6–3. Osmotic coefficients: 1, sucrose; 2, glycerol; 3, urea; 4, NaCl; 5, CaCl₂; 6, LaCl₃; 7, MgSO₄; 8, Al₂(SO₄)₃. The broken lines are corrected for the Debye-Hückel term.

the intercepts at $\bar{N}_1 = 1$ and $\bar{N}_1 = 0$ of the tangent to the curve of $\bar{G}^M$ vs. $\bar{N}_1$. So at each point we determine both $\bar{G}^M$ and $d\bar{G}^M/d\bar{N}_1$. If we subtract the ideal terms from the potentials, they determine $\bar{G}^E$ and $d\bar{G}^E/d\bar{N}_1$, and so for further deviation terms. If we determine the potentials from Eq. (6–1) instead of (6–2), the Gibbs-Duhem relation still holds approximately. Figure 6–4 shows

$$Q = \bar{N}_1 \log (p\bar{N}_1'/p_1^1\bar{N}_1) + \bar{N}_2 \log (p\bar{N}_2'/p_2^2\bar{N}_2) \qquad (6\text{--}8)$$

as a function of $\bar{N}_2$ and various deviation curves for chloroform(1)–ethanol(2) mixtures.[5] If the potentials were given by Eq. (6–1), Q would

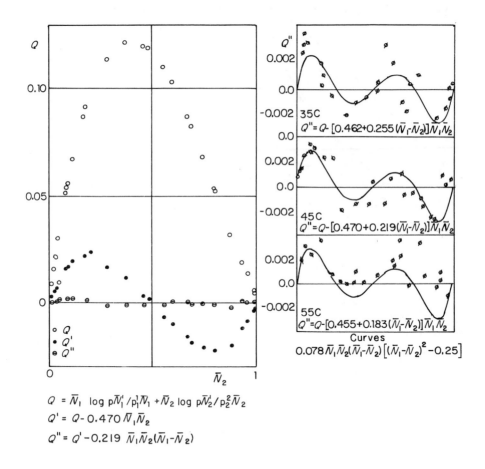

$$Q = \bar{N}_1 \log p\bar{N}_1'/p_1^1\bar{N}_1 + \bar{N}_2 \log p\bar{N}_2'/p_2^2\bar{N}_2$$
$$Q' = Q - 0.470\, \bar{N}_1\bar{N}_2$$
$$Q'' = Q' - 0.219\, \bar{N}_1\bar{N}_2(\bar{N}_1 - \bar{N}_2)$$

Figure 6–4. Vapor-liquid equilibrium of chloroform–ethanol mixtures.

[5]G. Scatchard, *Chem. Rev. 44*, 7 (1949), from measurements of G. Scatchard and C. L. Raymond, *J. Am. Chem. Soc. 60*, 1278 (1938).

be $G^E/2.3\ RT$. The difference between log $p\bar{N}_2'/p_2^2\bar{N}_2$ and log $p\bar{N}_1'/p_1^1\bar{N}_1$ should be very approximately the slope of Q. The left-hand part of Fig. 6–4 shows Q, Q', and Q'' at 45°C on a small scale. The right-hand part shows Q'' only, on a much larger scale, for 35°, 45°, and 55°C. The slopes are given only on the larger-scale figures. This is a much better method of checking consistency than calculating the potential of one component from that of the other by a Gibbs-Duhem integration. Each slope is compared with the slope determined from neighboring points, but there is no smoothing concealed in the integration. The method does share one disadvantage of the other, in that we have no criterion of how well the slopes should correspond. For a more definite comparison, we can calculate from our equation the pressure and vapor composition at each composition and compare these results with the measured values.

There is a group of chemical engineers who believe that the vapor composition can be measured much more accurately than the pressure. Using Eq. (6–1) we have, for a two component system,

$$(\bar{G}_A^E - \bar{G}_B^E)/RT = (\bar{G}_A^M - \bar{G}_B^M)/RT - \ln \bar{N}_A/\bar{N}_B$$
$$= \ln (\bar{N}_A'\bar{N}_B/\bar{N}_A\bar{N}_B') - \ln p_A^A/p_B^B. \qquad (6\text{–}9)$$

The various terms in the expansion are analogous to those in Fig. 5–3. If the vapor contains only A and B, Eq. (6–2) gives

$$(\bar{G}_A^E - \bar{G}_B^E)/RT = \ln (\bar{N}_A'\bar{N}_B/\bar{N}_A\bar{N}_B') - \ln p_A^A/p_B^B$$
$$+ [(p - p_A^A)(\beta_{AA} - \bar{V}_A) - (p - p_B^B)(\beta_{BB} - \bar{V}_B)$$
$$- (2\beta_{AB} - \beta_{AA} - \beta_{BB})\, p(\bar{N}_A' - \bar{N}_B')]R/T. \qquad (6\text{–}10)$$

The pressure must be known approximately to determine the corrections. It is possible to use the air-current method without determining the quantity of carrier gas to measure the difference in potentials. However, this will seldom be done precisely enough to use the equation corresponding to Eq. (6–2).

There are other groups who believe that the liquid composition and total pressure can be determined much more precisely than the vapor composition. We need only consider the case in which the vapor contains only components A and B. Then solution of Eq. (6–2) for $\ln p\bar{N}_A'$ gives

$$\ln p\, \bar{N}_A' = \ln p_A^A\bar{N}_A + \bar{G}_A^E/RT - (\beta_{AA}' - \bar{V}_A)(p - p_A)/RT$$
$$- (2\beta_{AB}' - \beta_{AA}' - \beta_{BB}')\, p\, \bar{N}_B^2/RT, \qquad (6\text{–}11)$$

and similarly for $\ln p\, \bar{N}_B'$. So

$$p = p\bar{N}_A' + p\bar{N}_B'$$
$$= p_A^A\bar{N}_A \exp (\bar{G}_A^E/RT) \exp [-(\beta_{AA}' - \bar{V}_A)(p - p_A^A)/RT] \exp [-(2\beta_{AB}'$$
$$- \beta_{AA}'\beta_{BB}')p\, \bar{N}_B'^2/RT] + p\, _B^B\bar{N}_B \exp (\bar{G}_B^E/RT) \exp [-(\beta_{BB}' - \bar{V}_B)(p$$
$$- p_B^B)/RT] \exp [-(2\beta_{AB}' - \beta_{AA}' - \beta_{BB}')p\bar{N}_A'^2/RT]. \qquad (6\text{–}12)$$

The last term in each exponential requires at least an approximate knowledge of the vapor composition.

The form corresponding to Eq. (6–1) is

$$p = p_A^A \bar{N}_A \exp{(\bar{G}_A^E/RT)} + p_B^B \bar{N}_B \exp{(\bar{G}_B^E/RT)}. \qquad (6\text{–}13)$$

If a single term is sufficient in the mole-fraction expansion, it may be determined readily from the pressure at $\bar{N}_A = 0.5$, for then

$$p = (p_A^A + p_B^B)\tfrac{1}{2} \exp{(\bar{G}_{AB}^0/4RT)}. \qquad (6\text{–}14)$$

If more terms are necessary, they must be determined by trial and error, and the pressure at $\bar{N}_A = 0.5$ does not depend upon G_{AB}^0 alone. If more than one term is necessary in the mole-fraction expansion, the volume-fraction expansion will usually be more economical. If corrections are made for deviations from the gas laws, the iteration becomes more tedious.

For a ternary system, we use the parameters obtained for the three component binaries and need one measurement of a ternary mixture to obtain γ_{123} and three more for δ_{1123}, δ_{1223}, and δ_{1233}. For measurements of both p and $\bar{N}_i'$, and for measurement of p only, γ_{123} should be determined with each fraction about one-third; for measurements of $\bar{N}_i'$ alone, the fractions must not all be equal. Redlich and Kister[6] have shown that the terms with three or more different subscripts are often unnecessary. This does not mean that there are no interactions involving three different species, but only that these interactions are related in a specific way to those that can occur in binary systems. Although this simplification holds exactly for the G_{ik}^0 terms, it cannot be general. For example, we found[7] that in systems of methanol(1) with carbon tetrachloride(2) and benzene(3) the fraction of methanol $\bar{N}_1$ is much more important than the distribution of $(1 - \bar{N}_1)$ between $\bar{N}_2$ and $\bar{N}_3$, and the properties of the ternary mixtures can be expressed much better by replacing $(\bar{N}_1 - \bar{N}_2)$ and $(\bar{N}_1 - \bar{N}_3)$ by $(2\bar{N}_1 - 1)$. This makes no difference, of course, in the binary mixtures or in the G_{ik}^0 terms.

6–4. Chemical Potentials and Freezing Points

A solid consisting of one pure component in equilibrium with a liquid mixture is the ideal, probably nonexistent, limit of extremely dilute solid solutions. For our purposes it is sufficient that the limit be approached closely enough that the potential of that component is

[6]O. Redlich and A. T. Kister, *Ind. Eng. Chem.* 40, 345 (1948).

[7]G Scatchard, L. B. Ticknor, J. R. Goates, and E. R. McCartney, *J. Am. Chem. Soc.* 74, 3721 (1952); G. Scatchard and L. B. Ticknor, *ibid.*, 3724.

the same in the equilibrium solid as it is in the pure component within the accuracy of our measurements. This close an approach is very often attained. Then the potential $\bar{G}_i$ of the component i in the solution is the same as that in the pure solid, or crystal, $\bar{G}_{ic}$, and it is possible to determine the potential in solution from the properties of the pure component:

$$\bar{G}_i^M = \bar{G}_i - \bar{G}_i^i = \bar{G}_i^c - \bar{G}_i^i = T \int_{1/T_i}^{1/T} (\bar{H}_i^c - \bar{H}_i^i) \, d \, (1/T),$$

$$(6\text{–}15)$$

in which T_i is the freezing point of the pure component i, T is the freezing point of the solution, and $\bar{H}_i^c$ and $\bar{H}_i^i$ are the molal enthalpies of component i in the solid and in the liquid states, so $\bar{H}_i^i - \bar{H}_i^c$ is the molal heat of fusion $\Delta\bar{H}_i$. Eq. (6–15) is exact. To integrate, we assume first that the heat of fusion is independent of the temperature. Then

$$\bar{G}_i^M/T = \Delta\bar{H}_i(1/T_i - 1/T) = - (\Delta\bar{H}_i/T_i^2)\theta/(1 - \theta/T_i), \quad (6\text{–}16)$$

if θ is $(T_i - T)$. For many substances the variation of the heat of fusion with the temperature is so slight that this expression may be used over a large range of temperature, in which case it is convenient to use $1/T$ as the independent variable. If only a small temperature range is considered, θ/T may be neglected relative to unity, the freezing-point depression θ is the convenient variable, and we consider component i to be the solvent S; then

$$- (\Delta\bar{H}_S/RT_S^2)\theta = \bar{G}_S^M/RT \quad (6\text{–}17)$$

and for the freezing point osmotic coefficient:

$$\phi_f = - 1000 \, \bar{G}_S^M/RT\bar{W}_S M = (1000 \, \Delta\bar{H}_S/RT_S^2\bar{W}_S)\theta/M. \quad (6\text{–}18)$$

If we assume that ϕ_f is unity, we have the usual law for the freezing-point depression:

$$\phi/M = (RT_S^2\bar{W}_S/1000 \, \Delta\bar{H}_S). \quad (6\text{–}19)$$

It should be noted, however, that the only assumption involved in Eq. (6–16) is that the heat of fusion is independent of the temperature. The usual elementary derivation through the equality of vapor pressures is needlessly complicated and makes the final expression appear much worse than it is.

For water solutions, $\Delta\bar{C}_p$ is large and it varies so rapidly with the temperature that even $d\Delta\bar{C}_p/dT$ must not be neglected in the most accurate work. If $\Delta\bar{H}_S = \Delta\bar{H}_S^0 + a(T - T_S) + b(T - T_S)^2/2$, so that $a = \Delta\bar{C}_{pS}$ and $b = d\Delta\bar{C}_{pS}/dT$, the integral becomes

$$\begin{aligned}
- \bar{G}_S^M/T &= \int_{1/T_S}^{1/T} [(\Delta\bar{H}_S^0 - aT_S + bT_S^2/2) \\
&\quad + (a - bT_S)T + bT^2/2]d(1/T) \\
&= (\Delta\bar{H}^0/T_S^2 - a/T_S + b/2)\,\theta/(1 - \theta/T_S) \\
&\quad - (a - bT_S)\ln(1 - \theta/T_S) + (b/2)\theta.
\end{aligned}$$

Expanding,

$$\begin{aligned}
- \bar{G}_S^M/T &\cong (\Delta\bar{H}_S^0/T_S^2)\theta + (\Delta\bar{H}_S^0/T_S^3 - a/2T_S^2)\theta^2 \\
&\quad + (\Delta\bar{H}_S^0/T_S^4 - 2a/3T_S^{\prime3} + b/6T_S^2)\theta^3 + \cdots, \quad (6\text{--}20)
\end{aligned}$$

For water, the "best" values of the properties are $T_S = 273.15°C$, $\Delta\bar{H}_S^0 = 6008$ J/mole, $\Delta\bar{C}_{pS} = 37.0$ J/mole deg., $d\Delta C_{pS}/dT = -0.209$ J/mole deg., $RT_i = 2270.7$ J, and $\bar{W}_{iS} = 18.0162$. The Δ's all refer to pure solvent. Then

$$\bar{G}_S^M/RT = -0.009686\,\theta - 0.0000056\,\theta^2 + 0.000000077\,\theta^3. \quad (6\text{--}21)$$

The second term becomes 0.1 percent of the whole when $\theta = 2°C$, and the third term does when $\theta = 12°C$. The measurement of the freezing-point depression is the most accurate method yet devised for the measurement of $\bar{G}^M$ for the solvent in very dilute solutions, except possibly for the osmotic pressure with solutes that cannot diffuse through a membrane permeable to the solvent. In favorable cases the depression can be determined to about 2×10^{-5} °C. For water this corresponds to a relative lowering of the vapor pressure of one part in 5×10^6, which would require a temperature control to 4×10^{-6} °C, or about one-fifth of the corresponding freezing-point depression, in addition to the difficulties of measuring such a small pressure difference. Measurements of the freezing-point depressions give the potentials at the freezing-point of the solutions. To obtain the value at the freezing-point of the solvent, or at any other temperature T'', it is necessary to know the heat of dilution, perhaps varying with the temperature, and

$$(\bar{G}_S^M/T)'' - (\bar{G}_S^M/T)' = - \int_{T'}^{T''} (\bar{H}_S^M/T^2)dT. \quad (6\text{--}22)$$

In the treatment of aqueous solutions it is customary to express the freezing-point depression as the osmotic coefficient, $\phi_f = \theta/1.860\,M$, in which M is the sum of the molalities of the solutes or the sum of the molalities of the ions for strong electrolytes, or as $j = 1 - \phi_f$. By Eq. (6–21), the isothermal osmotic coefficient of Eq. (C–43) is

$$\phi = - \bar{G}_S^M/0.0180162\,RTM = \phi_f[1 + 0.0011\,M\phi_f - 0.000026\,(M\phi_f)^2]. \quad (6\text{--}23)$$

Eq. (6–21) is based on the best values from the National Bureau of Standards for the various quantities involved.[8] Many of the calculations of freezing-point depressions are based on Lewis and Randall's computation from earlier values of these quantities, which led to 1.858 for the freezing-point constant and similar changes in Eqs. (6–21) and (6–23).

If the depression is so small that the difference in Eq. (6–22) may be neglected when T'' is the freezing-point of the solvent and T' is that of the solution, we may integrate by the Gibbs-Duhem equation to obtain

$$\bar{G}_2^{m_1}/RT - \ln M = \ln \gamma_2 = (\phi - 1) + \int_0^M (\phi - 1) \, dM/M.$$
(6–24)

Even when the change of ϕ with temperature cannot be neglected, it is sometimes convenient to make the integration in this way and to make the correction in a subsequent operation. In this case it is customary to denote the γ calculated from Eq. (6–24) by γ'.

The best experimental evidence concerning the variation of $\bar{G}^E$ of the solvent or of ϕ in very dilute solutions comes from measurements of the freezing-point depressions. Fig. 6–5 gives two illustrations, for

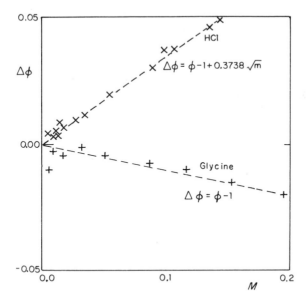

Figure 6–5. Freezing points of aqueous solutions.

[8]N. S. Osborne, H. F. Stimson, and D. C. Ginning, *J. Res. Nat. Bur. Stand.* 23, 197 (1939); N. S. Osborne, *ibid.* 643.

Table 6-2. Boiling-point elevations for water.

t	$\theta/M\phi_b$	$10^3 \times b$	$10^6 \times c$
60	0.3912	1.4	1.6
70	.4195	1.4	1.7
80	.4491	1.5	1.8
90	.4802	1.6	1.9
100	.5129	1.6	2.1

hydrochloric acid from the measurements of Randall and Vanselow,[9] $\Delta\phi = \phi_f - 1 + 0.3738\ m^{1/2}$, and for glycine from the measurements of Scatchard and Prentiss,[10] $\Delta\phi = \phi_f - 1$. Neither extends beyond 0.2 M ($m = 0.1$ for HCl) or a mole fraction of solute of 0.0036. The scattering in the most dilute solutions corresponds to the experimental error to be expected; it is larger for glycine because the determination of the concentration in a very dilute solution is much less accurate for a nonelectrolyte than for an electrolyte.

6-5. Chemical Potentials and Boiling Points

Eqs. (6–15) to (6–20) apply to the equilibrium between the liquid solution and the vapor if only one component is volatile and if $\bar{H}_i^c$ is replaced by $\bar{H}_i^g$, and if $\Delta\bar{H}_S$, $\Delta\bar{C}_p$, and $\Delta\bar{C}_p/dT$ are defined as differences between the enthalpy, heat capacity at constant pressure, and its temperature derivative of 1 mole of a volatile solvent in the liquid and in the gaseous state. Since the enthalpy difference is negative, the boiling point is raised by the addition of nonvolatile solutes. Since the enthalpy of evaporation is much larger than the enthalpy of fusion, θ/M is much smaller for boiling-point elevation than for freezing-point depression. Moreover, the boiling-point elevation cannot be measured with as great accuracy as the freezing-point depression. It has the great advantage, however, that measurements can be made at different pressures and therefore at different temperatures.

Table 6–2 gives values of $\theta/M\phi_b$, and the constants for the equation

$$\phi = \phi_b\,[1 + bM\phi_b + C(M\phi_b)^2] \tag{6-25}$$

for water are calculated from the parameters given by Smith,[11] who devised a differential boiling-point apparatus and measured the

[9]M. Randall and A. P. Vanselow, *J. Am. Chem. Soc.* 46, 2418 (1927).

[10]G. Scatchard and S. S. Prentiss, *J. Am. Chem. Soc.* 56, 1486 (1934).

[11]R. P. Smith, *J. Am. Chem. Soc.* 61, 497, 500 (1939); R. P. Smith and D. S. Hirtle, *ibid.*, 1123.

boiling-point elevations of aqueous sodium chloride solutions from 0.05 M to 4 M and from 60 to 100°C.

Figure 6–6 shows $\Delta\phi$ for sodium chloride at these temperatures and at 0 and 25°C calculated[12] for $\alpha = 1.55$ and $\beta = 0.03$ in the equation

$$\Delta\phi = \phi - AZ/\alpha - \beta M. \tag{6-26}$$

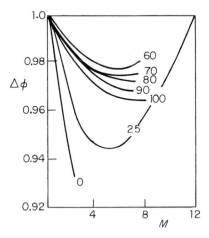

Figure 6–6. Boiling points of aqueous sodium chloride solutions: $\Delta\phi = \phi - AZ/1.55 - 0.03\ M$; the curve for $\Delta\phi$ vs. t must pass through a maximum below 70°C.

The differential-boiling-point apparatus requires a manostat. In the boiling-point determinations the pressure is kept constant at the vapor pressure of the solvent at its boiling point, and the measurement of this temperature is the simplest method of checking the pressure. The same apparatus may be used as a differential-vapor-pressure apparatus if the pressure is regulated to keep the equilibrium temperature of the solution constant, independent of the concentration. It is instructive to consider the difference between the properties of the solvent needed for the two treatments. For the differential-vapor-pressure method the vapor pressure of the pure solvent must be known as a function of the temperature, and the deviation from the ideal-gas laws as a function of the pressure. For the differential-boiling-point method we must know the enthalpy of evaporation of the solvent at its boiling-point and the difference between the heat capacities of the liquid and the vapor *at constant pressure* as a function of the temperature.

[12]G. Scatchard and L. F. Epstein, *Chem. Rev.* **30**, 211 (1942).

More precise measurements have been made by replacing the air connection between boiling solutions by an impervious diaphragm between two static systems,[13] using a null method to determine the temperatures at which the solvent and a solution of a nonvolatile solute have the same vapor pressure.

Mixtures of two or more volatile components are frequently studied by measurements of the boiling points at a constant pressure, often atmospheric, partly because this method is somewhat simpler experimentally but largely because distillations are made at nearly constant pressure. Unless the measurements are made at atmospheric pressure, any single measurement is the same as a vapor-pressure measurement, but a series of compositions is run at constant pressure instead of constant temperature. No refinements of calculations have been made with these measurements.

6-6.   Chemical Potentials and Osmotic Pressure; Light Scattering; Ultracentrifuge

The osmotic pressure is defined as the pressure at equilibrium across a membrane permeable to some but not all of the components of a solution if the concentration of components that cannot diffuse through the membrane is zero on one side. If more than one component can diffuse, the relations are more complicated. If only the solvent can diffuse, however, the relations are simple, particularly if the pressure on the solution is kept constant and there is a negative pressure on the solvent. Then

$$(\partial \bar{G}_S / \partial M_2)_p dM_2 = (\partial \bar{G}_S^S / \partial p) dp = - \bar{V}_S^S \, dP. \qquad (6\text{-}27)$$

Usually the molal volume of the solvent may be considered constant and

$$\phi_\pi = - \, 1000 \, \frac{\bar{G}_S - \bar{G}_S^S}{RT \widetilde{W}_S M}$$

$$= \frac{1000 \, \bar{V}_S^S P}{RT \widetilde{W}_S M} = \frac{1000 \, P}{RT \rho_S M} ; \qquad (6\text{-}28)$$

$1000 \, \bar{V}_S^S / \widetilde{W}_S M$ is the reciprocal of the number of moles of solute per liter of solvent. If the compressibility $\beta = - \, (\partial \ln V / \partial p)_T$ is taken into account,

$$\phi = \phi_\pi (1 + \beta P / 2). \qquad (6\text{-}29)$$

[13]O. L. I. Brown and C. M. Delaney, *J. Phys. Chem.* **58**, 255 (1954).

If the pressure on the solvent is kept constant, however,

$$\phi = \phi_\pi (\bar{V}_S/\bar{V}_S^S)(1 - \beta P/2). \tag{6-30}$$

For many years osmotic pressure was taught in physical chemistry courses because of its relation to the pressure of a gas at the same concentration but was almost never measured. Recently, the protein chemists and high polymer chemists have developed it into a very precise method. Since the molecular weight is one of the important unknowns, it is customary to use the concentration in grams of solute per kilogram of solvent, w, or per liter of solution, c, and to express the results as $P/w = RT\rho_s\phi_\pi/1000\ \bar{W}_2$ or as $P/c = RT\ \phi_\pi/1000\ \bar{W}_2$. In osmotic-pressure measurements, as in those of vapor pressure, the temperature is kept constant as the composition is changed and the temperature may be chosen at will. This makes it convenient for the comparison of solutions in different solvents, which may have very different freezing points and vapor pressures, and for the determination of the effect of molecular weight on the excess potential.

Fig. 6–7 shows P/c (P in grams/cm², c in grams/100 cm³) of polyisobutylene solutions in benzene and in cyclohexane.[14] The lines

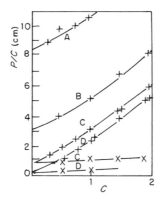

Figure 6–7. Osmotic pressure of polyisobutylene solutions: $+$ in cyclohexane; $\times$ in benzene. Values of $10^{-5}\ \bar{W}_2$ for the four curves are: A, 0.292; B, 0.781; C, 2.55; D, 7.7.

for benzene are nearly horizontal, showing that the solutions are nearly ideal. The lines for cyclohexane are drawn to have the same intercepts, that is, the same molecular weights, as for benzene, and they are drawn parallel to one another. We may write

[14]P. J. Flory, *J. Am. Chem. Soc.* **65**, 372 (1943).

$$P = (RT/1000)C_2 + (RT \ \beta_c/1000)C_2{}^2 + (RT \ \gamma_c/1000)C_2{}^3 + \cdots$$
$$= (RT/1000 \ \bar{W}_2)c_2 + (RT \ \beta_c/1000 \ \bar{W}_2{}^2)c_2{}^2$$
$$+ (RT \ \gamma_c/1000 \ \bar{W}_2{}^3)c_2{}^3 + \cdots,$$
$$P/c_2 = RT/1000 \ \bar{W}_2 + (RT \ \beta_c/1000 \ \bar{W}_2{}^2)c_2$$
$$+ (RT \ \gamma_c/1000 \ \bar{W}_2{}^3)c_2{}^2 + \cdots. \tag{6-31}$$

The agreement of the lines with the experimental points indicates that $\beta_c/\bar{W}_2{}^2$, $\gamma_c/\bar{W}_2{}^3$, ... are independent of the molecular weight, or that the interaction of a polymer unit (C_4H_8) in one molecule with a unit in another molecule is independent of the size of the molecule. The relations are not so simple with some other polymers.

In these measurements the polymer is treated as a single component, although it certainly is polydisperse, that is, it contains many different species, of different molecular weights. Since none of them can diffuse through the membrane, however, the ratio of the concentration of any species to that of any other species must remain constant. This fact justifies the treatment as a single component. In determining the molecular weight, the weight per unit volume is determined by analysis or by making up the solution to unit volume, and the number of molecules per unit volume is determined from the limit of the osmotic pressure–concentration ratio. The molecular weight so determined is the number–average molecular weight, $\bar{W}_{2N}$:

$$\bar{W}_{2N} = \Sigma_j C_j \bar{W}_j / \Sigma_j C_j = \Sigma_{cj} / \Sigma_j C_j / \bar{W}_j \tag{6-32}$$

if the separate species are represented by the subscript j.

If more than one component can diffuse through the membrane, it is convenient to keep the pressure and the composition constant on the side that does not contain the nondiffusible solute. Then this solution acts as a reservoir of all diffusible components at constant potential or activity. We shall use the subscript 1 for the solvent, 2 for the nondiffusible component, which may be complex, and $I, J, K, \ldots$ for diffusible components; the subscripts i and j will include solvent and all diffusible solutes. Then

$$d\bar{G}_i/dm_2 = (\partial \bar{G}_i/\partial m_2)_{m,p} + \Sigma_j(\partial \bar{G}_i/\partial m_j)_{m,p}dm_j/dm_2$$
$$+ (\partial \bar{G}_i/\partial P)_m \ dP/dm_2 = 0. \tag{6-33}$$

There are as many equations of this type as there are diffusible components. Multiplying each by m_i and adding for all diffusible components give

$$(\Sigma_i m_i d\bar{G}_i/dm_2)_{\bar{G}} = \Sigma_i m_i(\partial \bar{G}_i/\partial m_2)_{m,p}$$
$$+ \Sigma_{ij}m_i(\partial \bar{G}_i/\partial m_j)_{m,p}(dm_j/dm_2)_{\bar{G}}$$
$$+ \Sigma_i m_i(d\bar{G}_i/\partial P)_m(dP/dm_2)_{\bar{G}} = 0 \tag{6-34}$$

and

$$\Sigma_i m_i(d\bar{G}_i/dP)_m = \Sigma_i m_i \bar{V}_i. \tag{6-35}$$

Here $\Sigma_i m_i \bar{V}_i$ is the volume containing 1 kg of solvent minus $m_2 \bar{V}_2$. Making this substitution and applying the Gibbs-Duhem operation to the first two terms, together with the relation $(\partial \bar{G}_2/\partial m_j)_m = (\partial \bar{G}_j/\partial m_2)_m$, we obtain

$$\Sigma_i m_i \bar{V}_i (dP/dm_2)_{\bar{G}} = [(\partial \bar{G}_2/\partial m_2)_{m,p} + \Sigma_j(\partial \bar{G}_j/\partial m_2)_{m,p}(dm_j/dm_2)_{\bar{G},p}]m_2. \tag{6-36}$$

For the diffusible components, $(\partial \bar{G}_i/\partial m_j)_{m,p}$ and $\bar{V}_i$ are usually known from other measurements. Then the values of $(\partial \bar{G}_i/\partial m_2)_{m,p}$ are determined from the set of simultaneous equations (6–33). Then the measurement of the osmotic pressure determines $(\partial \bar{G}_2/\partial m_2)_{m,p}$.

Light is scattered by any material medium. Much can be learned about the shapes of particles that have dimensions of the same order as the wavelength of the light from the relation of the scattering to the angle between the incident beam and the scattered beam and to the wavelength, and from the polarization of the scattered light. We shall confine our attention here to the fraction of the light beam scattered in all directions in going unit distance through the medium, which is called the turbidity, τ. We shall also confine our attention to two-component systems, except that, as in the treatment of osmotic pressure, we may consider a polydisperse polymer as a single component.

Einstein[15] showed that the turbidity is related to the fluctuations in refractive index. The fluctuations in the refractive index due to fluctuations in density may be corrected for by comparison of the turbidity of a dilute solution with that of the solvent. There is left the effect due to fluctuations in the concentration of the solute, which Einstein related to the osmotic pressure. We shall follow Stockmayer[16] in relating it rather to the potential of the solute, since this treatment can be extended naturally to the treatment of polycomponent solutions. Applying the Gibbs-Duhem operation to Eq. (6–27) gives

$$\bar{V}_S^S \, dP = -(\partial \bar{G}_S/\partial m_2)_p \, dm_2 = \frac{m_2}{m_1}(\partial \bar{G}_2/\partial m_2)_p \, dm_2, \tag{6-37}$$

$$(\partial \bar{G}_2/\partial m_2)_{T,p} = HV_0\psi_2^2/(\tau - \tau_s), \tag{6-38}$$

$$H = 32\, \pi^3 kT\, n^2/3\, \lambda^4, \tag{6-39}$$

$$\psi_2 = (\partial n/\partial m_2)_{T,p}, \tag{6-40}$$

in which λ is the wavelength, in vacuum, of the light, and V_0 is the volume of 1 kg of solvent.

[15]A. Einstein, *Ann. Phys. 33*, 1275 (1910).
[16]W. H. Stockmayer, *J. Chem. Phys. 18*, 58 (1950).

If $\bar{G}_2$ is given by Eq. (6–3),

$$\frac{d\bar{G}_2}{dm_2} = RT\left(\frac{1}{m_2} + 2\beta + \frac{3}{2}\gamma m_2 + \cdots\right),$$

(6–41)

so

$$\frac{HV_0\psi_i^2 m_2}{RT(\tau - \tau_s)} = 1 + 2\beta m_2 + \frac{3}{2}\gamma m_2^2 + \cdots.$$

(6–42)

The quantities m_2 and ψ_2 are not measured directly. Instead we determine $w_2 = \bar{W}_2 m_2$ and $(\partial n/\partial c_2)_{T,p} = \psi_2/\bar{W}_2$, so the product $\psi_2^2 m_2$ corresponds to $(\partial n/\partial w_2)_{T,p}^2 \bar{W}_2 c_2$. Then the extrapolated value of Eq. (6–42) at zero concentration gives the relation[17].

$$\bar{W}_2 = \Sigma_j(\partial n/\partial c_j)_{T,p}^2 \bar{W}_j c_j/\Sigma_j(\partial n/\partial c_j)_{T,p}^2 c_j.$$

(6–43)

In most cases $(\partial n/\partial c_j)_{T,p}$ is nearly independent of the concentration. For many high polymers it is independent of the molecular weight. If it is the same for all species it may be taken outside the sums in Eq. (6–43) and canceled, to give

$$\bar{W}_{2W} = \Sigma_j\bar{W}_j c_j/\Sigma_j c_j = \Sigma_j\bar{W}_j^2 m_j/\Sigma_j\bar{W}_j m_j.$$

(6–44)

If the system is polydisperse, $\bar{W}_{2W}$ obtained from light scattering is larger than $\bar{W}_{2N}$ obtained from osmotic pressure. The ratio may be taken as a measure of the heterogeneity.

Stockmayer's treatment of systems of C components is to take $(\partial\bar{G}_i/\partial m_k)_{T,p,m} = a_{ik}$; $|a_{ik}|$ is the determinant of all the a_{ik} excluding the first component, the solvent, and $A_{ij} = \partial |a_{ij}|/\partial a_{ij}$ is the cofactor of element a_{ij} in $|a_{ik}|$. Then

$$(\tau - \tau_s) = HV_0 \Sigma_i \Sigma_j \psi_i\psi_j A_{ij}/|a_{ij}|.$$

(6–45)

The light-scattering measurements are made at constant pressure, so it is not possible to change the concentration of component 2 and keep the potentials of all other components constant as in the osmotic-pressure measurements. There is a great advantage, however, in keeping constant the potentials of all solutes other than 2. This requires that their concentrations change with that of component 2.

With a two-component system in an equilibrium centrifuge, if the densities and the activity coefficient are independent of the distance x from the axis,

$$d(\ln c_2)/d(x^2) = [\bar{W}_2(1 - \rho_1/\rho_2)\omega^2/2RT] = A\bar{W}_2,$$

(6–46)

$$dc_2/d(x^2) = A\bar{W}_2 c_2,$$

(6–47)

---

[17]This simplification for polycomponent systems is possible at the limit because all terms involving two or more molecules disappear there.

$$d \ln [dc_2/d(x^2)] = d \ln c_2, \tag{6-48}$$

$$d\, n/d(x^2) = \psi_2 dc_2/d(x^2), \tag{6-49}$$

$$\frac{d \ln [dn/d(x^2)]}{d(x^2)} = \frac{d \ln [dc_2/d(x^2)]}{d(x^2)} = \frac{d \ln c_2}{d(x^2)} = A\bar{W}_2. \tag{6-50}$$

These equations give a means of determining the molecular weight. For a polycomponent system in which $\rho_i = \rho_j = \rho_2$ and $\psi_i = \psi_j = \psi_2$,

$$\bar{W}_2 \frac{dn}{d(x^2)} = \psi_2 \Sigma_i dw_i/d(x^2) = \psi_2 \Sigma_i A\bar{W}_i c_i, \tag{6-51}$$

$$\ln [dn/d(x^2)] = \ln \psi_2 + \ln A\, \Sigma_i \bar{W}_i c_i, \tag{6-52}$$

$$\frac{d \ln [dn/d(x^2)]}{d(x^2)} = \frac{d\, A\, \Sigma_i \bar{W}_i c_i}{A\, \Sigma_i \bar{W}_i c_i d(x^2)} = \frac{A\, \Sigma_i \bar{W}_i^2 c_i}{\Sigma_i \bar{W}_i c_i}. \tag{6-53}$$

The molecular weight $\Sigma_i \bar{W}_i^2 c_i/\Sigma_i \bar{W}_i c_i = \bar{W}_{2Z}$ is called the Z-average molecular weight. In a polydisperse system it is larger than the weight-average molecular weight. If there is random distribution in the degree of polymerization, the relations between the number-average, weight-average, and Z-average molecular weights is

$$\bar{W}_{2N} = \bar{W}_{2W}/2 = \bar{W}_{2Z}/3. \tag{6-54}$$

The consideration of variation in activity coefficients in the centrifuge is a complicated one.

6-7. Chemical Potentials and Solubility

The effect of the addition of one solute upon the activity of another can sometimes be measured by determining its effect upon the solubility of a crystalline solid. The potential of the second component is maintained constant at the value of the potential of the solid but its concentration will change if its excess potential is altered. For a nonelectrolyte,

$$\bar{G}_i/RT = \bar{G}_i^c/RT = \bar{G}_i^s/RT + \ln m_i + 2\Sigma_j \beta_{ij} m_j,$$

and

$$\ln (m_i/m_i^0) = -2\Sigma_j \beta_{ij} m_j + 2\beta_{ii} m_i^0, \tag{6-55}$$

in which m_i^0 is the solubility when the component i is the only solute.

If the solid is a compound that dissociates in solution into ν_k moles of component k and ν_ℓ moles of component ℓ,

$$\ln (m_k^{\nu_k} m_\ell^{\nu_\ell}/m_k^{0\nu_k} m_\ell^{0\nu_\ell}) = -2\Sigma_j(\nu_k \beta_{jk} + \nu_\ell \beta_{j\ell}) m_j$$
$$+ 2(\nu_k^2 \beta_{kk} + 2\nu_k^{\nu_\ell} \beta_{k\ell} + \nu_\ell^2 \beta_{\ell\ell}) m_i^0. \tag{6-56}$$

Most of the accurate measurements have been made with electrolytes for which it is necessary to add to Eq. (6–56) a term for the electrostatic interaction:

$$
\begin{aligned}
\ln \left(m_k^{\nu_k} m_\ell^{\nu_\ell}/m_k^{0\nu_k} m_\ell^{0\nu_\ell}\right) = & - (\nu_k + \nu_\ell)\, z_k z_\ell\, A\{(I/2)^{1/2}/[1 + \alpha(I/2)^{1/2}] \\
& - (I^0/2)^{1/2}/[1 + \alpha(I^0/2)^{1/2}]\} \\
& - 2\Sigma_j(\nu_k\beta_{jk} + \nu_\ell\beta_{j\ell})m_j \\
& + 2(\nu_k^2\beta_{kk} + 2\nu_k\nu_\ell\beta_{k\ell} + \nu_\ell^2\beta_{\ell\ell})m_i^0.
\end{aligned}
\tag{6–57}
$$

If the whole right-hand side of Eq. (6–57) is zero, the solubility-product constant results. For dilute solutions the β terms may often be ignored. The effect of electrostatic interaction is greater than is usually recognized, and may even swamp out the common-ion effect. We should expect it to preponderate if the salt is not extremely insoluble, if the valence type is unsymmetric, and if it is the ion of higher valence, and therefore with lower exponent, that is added. Thus the concentration of hydroxyl ion in equilibrium with solid strontium hydroxide is increased on the addition of strontium nitrate, and the concentration of nitrate ion in equilibrium with solid strontium nitrate is increased on the addition of strontium hydroxide.[18]

This method gives an interesting check on the electrostatic interaction because the valences of the solute salt and of the solvent salt can be varied independently. Fig. 6–8 shows some measurements of the solubilities of some complex cobalt compounds for which the β and higher terms compensate or slightly overcompensate for the denominator of the electrostatic term, so that the solubility follows the Debye-Hückel limiting law to moderately high concentrations.[19]

6–8. Chemical Potentials and Electromotive Force

The *electromotive force* of an electric cell is the difference in potential between two pieces of the same metal, usually copper, one of which is attached to each electrode of the cell. Usually the whole cell is at the same temperature and pressure. The simplest way to treat this electromotive force is to consider it as the measure of the reversible work at constant temperature and pressure of transferring unit, usually infinitesimal, quantity of electricity from one piece of metal to the other and therefore as a measure of the increase in free energy in the change of state that accompanies this transfer of electricity. The most important step in the understanding of electromotive force

[18]C. L. Parsons and C. L. Perkins, *J. Am. Chem. Soc.* 32, 1387 (1910).
[19]J. N. Brönsted and V. K. La Mer, *J. Am. Chem. Soc.* 46, 555 (1924).

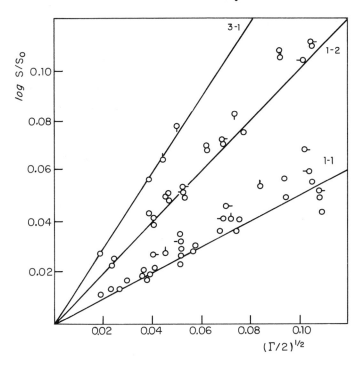

Figure 6–8. Salting in of electrolytes of different valence types. Solvent salts are: ○ 1–1, NaCl, KNO₃; ○ 2–1, BaCl₂; ○ 1–2, K₂SO₄; ○ 2–2, MgSO₄; ○ 1–3, K₃CO(CN)₆. Dissolved salts are: 1–1, [Co(NH₃)₄C₂O₄] [Co(NH₃)₂(NO₂)₂C₂O₄], [Co(NH₃)₄NO₂CNS] [Co(NH₃)₂(NO₂)₂C₂O₄]; 1–2, [CO(NH₃)₄C₂O₄]₂ [S₂O₆]; 3–1, [Co(NH₃)₆] [Co(NH₃)₂(NO₂)₂C₂O₄]₃.

is the realization of the necessity of determining this change of state.

It is the usage of American physical chemists to call the electromotive force positive if positive charge tends to pass through the cell from left to right. Then for the change of state when $N\mathfrak{F}$ coulombs of positive electricity pass through the cell from left to right,

$$\Delta G = - E N \mathfrak{F}, \tag{6–58}$$

in which E is the electromotive force. A cell consists of two electrodes and one or more solutions of ions. The change of state is the sum of the changes at the electrodes and those at each boundary between two solutions. The changes at solution boundaries, usually called *liquid junctions*, are often complicated, and we shall postpone discussion of them to Sec. 8–9.

Here we consider only one simple cell containing hydrochloric acid at concentration m, with a platinized platinum electrode saturated with hydrogen at a pressure p_{H_2} and a silver–silver chloride electrode. We write such a cell

$$H_2(p_{H_2}), \; H^+Cl^-(m), \; AgCl(s), \; Ag(s).$$

The changes of state at the two electrodes are

$$\tfrac{1}{2}H_2(p_{H_2}) = H^+(m) + \epsilon^-,$$
$$AgCl(s) + \epsilon^- = Ag(s) + Cl^-,$$

in which ϵ^- represents the electron, and the total change of state for one Faraday is

$$AgCl(s) + \tfrac{1}{2}H_2(p_{H_2}) = Ag(s) + H^+Cl^-(m).$$

Therefore

$$E\mathfrak{F} = -\bar{G}_{Ag(s)} - \bar{G}_{H^+Cl^-(m)} + \bar{G}_{AgCl(s)} + \tfrac{1}{2}\bar{G}_{H_2(p_{H_2})}.$$

It is customary to define *molal electrode potentials*, E_H^0, E_{AgCl}^0, and so on, by the relation

$$(E_H^0 - E_{AgCl}^0)\,\mathfrak{F} = -\bar{G}_{Ag(s)} + \bar{G}_{AgCl(s)} - \bar{G}_{H^+Cl^-}^w + \tfrac{1}{2}\bar{G}_{H_2}^0,$$

and values of E^0 for various electrodes on this basis are usually tabulated as differences from the hydrogen electrode, that is, with the arbitrary definition that the potential of the hydrogen electrode is zero. If we stick to the definition of electric potential given above, only differences in potential are measurable so we can make one arbitrary assumption without any loss of generality. Then

$$E\mathfrak{F} = (E_H^0 - E_{AgCl}^0)\,\mathfrak{F} - \bar{G}_{HCl}^{mw} + \tfrac{1}{2}(\bar{G}_{H_2} - \bar{G}_{H_2}^0).$$

When hydrogen electrodes are used to study liquid solutions, it is customary to correct the hydrogen pressure to 1 atm partial pressure so that the last term is zero. Lewis and Randall define E_0' by the relation

$$E + (2RT/\mathfrak{F})\ln m_2 = E_0',$$

so that

$$E_0' = (E_H^0 - E_{AgCl}^0) - \bar{G}_{HCl}^{ew}/\mathfrak{F}.$$

It is convenient to go still further in taking deviation functions. Fig. 6–2 shows the results of several observers for $E_0' - 59.62\,m^{1/2} + 90m$. The first term gives the Debye-Hückel limiting law, and the second is a purely empirical linear correction which is approximately equivalent to multiplying the limiting law by $1/(1 + 1.5m^{1/2})$.

This cell is not strictly without liquid junction. The solution around the left-hand electrode is saturated with hydrogen and that around the right-hand electrode is saturated with silver chloride. If there were no liquid junction, the cell reaction would usually occur without passage of current. For this cell it appears that the disturbance is not serious, for Nonhebel[20] made measurements with all the solution saturated with both hydrogen and silver chloride, with results that differ very little from those of other observers. With the corresponding cell with mercury-calomel electrode, however, mixing the solutions is very disturbing.

It is necessary to correct the electromotive force for the chloride ion coming from the dissolved silver chloride. The solubility product is about 2.4×10^{-10}. So the concentration of chloride ion is related to that of hydrogen ion approximately by

$$(Cl^-) = (H^+)[1 + 2.4 \times 10^{-10}/(H^+)^2].$$

In 10^{-4} molal acid the concentration excess is 2.4 percent, which corresponds to 0.6 mV, but at 10^{-3} molal acid, it is only 0.006 mV.

6-9. Chemical Potentials and Chemical Equilibria

In liquid solutions, as in gases, there are compensations between the deviations of the reactants and those of the products so that the change in the equilibrium constant of a chemical reaction is often less than the change in the activity of one of the components.

We repeat the Gibbs equilibrium condition

$$\Sigma_i \nu_i \bar{G}_i = 0 \qquad (1\text{--}1)$$

and define the equilibrium constant K_m by the first equality of Eq. (6–59):

$$\ln K_m = \Sigma_i \nu_i \ln m_i$$
$$= \ln K_m^0 - \Sigma_i \nu_i \left(2\Sigma_j \beta_{ij} m_j + \Sigma_{jk} \tfrac{3}{2} \gamma_{ijk} m_j m_k + \cdots \right). \qquad (6\text{--}59)$$

If the reaction involves ions,

$$\ln K_m = \ln K_{m0} + [A(I/2)^{1/2}/(1 + \alpha(I/2)^{1/2}] \Sigma_i \nu_i z_i^2$$
$$- \Sigma_i \nu_i (2\Sigma_j \beta_{ij} m_j + \Sigma_{jk} \tfrac{3}{2} \gamma_{ijk} m_j m_k + \cdots). \qquad (6\text{--}60)$$

Similar expressions may be written for $\ln K_c$ and $\ln K_x$. The latter will be complicated by the fact that addition of one component affects the mole fraction of every component. Many chemical equilibria in dilute solutions have been carefully studied. Reactions that

---

[20]G. Nonhebel, *Phil. Mag. 2*, 1085 (1926).

create or destroy a highly polar group are much more important in solutions than in gaseous reactions. Even in solutions, however, the reactions in which corrections to the ideal laws are important are those in which two or more ions react.

If an ion combines with a neutral molecule, the product is another ion with the same charge as the first, so that $\Sigma_i \nu_i z_i^2$, and therefore the approximate correction for electrostatic interaction, is zero. If two ions of valence z_A and z_B combine to form a molecule of valence $z_A + z_B$, $\Sigma_i \nu_i z_i^2$ is $2z_A z_B$, so that the electrostatic interaction makes an increase of the ionic strength favor association if the sign of the charges is the same, and favor dissociation if the charges are different.

6-10. Immiscible Liquids and Solid Solutions

If a binary liquid mixture separates into two phases in equilibrium, the potentials of the two components must be the same in each. The determination of the composition of the two saturated phases at each temperature enables the calculation of two parameters as functions of the temperature. They may be expressed in either the mole-fraction or the volume-fraction system.[21] We shall use the simpler mole-fraction system to illustrate, and the subscripts α and β to indicate the two phases. Then

$$\begin{aligned}
\bar{G}_1^M/RT &= \ln \bar{N}_1^\alpha + (1 - \bar{N}_1^\alpha)^2[g_{12}^0 - g_{12}'(1 - 4\bar{N}_1^\alpha)] \\
&= \ln \bar{N}_1^\beta + (1 - \bar{N}_1^\beta)^2[g_{12}^0 - g_{12}'(1 - 4\bar{N}_1^\beta)], \\
\bar{G}_2^M/RT &= \ln (1 - \bar{N}_1^\alpha) + (\bar{N}_1^\alpha)^2[g_{12}^0 - g_{12}'(3 - 4\bar{N}_1^\alpha)] \\
&= \ln (1 - \bar{N}_1^\beta) + (\bar{N}_1^\beta)^2[g_{12}^0 - g_{12}'(3 - 4\bar{N}_1^\beta)].
\end{aligned} \quad (6\text{-}61)$$

When there is separation into two phases, the curve for the free energy of mixing has two points with a common tangent. If the curve is continuous through the metastable region, there is another point between them with a tangent that has the same slope, and two inflections. At the critical mixing point the three points of tangency and the inflection points merge to a single point. We shall consider only the case in which g_{12}' is zero. Then

$$\begin{aligned}
\bar{G}/RT &= \bar{N}_1 \ln \bar{N}_1 + (1 - \bar{N}_1) \ln (1 - \bar{N}_1) + g_{12}^0 \bar{N}_1(1 - \bar{N}_1) \\
&\quad + g_{12}' \bar{N}_1(1 - \bar{N}_1)(1 - 2\bar{N}_1),
\end{aligned} \quad (6\text{-}62)$$

$$\begin{aligned}
\frac{d(\bar{G}/RT)}{d\bar{N}_1} &= \ln \bar{N}_1 - \ln (1 - \bar{N}_1) + g_{12}^0(1 - 2\bar{N}_1) \\
&\quad + g_{12}'(1 - 6\bar{N}_1 + 6\bar{N}_1^2),
\end{aligned} \quad (6\text{-}63)$$

$$\frac{d^2(\bar{G}/RT)}{d\bar{N}_1^2} = \frac{1}{\bar{N}_1} + \frac{1}{1 - \bar{N}_1} - 2g_{12}^0 - 6g_{12}'(1 - 2\bar{N}_1). \quad (6\text{-}64)$$

---

[21] G. Scatchard and W. J. Hamer, *J. Am. Chem. Soc.* 57, 1805 (1935).

From symmetry considerations we see that the critical mixing composition must be $\bar{N}_1 = \bar{N}_2 = 0.5$, and therefore $g_{12}^0 = 2$.

For the inverse problem of determining the composition of the phases in equilibrium, when the parameters of the excess free energy of mixing are known, there is no restriction as to the number of parameters. We may use the Gibbs method of determining the composition of the two points on the curve for the free energy of mixing that have a common tangent. This method is not capable of great accuracy, however, particularly when one of the compositions is near one side of the diagram. Any degree of precision may be obtained with sufficient patience by use of a method similar to that of Fig. 3–3.[22] The potential of component 2 is plotted against that of component 1. The intersection of the two stable branches will give the equilibrium potentials and compositions.

If a liquid solution and a solid solution are in equilibrium, the compositions of the two saturated phases at each temperature permit the calculation of two parameters as functions of temperature, provided that the change in potential on melting is known for each component.[23] These may be determined as in Sec. 6–4. If the parameters are known in one phase, the two determined from the equilibrium compositions may be attributed to the other phase; otherwise one parameter must be attributed to each phase. The determination of the composition of a saturated solid solution is extremely difficult. If the parameters are known, the potentials and composition of the saturated phases are obtained by plotting the potential of component 2 against that of component 1 for each phase.

Even one parameter for each phase is sufficient to explain a wide variation in behavior. We shall call the liquid phase α and the solid phase β. If both phases are ideal, the liquidus and solidus on the temperature–mole fraction diagram form a lens-shaped two-phase region almost symmetric around the straight line between the melting points. If $g_{12}^\alpha = g_{12}^\beta$, the curves approach the ideal curves asymptotically at each end. If the g's are positive, they spread more in the middle; if the g's are negative, the curves spread less. If g_{12}^β is less positive than g_{12}^α, the curves are raised above the ideal curves. The usual case is that g_{12}^β is more positive than g_{12}^α, in which case the curves are lower than the ideal. If g_{12}^β is sufficiently positive to give separation into two solid phases, we have a eutectic or a peritectic point.

The method is general enough to cover a very large range of phase equilibria. In the $\bar{G}_2$ vs. $\bar{G}_1$ plot, a phase consisting of a pure component 1 gives a vertical line, a pure component 2 gives a horizontal line, a

[22]G. Scatchard, *J. Am. Chem. Soc. 62*, 2426 (1940).
[23]G. Scatchard and W. J. Hamer, *J. Am. Chem. Soc. 57*, 1809 (1935).

compound of constant composition gives a straight line with slope $-\nu_1/\nu_2$, and an ideal solution gives a rectangular hyperbola.

We noted in Sec. 6–4 that a solid consisting of one pure component in equilibrium with a liquid mixture is the ideal, probably nonexistent, limit of extremely dilute solid solutions. This is equally true of a pure solid compound in equilibrium with a liquid containing an excess of one of its components, and this case is easier to illustrate on a diagram of free energy vs. mole fraction. Fig. 6–9 shows $\bar{G}/T$ vs. $\bar{N}_B$ for a

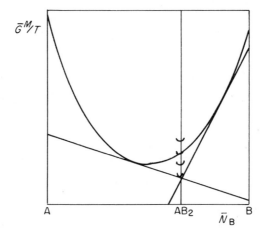

Figure 6–9. Equilibrium between ideal solution and solid compound.

liquid mixture of A and B and for a solid compound AB_2 that forms solid solutions with each component. The shapes of the curves are assumed to be independent of the temperature, and the standard state is the unmixed liquid phase. So the liquid curve is unchanged with changing temperature. The four solid curves represent four different temperatures. The top one is so high that the solid is never stable, the others decrease regularly, and the second is just below the melting point. The tangents are drawn only for the lowest curve. The practical criteria for a pure compound are that the difference between the composition of the saturated solid phases and the stoichiometric composition is less than the analytic precision and that the difference between the potential of the compound in the saturated phases and that of the pure compound, that is, the difference between the intersections with the perpendicular of the tangents and the solid curve, is smaller than the error of the determination. The practical criteria for a pure component are the same, but in this case the solid curve starts with a slope of $-\infty$ but must pass through a minimum within the error of measurement, so even the approximate case is difficult to illustrate graphically.

Fig. 6–9 also illustrates the fact that the top of the melting-point curve of a compound must be horizontal unless the compound is completely undissociated in the liquid solution. This is best shown analytically. In the liquid phase,

$$\bar{G}_{A_a B_b} = a\bar{G}_A + b\bar{G}_B, \qquad (6\text{–}65)$$

$$d\bar{G}_{A_a B_b}/d\bar{N}_a = ad\bar{G}_A/dN_a + bd\bar{G}_B/dN_a. \qquad (6\text{–}66)$$

At the stoichiometric composition, $\bar{N}_A = a/(a + b)$ and $\bar{N}_B = b/(a + b)$, so that

$$d\bar{G}_{A_a B_b}/dN_a = (a + b)(\bar{N}_A d\bar{G}_A/d\bar{N}_A + \bar{N}_B d\bar{G}_B/d\bar{N}_A) = 0 \qquad (6\text{–}67)$$

by the Gibbs-Duhem relation. However, $d\bar{G}_{A_a B_b}/dT$ is finite, so $dT/d\bar{N}_a$ is zero.

The equilibrium between two liquid phases in polycomponent systems has many applications. If two solvents are nearly immiscible, the distribution of a solute that gives an ideal solution in one of them may give a means of studying the deviations from ideality in the other or the effect on its potential of the addition of another solute soluble only in the second solvent.

If the two liquids are slightly miscible, the miscibility is increased by the addition of a solute that forms nearly ideal solutions with both, and it is decreased by the addition of a solute that has large positive deviations with one but not with the other. An example of the latter is the salting out of nonelectrolytes from aqueous solutions.

The distribution of solutes between two liquids is made the basis of many important separation processes.

7. Electrolytes. The Ion Atmosphere and the Electrostatic Potential

7–1. Historical

The most obvious characteristic of electrolytes, which distinguishes them from other substances, is the ability to exist as ions that conduct the electric current. This property makes it possible to study electrolyte solutions by two methods that are not applicable to nonelectrolyte solutions. In the first place, it is often possible to construct an electric cell the electromotive force of which can be measured easily and accurately, and in which the net material effect of passing current is the transfer of an electrolyte from one solution to another. Since the equilibrium electromotive force multiplied by the faraday is the reversible electric work of transferring one equivalent at constant temperature and pressure, it is also the difference in chemical potential per equivalent.

The second method is the study of conductance itself. The conductance has no direct thermodynamic significance, but it is nevertheless very useful as a means of studying the interaction between ions. The specific conductance L' of a solution is the reciprocal of the resistance of a cube 1 cm on an edge. It is made up of two parts: the specific conductance L of the ions of the solute and the conductance L_0 of the ions of the solvent and of small amounts of impurities that cannot be removed from it. The molal conductance Λ is the conductance of 1 mole of the solute between electrodes 1 cm apart and of sufficient area to contain the mole of solute and the accompanying solvent between them. It follows that

$$\Lambda = 1000 \, L/C = 1000(L' - L_0)/C. \qquad (7–1)$$

By the study of transference, that is, the movement of the electrolyte as a whole due to the passage of the current, combined with a study of conductance, it is possible to determine the conductances of the different ion species Λ_+ and Λ_-.

The specific conductance of a typical salt solution is very roughly proportional to the concentration of the salt. The first picture was that of Grotthuss,[1] that the ions are formed by the electric field first orienting the dipole salt molecules and then forcing away the ends of the chain, with a resultant shift of partners all along the line. The great achievement of Arrhenius was the realization that the molecules of electrolytes break into ions independently of any external field.[2] This much of Arrhenius' theory—and it is by far the most important part—has been established by many kinds of experimental verification. In fact, it is only within recent years that experimental technique has been developed sufficiently to measure an increased dissociation of weak electrolytes due to an external field.

Arrhenius announced his theory at almost the same time that van't Hoff announced his theory of osmotic pressures and Raoult announced his law of vapor pressures. The rapid development in the next few years due to the interplay of these theories marks the birth of physical chemistry as a separate branch of science. Arrhenius assumed that ions behave as perfect solutes and that the conductance is proportional to the concentration of the ions, or that the molal conductance is proportional to the fraction of the electrolyte that exists as ions. These supplementary assumptions have since been shown to be inexact, and sometimes grossly so. It is unfortunate that the name of Arrhenius is today associated with them rather than with the more fundamental assumption of the existence of ions independently of the external field.

The theory of Arrhenius was immediately tested by Ostwald[3] with solutions of slightly ionized organic acids and bases, and the agreement was so satisfactory that the special case of the law of mass action applied to a dissociating substance at different concentrations is known as the Ostwald dilution law. The theory was also tested with salt solutions by Planck,[4] who found that the law of mass action does not apply to highly ionized substances. Much time was spent during the succeeding years in attempts to express the behavior of salt solutions by more or less empirical relations. Much of this time could have been saved had it clearly recognized that, if the law of mass action does not apply, it is a thermodynamic necessity that neither the osmotic coefficient nor the activity coefficient is a measure

[1]C. J. T. de Grotthuss, *Ann. Chim. 58*, 54 (1806).

[2]S. Arrhenius, *Bijhanj till K. Svenska Vet. Akad. Handl.* No. 13 (1884); *Sixth Circular of the British Association Committee for Electrolysis* (May 1887); *Z. phys. Chem. 1*, 631 (1887).

[3]W. Ostwald, *Z. phys. Chem. 2*, 36 (1888).

[4]M. Planck, *Ann. Phys. 34*, 139 (1888).

of the degree of dissociation, and it is highly probable therefore that the molal conductance ratio is not.

The experimental evidence against the applicability of the laws of ideal solutions to ions was, in fact, ready at hand when the theory of Arrhenius was proposed. We may formulate the law of mass action for the formation of a molecule from ν_+ cations and ν_- anions without the formation of intermediate ions (setting $\nu = \nu_+ + \nu_-$) as

$$\alpha/(1 - \alpha)^\nu = K\nu_+{}^{\nu_+}\nu_-{}^{\nu_-}C^{\nu-1}, \tag{7-2}$$

in which α is the fraction existing as undissociated molecule and $(1 - \alpha)$ is therefore the fraction ionized, C is the total concentration, and K is the reciprocal of the ionization constant, or the association constant. Since ν is a positive integer greater than 1, α must be proportional to an integral power of C when α is so small that $(1 - \alpha)$ does not differ essentially from 1. Arrhenius' assumption that the conductance measures the concentrations of the ions is equivalent to

$$(\Lambda_0 - \Lambda)/\Lambda_0 = \alpha. \tag{7-3}$$

In dilute solutions, therefore, the assumptions underlying Eqs. (7–2) and (7–3) require that $(\Lambda_0 - \Lambda)/\Lambda_0$ be proportional to an integral power of the concentration. At that time Kohlrausch[5] had already made accurate measurements of the conductance of salt solutions and had stated the law for moderately concentrated solutions,

$$(\Lambda_0 - \Lambda)/\Lambda_0 = kC^{1/3}, \tag{7-4}$$

in which Λ is the molal conductance at the concentration C, Λ_0 is the molal conductance at infinite dilution, and k is a constant. A little later Kohlrausch[6] announced the law for very dilute solutions,

$$(\Lambda_0 - \Lambda)/\Lambda_0 = k'C^{1/2}. \tag{7-5}$$

Eqs. (7–2) and (7–3) are quite irreconcilable with Eq. (7–4) or (7–5).

It has since been found that for strong electrolytes 1 minus the osmotic coefficient and 1 minus the activity coefficient, usually expressed as the negative of the logarithm of the activity coefficient, are also generally proportional to the square root of the concentration in very dilute solutions. Either of these quantities should be a measure of the degree of association if the solutions were species-ideal. The constant of proportionality depends only upon the valence type of the electrolyte and upon the dielectric constant of the solvent and the temperature. For the conductance function the constant depends also upon the limiting conductance at zero concentration. The variation

[5] F. Kohlrausch, *Ann. Phys.* 26, 1661 (1885).
[6] F. Kohlrausch, *Z. Elektrochem.* 13, 333 (1907).

with the square root of the concentration, which has also been found with many other properties of electrolyte solutions, shows that there is something quite different from anything found with nonelectrolyte solutions, even when their deviations from the ideal-solution laws are very large. The dependence upon the ionic type and upon the dielectric constant of the solvent suggests immediately that the additional effect is due to the long-range interaction of the electric charges.

Even for weak acids and bases, for which Ostwald found good agreement with the simple assumptions of Arrhenius, the agreement is improved if the square-root variations of the activity coefficient and of conductance functions are taken to be the same as for other ions of the same valence type.[7,8] The neglect of these factors is not so important with weak electrolytes as with strong, for the concentration of ions is always small.

Van Laar[9] was the first to suggest that the interaction of the electric charges is responsible for the unique behavior of electrolyte solutions. Apparently Noyes[10] was the first to consider the evidence from the color and optical activity of salt solutions that strong electrolytes are completely ionized. He rejected the idea because of the approximate agreement of the degree of ionization determined from conductance, from freezing points, and from electromotive force by the species-ideal theory of Arrhenius. Sutherland[11] and Bjerrum[12] accepted the idea of complete ionization and considered the deviations of these properties from those of completely ionized ideal solutes as due to the electric interactions of the ionic charges. We have seen from the discussion of association and compound formation in nonelectrolyte solutions that the degree of association cannot be measured exactly or even defined accurately if the deviations from the ideal solution laws are large. The longer the range of attraction, the greater is the difficulty; and it is therefore even greater with electrolyte solutions than with nonelectrolytes. What we need to retain of the theory of complete ionization, however, is that there is no advantage in assuming the existence of neutral molecules of a typical salt.

Milner[13] was the first to calculate the effect of interionic attraction

[7]D. A. MacInnes, *J. Am. Chem. Soc. 48*, 2068 (1926).

[8]M. S. Sherrill and A. A. Noyes, *J. Am. Chem. Soc. 48*, 1861 (1926).

[9]J. J. van Laar, *Z. phys. Chem. 18*, 274 (1895); *Arch. Musée Teyler*, Series II, 7, 59 (1902).

[10]A. A. Noyes, *Science 20*, 584 (1904).

[11]W. Sutherland, *Phil. Mag.* (6) *12*, 1 (1907).

[12]N. Bjerrum, *Proc. VII Internat. Cong. App. Chem. (London) 10*, 55 (1909). This is included with other papers, all translated into English, in N. Bjerrum, *Selected Papers* (Copenhagen, 1949), p. 56.

[13]S. R. Milner, *Phil. Mag.* (6) *23*, 551 (1912).

on the conductance and on the osmotic coefficient. Assuming that the interaction is given by Coulomb's law and that the distribution of ions around any one taken as the central ion is given by Boltzmann's distribution law, he applied the theory of probability to the calculation of the distribution and the energy involved. Except for minor errors in his very laborious calculations, his results agree with those obtained later by other methods. We may even say that his method has the best theoretical foundation of any yet devised. However, the complexities of his theory and particularly of his calculations, together with the fact that he failed to derive an analytic expression for his results, prevented his work from receiving the recognition it deserved.

Ghosh[14] reattacked the problem, considering the ions to form a sort of expanded crystal lattice with solvent in the empty spaces, and derived a cube-root law for the change of conductance and of osmotic coefficient with concentration. His work was certainly not ignored, but it was criticized by many workers both because of the theory and because of the treatment of the experimental data.

The next important theoretical development was that of Debye and Hückel.[15] They used the Boltzmann law as Milner did, but succeeded in obtaining a simple analytic expression by using Poisson's equation in place of the equivalent Coulomb's law. Their theory has been so successful that subsequent theoretical work has consisted either of attempts to avoid or to justify some of the unorthodox statistical mechanics of their simple treatment or of extending that treatment to include other effects and to determine other properties. The rest of this chapter and the next will be devoted to a development of the Debye picture of electrolyte solutions and his method of treating them.

7-2. The Problem

We have seen that it is possible to obtain an approximate picture of the behavior of nonelectrolyte solutions by considering the distribution of molecules to be random, and also that this treatment, like that which attributes all deviations from ideality to chemical reactions, leads to the nonideal part of the potential of the solute in dilute solutions being a linear function of the solute concentration. For electrolyte solutions there is a term proportional to the square root of the concentration that cannot be treated without taking into account the variations from random distribution. The problem of the treatment of

---

[14]C. Ghosh, *J. Chem. Soc. 113*, 449, 627, 707, 790 (1918).
[15]P. Debye and E. Hückel, *Phys. Z. 24*, 185, 305 (1923).

electrolyte solutions might be defined as the determination of the distribution of the ions.

On account of the electric charges of the ions, electrolyte solutions differ from nonelectrolyte mixtures in two important respects. The mutual energy of two ions decreases with increasing distance between the ions much less rapidly than the mutual energy of two neutral molecules—as the inverse first power instead of as the inverse sixth power of the distance of separation. There is also a duality introduced by the fact that positive and negative ions must always exist together and in such a ratio that the total charge of the negative ions is equal, but opposite in sign, to the total charge of the positive ions.

The Coulomb forces between the ionic charges are very large. To illustrate their magnitude, Debye makes the following computations. If 1 mole of sodium chloride were completely ionized and the sodium ions were removed from the chloride ions as far as the north pole of the Earth is removed from the south pole and they were separated by a vacuum or by air, they would still attract each other with a force of 50 tons. A distance of 10^{-7} cm is fairly large relative to the ionic dimensions. The electric force at this distance from the center of a univalent ion in water solution is 2×10^6 V/cm if the dielectric constant of the water is not altered by this field. If the dielectric constant is changed, the force has some value between that given above and the corresponding value in a vacuum, which is 160×10^6 V/cm.

Let us consider a solution containing a small concentration of univalent ions, and look at a unit volume that contains, on the average, one ion. Let us also assume that it is possible to locate the ions so precisely, perhaps by considering their centers of gravity, that we can say definitely of each ion that it is either inside or outside this volume element. We shall be interested in the total charge within this element —the excess of positive over negative ions—and we shall consider the observations to be made instantaneously and at equal time intervals.

First let the volume element be fixed in space. Since the number of positive ions is equal to the number of negative ions, and since there is no external electric field, the average charge will be zero. The instantaneous charge may be either positive or negative, or it may be zero if no ions are present or if there are equal numbers of positive and negative ions. If the solution is so dilute that the distribution is random, that is, if the ions are so far apart on the average that even their electrostatic interactions may be neglected, the probabilities of any given total number of ions and of any total charge in unit volume are listed in Table 7-1. The probability of any number n is $1/en!$. If there are two ions, the charge will be ± 2 half the time and 0 half the time. If there are three ions, then the charge will be ± 3 one-quarter

Table 7-1. Random distribution of ions.

Number	Total ions	Net charge (absolute)
0	0.368	0.465
1	.368	.416
2	.184	.100
3	.061	.016
4	.015	.002
5	.003	.000
6	.001	.000

of the time, and ± 1 three-quarters of the time. If the concentration increases, the fractions of higher charges will decrease. The picture of the charge might be represented by Fig. 7-1 (a), in which the arrows represent the average charge.

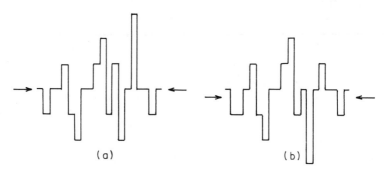

(a)　　　　　　　　　　　　(b)

Figure 7-1. Instantaneous charge in volume elements (a) fixed in space (average, 0.0), (b) fixed relative to central ion (average, −0.25).

Next let the volume element be fixed at a given distance and in a given direction from the center of a positive ion. Since this ion cannot get into the volume element, there is a possibility of one more negative ion than positive ions entering, and the average charge will be negative. Since the fixed ion will repel other positive ions and attract negative ions, the magnitude of the charge will depend upon the distance between the volume element and the fixed charge. The series of instantaneous pictures of the charge might now look like Fig. 7-1 (b). Our problem might be more definitely stated as the determination of the average distribution of the excess negative charge as a function of the distance from the positive central ion. The effect on the measurable properties of the solutions, however, is more conveniently determined in terms of other quantities that are related to this distribution. The excess charge cannot be evenly distributed throughout because the total charge, including that of the central ion, must be so distributed. On the other hand, it is not usually packed close around the central ion, but extends to great distances from it. Debye calls this distribution of the charge of opposite sign around the central ion the "ion atmosphere." It is important to remember that the total charge of the ion atmosphere is equal in magnitude, though opposite in sign, to that of the central ion.

7-3. The Solution Model

As with nonelectrolyte solutions, it is obviously impossible to make any theory that applies quantitatively and rigorously to any real solution of ions. It is necessary to take a model that is very much simpler than a real solution, but we shall make our model as nearly like the real solutions as possible. We shall assume that each ion is a sphere of radius b_i within which the distribution of the charge and the dielectric constant are spherically symmetric, independent of the medium, and not deformed by any electric field, and outside of which the dielectric constant is uniform. Two ions cannot approach more closely than the distance a_{ij}, measured from center to center, but at greater distances the interaction is given completely by the Coulomb forces between the ionic charges. Moreover, a_{ij} is not smaller than b_i or b_j. For simplicity we shall assume that the distance a_{ij} is the same for all the ions in the solution. Some of these assumptions will be made more general in subsequent treatments.

To calculate the chemical potentials we shall assume that it is possible to charge and discharge the ions reversibly, that is, to add or subtract infinitesimal fractions of electrons, and that it is possible to carry out this charging process so that, at each stage of the process, each ion bears the same fraction of its final charge. We shall assume further that during this charging process the dielectric constant and the volume of the solution remain constant. The validity of the assumptions will be discussed later.

7-4. The Electric Potential and the Distribution of Ions

Boltzmann's distribution law may be written

$$c_i = c_{i0}e^{-\psi\epsilon z_i/kT}, \tag{7-6}$$

n which c_i is the average concentration in molecules per cubic centimeter of the ith species of ions at a point where the electric potential is ψ, c_{i0} is the average concentration at a point where the potential is zero, ϵ is the protonic charge, z_i is the valence of the ith species, to be taken negative for a negative ion, T is the absolute temperature, and k is Boltzmann's constant, the gas constant R divided by Avogadro's number. Since the average electric potential differs only very slightly from zero over most of space, c_{i0} may be taken as the average concentration of the ith species over all the solution. The average electric density ρ is then given by

$$\rho = \Sigma_i c_i \epsilon z_i = \epsilon \Sigma_i c_{i0} z_i e^{-\psi \epsilon z_i/kT}. \tag{7-7}$$

Poisson's equation may be written

$$\nabla^2 \psi = - 4\pi\rho/D, \tag{7-8}$$

in which ∇^2 is the Laplacian operator, D is the dielectric constant, and π has its usual geometric significance. Eqs. (7-7) and (7-8) may be combined to eliminate ρ:

$$\nabla^2 \psi = - (4\pi\epsilon/D) \, \Sigma_i c_{i0} z_i e^{-\psi \epsilon z_i/kT}. \tag{7-9}$$

This expression cannot be integrated analytically (see, however, Sec. 8-7). To obtain a solution we assume that $\psi \epsilon z_i/kT$ is very small and expand the exponential in a power series, neglecting all but the first two terms, to give

$$\rho = \epsilon \Sigma_i c_{i0} z_i - (\epsilon^2 \psi/kT) \, \Sigma_i c_{i0} z_i^2 + \cdots. \tag{7-10}$$

The first term is zero because the total charge of the solution must be zero, so

$$\nabla^2 \psi = (\psi 4\pi\epsilon^2/DkT) \, \Sigma_i c_{i0} z_i^2 = \kappa^2 \psi, \tag{7-11}$$

in which

$$\kappa = [(4\pi\epsilon^2/DkT) \, \Sigma_i c_{i0} z_i^2]^{1/2} = [(4\pi\epsilon^2/DkTV) \, \Sigma_i n_{i0} z_i^2]^{1/2}$$
$$= [(4\pi N_A \epsilon^2/1000 DkT) \, \Sigma_i C_{i0} z_i^2]^{1/2} = [(4\pi N_A \epsilon^2/1000 DkT)\Gamma]^{1/2}, \tag{7-12}$$

where n_{i0} is the total number of ions of the ith species in the system, V is the total volume, C_{i0} is the concentration of the ith species in moles per liter, N_A is Avogadro's number, and Γ is the ional concentration in moles per liter, defined as

$$\Gamma = \Sigma_i C_{i0} z_i^2. \tag{7-13}$$

For uniunivalent electrolytes the ional concentration is equal to twice the molal volume concentration, for a 2–1 or a 1–2 electrolyte it is six times the molal concentration, for a 2–2 electrolyte it is eight times the molal concentration, and so on. Another definition of the ional concentration, which is convenient for some purposes, is the sum of the equivalent concentration of each ion multiplied by the absolute value of its valence. The ionic strength, except that it is usually expressed in moles per kilogram of solvent, is half the ional concentration. When the ional concentration is expressed in weight-concentration units we shall give it the symbol I. If the dielectric constant is independent of the electrolyte concentration for a given solvent, temperature, and pressure, the quantity κ is proportional to the square root of the ional concentration in moles per unit volume of solution. If, on the other hand, the dielectric constant is proportional to the concentration of the solvent, then κ is proportional to $I^{1/2}$, the square root of the ional concentration in moles per unit volume of solvent, and thus equally well to the square root of the ional concentration in moles per kilogram of solvent. Adopting the latter, we obtain for water

$$\kappa/(I/2)^{1/2} = 0.3240 \times 10^8[1 + 0.05217(t/100) - 0.00916(t/100)^2$$
$$+ 0.00888(t/100)^3] \text{ cm}^{-1}, \tag{7–14}$$

in which t is the Celsius temperature and I is in moles per kilogram of solvent; see Eq. (5–54). The quantity κ will be found to be very important in the theory of electrolytes. It is not a function of position in the solution and it has the dimensions of a reciprocal length. For reasons that will be brought out later in this section, Debye calls $1/\kappa$ the "thickness of the ion atmosphere."

Before considering the distribution of ions about a central ion, we shall determine the distribution about a plane surface at which there is a change in the electric potential. We shall let the surface have infinite extent and have a constant potential throughout. The origin of our rectangular coordinate system will be taken in the surface, and the x-axis is perpendicular to this surface with the solution on the positive side. The change of the potential ψ must be continuous, since $-d\psi/dx$ is the electric force and $-d^2\psi/dx^2$ is proportional to the electric density, and both of these quantities are finite. These three quantities may be something like Fig. 7–2. The electric force passes through a maximum, and the electric density passes through a maximum and a minimum. There is an excess concentration of positive charges on the left-hand side of the surface and of negative charge on the right. Such a distribution of charge is known as a Helmholtz double layer. It is important in electroosmosis and cataphoresis, as well as in the physics of colloidal and molecular ions. It has long been known that the thickness of the double layer is unimportant provided that it is small

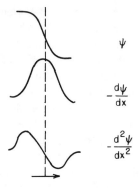

Figure 7–2. Distribution of charge about a plane surface.

relative to the radius of curvature of the surface. So the actual thickness of the surface is not important in the plane-surface problem. For ions, on the other hand, particularly in dilute solutions, the thickness is very large relative to ionic dimensions, and it therefore becomes the important factor.

In rectangular coordinates,

$$\nabla^2 \psi = d^2\psi/dx^2 + d^2\psi/dy^2 + d^2\psi/dz^2, \qquad (7\text{--}15)$$

and in our case, since the y,z planes are equipotential surfaces,

$$\nabla^2 \psi = d^2\psi/dx^2 = \kappa^2\psi. \qquad (7\text{--}16)$$

The solution, most easily verified by differentiation, is

$$\psi = Ce^{\kappa x} + C'e^{-\kappa x}. \qquad (7\text{--}17)$$

The parameters C and C' may be determined from the boundary conditions that $\psi = 0$ when $x = \infty$ and $\psi = \psi_0$ when $x = 0$. So

$$\psi = \psi_0 e^{-\kappa x}. \qquad (7\text{--}18)$$

The reciprocal of κ is the distance at which the potential falls to $1/e$ times its value at the surface. It may also be shown that it is the distance at which the electric density has fallen to $1/e$ times its value at the surface. With changed boundary conditions, Eq. (7–17) may be used to determine the variation of the potential or of the charge between the plates of a parallel-plate condenser.

For the potential around an ion we express the Laplacian in spherical coordinates:

$$\nabla^2 \psi = \frac{1}{r^2} \frac{d}{dr}\left(r^2 \frac{d\psi}{dr}\right) + \frac{1}{r^2\sin\theta} \frac{d}{d\theta}\left(\sin\theta \frac{d\psi}{d\theta}\right) + \frac{1}{r^2\sin^2\theta} \frac{d^2\psi}{d\phi^2},$$

$$(7\text{--}19)$$

and, by our assumption of spherical symmetry,

$$\nabla^2 \psi = \frac{1}{r^2} \frac{d}{dr} \left(r^2 \frac{d\psi}{dr} \right) = y^4 \frac{d^2\psi}{dy^2} = \kappa^2 \psi, \tag{7-20}$$

in which $y = 1/r$. The solution, which is again most easily verified by differentiation, is

$$\psi = Cye^{-\kappa/y} + C'ye^{+\kappa/y} = Ce^{-\kappa r}/r + C'e^{+\kappa r}/r. \tag{7-21}$$

The two constants are again to be determined from the boundary conditions. As r approaches infinity, $e^{+\kappa r}/r$ also approaches infinity, but ψ approaches zero; so C' must be zero. To determine C we make use of the fact that both ψ and $d\psi/dr$ must be continuous when $r = a$. When $r \geq a$,

$$\psi = Ce^{-\kappa r}/r \tag{7-22}$$

and

$$\frac{d\psi}{dr} = -Ce^{-\kappa r}(1 + \kappa r)/r^2, \tag{7-23}$$

and when $b \leq r \leq a$, since there is no charge within this shell, Coulomb's law may be used, to give

$$d\psi/dr = -\epsilon z_i/Dr^2, \tag{7-24}$$

$$\psi = \epsilon z_i/Dr + K. \tag{7-25}$$

From Eqs. (7–23) and (7–24), when $r = a$,

$$Ce^{-\kappa a}(1 + \kappa a) = \epsilon z_i/D, \tag{7-26}$$

$$C = \epsilon z_i e^{\kappa a}/D(1 + \kappa a), \tag{7-27}$$

and when $r \geq a$,

$$\psi = \epsilon z_i e^{-\kappa(r-a)}/Dr(1 + \kappa a). \tag{7-28}$$

When $r = a$,

$$\psi_a = \epsilon z_i/Da(1 + \kappa a) = \epsilon z_i/Da - \epsilon z_i \kappa/D(1 + \kappa a) = \epsilon z_i/Da + K, \tag{7-29}$$

$$K = -\epsilon z_i \kappa/D(1 + \kappa a); \tag{7-30}$$

and when $r = b_i$, from Eqs. (7–25) and (7–30),

$$\psi_b = (\epsilon z_i/D)[1/b_i - \kappa/(1 + \kappa a)]. \tag{7-31}$$

Evidently the first term in Eq. (7–29) or (7–31) is the contribution of the ion itself to the potential, and the second term is the contribution of the ion atmosphere. Let us stop a moment to consider the distribution of the ions in this atmosphere using the approximate equation that corresponds to our solution, that is, Eq. (7–10) split for the separate species:

$$c_j = c_{j0}[1 - \epsilon z_j \psi/kT] = c_{j0}[1 - \epsilon^2 z_i z_j e^{-\kappa(r-a)}/DkTr(1 + \kappa a)].$$
(7–32)

When $r = a$, this becomes

$$c_{ja} = c_{j0}[1 - \epsilon^2 z_i z_j/DkTa(1 + \kappa a)].$$
(7–33)

Since every factor in the second term of Eq. (7–32) except $z_i z_j$ is intrinsically positive, it is obvious that the concentration of an ion with the same sign as the central ion is always smaller than the average concentration, whereas the concentration of an ion of the opposite sign is always larger than the average. According to this approximation, the total concentration will be independent of the distance from the central ion but the ratio of ions of sign opposite to that of the central ion to those with the same sign as the central ion will increase as the distance from the central ion decreases. Eq. (7–33) shows that at the inner edge of the ion atmosphere, where $r = a$, the ratio of the concentration to the average concentration becomes nearer unity as the concentration increases. Returning to Eq. (7–32), we see that the greater the concentration the more rapidly the average concentration approaches unity with increasing distance from the central ion.

The electric density at any point is given by

$$\rho = \Sigma_j c_j \epsilon z_j = -z_i \epsilon^3 e^{-\kappa(r-a)} \Sigma_j c_{j0} z_j^2/kTDr(1 + \kappa a).$$
(7–34)

Of more interest than the density itself is dq/dr, the ratio of the quantity of charge dq in a shell of thickness dr to that thickness:

$$dq/dr = -\epsilon z_i 4\pi\epsilon^2 r e^{-\kappa(r-a)} \Sigma_j c_{j0} z_j^2/kTD(1 + \kappa a)$$
$$= -\epsilon z_i \kappa e^{\kappa a} \kappa r e^{-\kappa r}/(1 + \kappa a),$$
(7–35)

or, taking κr as the unit of distance,

$$dq/d\kappa r = [-\epsilon z_i e^{\kappa a}/(1 + \kappa a)]\kappa r e^{-\kappa r}.$$
(7–36)

Then

$$\int_{r=R}^{r=\infty} dq = [-\epsilon z_i e^{\kappa a}/(1 + \kappa a)](1 + \kappa R)e^{-\kappa R}$$
(7–37)

or

$$\int_{r=a}^{r=\infty} dq = -\epsilon z_i.$$
(7–38)

MacInnes[16] uses an integration like that in Eqs. (7–34) to (7–38)

---

[16]D. A. MacInnes, *The Principles of Electrochemistry* (Reinhold, New York, 1939).

and the fact that the charge outside a must be $- \epsilon z_i$ to determine the value of the constant C, starting with

$$\rho = - (\epsilon^2 \Sigma_j c_{j0} z_j^2 / kT) \psi = - (\epsilon^2 \Sigma_j c_{j0} z_j^2 / kT) C e^{-\kappa r} / r. \qquad (7\text{--}39)$$

His method is simpler than the one we have used for the case of spherical symmetry. In more complicated cases, however, it is necessary to use an extension of the method of Eqs. (7–23) to (7–27).

Fig. 7–3 shows the fraction of the charge in a spherical shell at a

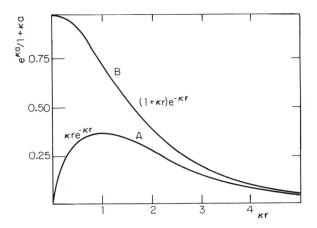

Figure 7–3. Distribution of charge about a central ion: (A) fraction of charge of ion atmosphere between κr and $(\kappa r + d\kappa r)$ divided by $d\kappa r$; (B) fraction of charge outside κr. The ordinate of either curve is to be divided by the ordinate of curve B at $\kappa r = \kappa a$ to normalize.

distance κr from the central ion (curve A), and also the fraction of the charge further from the central ion than κr (curve B). The units of the abscissas are κr; the units of the ordinates are $e^{\kappa a} / (1 + \kappa a)$. To normalize it is necessary to divide the ordinate of either curve by the value of the B curve at $\kappa r = \kappa a$. With κr as the unit of distance, the distribution is independent of the concentration except that the inner edge of the atmosphere expands with increasing κ. Moreover, more of the atmosphere is at the distance $1/\kappa$ than at any other distance. This is the best justification of the choice of $1/\kappa$ as the thickness of the ion atmosphere. It should be noted, however, that almost three-quarters of the ion atmosphere is at a greater distance from the central ion. Since κ is proportional to the square root of the concentration, the thickness of the ion atmosphere, $1/\kappa$, approaches infinity as the ion concentration approaches zero.

The most accurate tests of the Debye theory are the measurements of transport properties of electrolyte solutions, such as conductance and its relation to field strength or the frequency, transference, viscosity, and diffusion of ions. The fact that there is neither equilibrium nor spherical symmetry makes the mathematics very much more complicated than for the study of chemical potentials, so there has been very little effort to determine theoretically more than the limiting law for the effect of concentration on these properties. Even this is so complicated that we shall not attempt to follow the mathematical development, but content ourselves with presenting the physical picture upon which the theory rests and the final results as compared with experimental measurements. The detailed treatment can be found in the literature.[17] Except for the relation of transference to the electromotive force of concentration cells, these transport problems are not directly related to the study of equilibria.

7-5. The Time of Relaxation

The potential at the central ion is made up of two parts, that due directly to its own charge and that due to the ion atmosphere which is held by the charge on the central ion. If the charge on the central ion were suddenly obliterated, its direct share in the potential would also disappear instantaneously and with it the force holding the ion atmosphere. But the ion atmosphere does not disappear instantaneously because it is removed only by the Brownian movement of the ions composing the atmosphere, which slowly resume a random distribution. The time required for this dissipation depends upon the mobility of the ions and also upon the distance they must travel. Owing to this latter cause, the time is greater the smaller the concentration, for the ion atmosphere is scattered over a larger volume at low concentrations. A constant θ, with the dimension of time, introduced by Debye to make the equations dimensionless just as κ, with the dimension of reciprocal length, does in the equilibrium equations, is closely related to the time necessary for the potential at the position of the central ion due to the ion atmosphere to be reduced to $1/e$ times the value when the central ion was present, and is called the *time of relaxation*. It is defined by the relation

$$\theta = 15.34 \times 10^{-8}(z_+ + |z_-|)/(\lambda_+ + \lambda_-)kT\kappa^2, \qquad (7\text{-}40)$$

in which z_+ and z_- are the valences of the ions and λ_+ and λ_- their

[17]D. A. MacInnes, *Principles of Electrochemistry;* L. Onsager and R. M. Fuoss, *J. Phys. Chem. 34,* 2689 (1932); H. S. Harned and B. B. Owen, *The Physical Chemistry of Electrolytic Solutions* (Reinhold, New York, 1943).

equivalent conductances. For aqueous solutions at room temperature, θ lies between $10^{-11}/C$ and $10^{-10}/C$ sec when C is expressed in moles per liter.

For ordinary conductance measurements, at low field strengths and low frequencies, it is possible to treat the time-of-relaxation effect as an infinitesimal displacement from equilibrium conditions. The potential due to the ion atmosphere decreases as an inverse exponential of the distance from the central ion. Since the sign of this potential is opposite to that of the central ion, work is required to move the central ion from its equilibrium position. Expressing the effect in other terms, we may say that, on account of the finite time necessary for the formation of the ion atmosphere, the atmosphere always lags behind a moving ion, so that there is always more of the charge of opposite sign behind the ion than corresponds to the equilibrium atmosphere, whereas ahead of it there is always less. Both of these excess charges— that of the opposite sign in the rear and that of the same sign in front —tend to hold back the ion. Debye and Hückel,[18] who named this unsymmetric distribution of charge the "ion cloud," calculated for this effect a decrease in mobility proportional to the square root of the concentration, and somewhat less than half as great as the total measured effect of increasing concentration. Their calculations assumed that the ion moves in a straight line parallel to the field. Onsager[19] corrected this result for the Brownian movement of the central ion, introducing a factor that is $2 - 2^{1/2} = 0.586$ for symmetric salts. For other cases the relation is somewhat more complicated.

7–6. Electrophoresis

The effect of the time of relaxation accounts for only about a third of the decrease of mobility with concentration. The rest is explained by the Debye theory as arising from the effect of electrophoresis. Let us go back to Stokes's treatment of a sphere moving in a continuous, incompressible, viscous fluid. Stokes's method is to assume the sphere at rest, with the fluid moving past it under constant hydrostatic pressure. At the surface of the sphere the velocity of the fluid is zero, that is, there is no slip; at an infinite distance from the sphere the velocity is independent of direction from the sphere; and the lines of flow bulge around the sphere. Assuming that there is no force acting on the fluid other than the hydrostatic pressure, one obtains the familiar Stokes's law,

---

[18]P. Debye and E. Hückel, *Phys. Z. 24*, 305 (1923).
[19]L. Onsager, *Phys. Z. 27*, 388 (1926); *28*, 277 (1927).

$$u = F/6\pi\eta r, \tag{7-41}$$

in which u is the velocity of the particle, F the force acting on it, and r its radius, while η is the viscosity of the fluid. The equation should be applicable in the limit of extremely low concentration except for the fact that the ions are so small that it is hardly legitimate to consider the solvent continuous and of uniform viscosity. The equation may be used, in fact, for the larger ions to relate the mobility at zero concentration to the size of the ion or to the viscosity of the medium.

At higher concentrations there is a correction due to the fact that an ion is surrounded by its ion atmosphere and therefore the medium around it bears the opposite charge and is acted on by a force tending to move it in the direction opposite to the motion of the central ion when in an electric field. This introduces a second retarding action which, since it depends upon the thickness of the ion atmosphere, is also proportional to the square root of the concentration. The method of treatment is similar to that of Smoluchowski[20] for the electrophoresis of charged colloid particles.

For electrophoresis, we take

$$F = \psi_{ib}DbX, \tag{7-42}$$

in which X is the field strength due to the external field. Then

$$\frac{u_i}{X} = \frac{\psi_{ib}Db}{6\pi\eta b} - f(\kappa) = \frac{Db}{6\pi\eta b} \frac{\epsilon z_i}{Db}\left(1 - \frac{\kappa b}{1 + \kappa a}\right) - f(\kappa)$$

$$= \frac{\epsilon z_i}{6\pi\eta b}\left(1 - \frac{\kappa b}{1 + \kappa a}\right) - f(\kappa), \tag{7-43}$$

in which $f(\kappa)$ is the contribution of the time-of-relaxation effect to the decrease in mobility. For very small concentrations it is proportional to κ. Then

$$\left(\frac{u_i}{X}\right)_0 = \frac{\epsilon z_i}{6\pi\eta b}, \tag{7-44}$$

$$\left(\frac{u_i}{X}\right)_0 - \left(\frac{u_i}{X}\right) = \left(\frac{\epsilon z_i}{6\pi\eta}\right)\frac{\kappa}{1 + \kappa a} + f(\kappa), \tag{7-45}$$

$$1 - \left(\frac{u_i}{X}\right)\left(\frac{X}{u_i}\right)_0 = \frac{\kappa b}{1 + \kappa a} + \frac{6\pi\eta b}{\epsilon z_i}f(\kappa). \tag{7-46}$$

Eq. (7–46) is the form derived by Debye and Hückel, and Eq. (7–45) is the modification of Onsager, who succeeded in eliminating the size b by relating it to the mobility at zero concentration. We shall use

---

[20]M. von Smoluchowski in L. Graetz, ed., *Handbuch der Elektrizität und des Magnetismus* (Barth, Leipzig, 1918–28), vol. II, p. 384.

Eq. (7–45), and only as the limiting law in which κa is neglected relative to unity and $f(\kappa)$ is proportional to κ.

7-7. Conductance Equations

The final equations for the conductance apply only to solutions containing but two kinds of ions. They are

$$\Lambda_i = \Lambda_{i0} - \left[\frac{1.394 \times 10^6 w \Lambda_{i0}}{(D_0 T)^{3/2}} + \frac{41.0 z_i}{(D_0 T)^{1/2} \eta} \right] (\Gamma/2)^{1/2} \qquad (7\text{–}47)$$

$$\Lambda = \Lambda_+ + \Lambda_- = \Lambda_0 - \left[\frac{1.394 \times 10^6 w \Lambda_0}{(D_0 T)^{3/2}} + \frac{41.0(z_+ - z_-)}{\eta (D_0 T)^{1/2}} \right] (\Gamma/2)^{1/2}, \qquad (7\text{–}48)$$

in which D_0 is the dielectric constant of the medium, Λ_i is the equivalent conductance of the ith ion species and z_i its valence, reckoned negative for a negative ion, and Λ is the equivalent conductance of the electrolyte; Λ_{i0} and Λ_0 are the corresponding quantities at the limit of zero concentration,

$$w = -2q z_+ z_- / (1 + q^{1/2}), \qquad (7\text{–}49)$$

and

$$q = -z_+ z_- (\Lambda_{+0} + \Lambda_{-0}) / (z_+ - z_-)(z_+ \Lambda_{-0} - z_- \Lambda_{+0}). \qquad (7\text{–}50)$$

If $z_+ = -z_- = z$, then $q = 1/2$ and $w = z^2/(1 + (1/2)^{1/2}$, so that the equations become

$$\Lambda_i = \Lambda_{i0} - \left[\frac{0.817 \times 10^6 z^2 \Lambda_{i0}}{(D_0 T)^{3/2}} + \frac{41.0 z}{(D_0 T)^{1/2} \eta} \right] (\Gamma/2)^{1/2} \qquad (7\text{–}51)$$

and

$$\Lambda = \Lambda_0 - \left[\frac{0.817 \times 10^6 z^2 \Lambda_0}{(D_0 T)^{3/2}} + \frac{82.0 z}{(D_0 T)^{1/2} \eta} \right] (\Gamma/2)^{1/2}. \qquad (7\text{–}52)$$

We note, first of all, that in the limiting law $(\Lambda_0 - \Lambda)/\Lambda_0$ is proportional to the square root of the ional concentration, or of the molal or equivalent concentration. Such a relation was given in the empirical modifications of the law of mass action introduced by Rudolphi and van't Hoff, but was first shown to give an exact expression of the behavior of very dilute solutions by Kohlrausch[21] in 1907. The slope is specific for the different electrolytes; as Λ_0 increases, $(\Lambda_0 - \Lambda)/\Gamma^{1/2}$ also increases, but $(\Lambda_0 - \Lambda)/\Lambda_0 \Gamma^{1/2}$ decreases. Also, the conductance of any ion is independent of the ion of opposite sign with which it is associated if $z_+ = -z_-$, and depends only upon the

---

[21] F. Kohlrausch, Z. Elektrochem. 13, 333 (1907).

ional concentration. This relation was first noted by Kohlrausch and was emphasized, always empirically, by MacInnes[22] and by Lewis and Randall.[23] It applies strictly, however, only to the limiting law.

The dielectric constant occurs in the first term as the inverse three-halves power and in the second as the inverse square root. Since the two terms are approximately equal, this gives an approximate agreement with the empirical rule found by Walden that, at the same concentration in different solvents, $(\Lambda_0 - \Lambda)/\Lambda_0$ is inversely proportional to the dielectric constant.

At finite concentrations, Eq. (7–52) is not exact. Since Λ, not Λ_0, is the measured quantity, it is convenient to solve for Λ_0', the value of Λ_0 given by Eq. (7–48):

$$\Lambda'_0 = \frac{\left[\dfrac{\Lambda + 41.0(z_+ - z_-)(\Gamma/2)^{1/2}}{\eta(D_0 T)^{1/2}} \right]}{\left[\dfrac{1 - 1.394 \times 10^6 w(\Gamma/2)^{1/2}}{(D_0 T)^{3/2}} \right]}.$$

$$(7\text{–}53)$$

If Λ can be expressed as a power series in $C^{1/2}$, Λ'_0 should begin to change with the concentration as a linear function of C, although higher terms might be necessary to express the change at higher concentrations. Then Eq. (7–53) should give a convenient method of extrapolation for Λ_0. The plot of Λ'_0 against C is often called a Shedlovsky plot.[24] If Eq. (7–52) does not give the correct limiting law, however, the slope of Λ'_0 vs. C will be infinite when $C =$ zero. It is also possible, and theoretically probable in some cases, that Λ'_0 may contain a term in $C \ln C$, which also leads to curvature in very dilute solutions. Shedlovsky has found that for the alkali chlorides in water a linear term is sufficient up to about 0.1 M, whereas for nitrates there is curvature. Fig. 7–4 contains a plot of some of his data on uni univalent salts in water at 25°C, for which Eq. (7–53) becomes

$$\Lambda'_0 = \frac{\Lambda + 59.79\ C^{1/2}}{1 - 0.2274\ C^{1/2}}.$$

HCl, LiCl, and NaCl behave like KCl; $AgNO_3$ behaves like KNO_3 but the curvature is not so great.

Most of the measurements have not been treated in this way. Debye and Hückel computed Λ_0 and the slope of the limiting-law term from Kohlrausch's data by using the method of least squares

[22]D. A. MacInnes, *J. Am. Chem. Soc. 41*, 1086 (1919); *43*, 1217 (1921).
[23]G. N. Lewis and M. Randall, *Thermodynamics* (McGraw-Hill, New York, 1923), p. 32; G. N. Lewis, *J. Am. Chem. Soc. 34*, 1631 (1912).
[24]T. Shedlovsky, *J. Am. Chem. Soc. 54*, 1405 (1932).

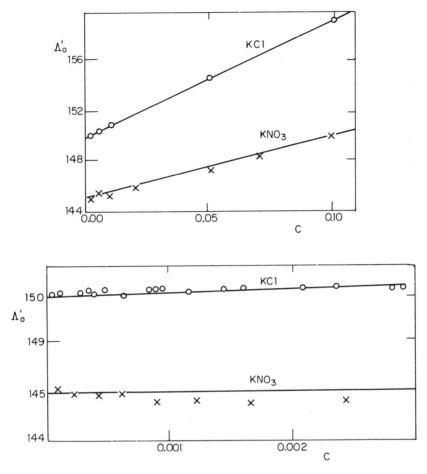

Figure 7–4. Conductance of potassium chloride and potassium nitrate in water at 25 C: (above) smoothed values at rounded concentrations; (below) actual experimental points in very dilute solutions. In both cases the straight lines are determined by the values at $C = 0.0$ and $C = 0.1$.

with a quadratic equation in $\Gamma^{1/2}$ for the measurements from $\Gamma = 0.0002$ to 0.016. Table 7–2 contains the comparisons of their results with Eq. (7–48) for water at 18°C and a similar comparison for methyl alcohol at 25°C with the limiting slope obtained graphically from the measurements in the range $\Gamma = 0.0002$ to 0.004. The measurements of Kraus and Parker[25] agree very well with Eq. (7–53), but those with iodic acid require a very large term linear in the concentration, in-

_____

[25]C. A. Kraus and H. C. Parker, *J. Am. Chem. Soc.* **44**, 2429 (1922).

Table 7-2. Tests of limiting conductance law.

| | Water at 18° C[a] | | | | Methyl alcohol at 25° C[b] | | |
| | | Limiting slope | | | | Limiting slope | |
Salt	Λ_0	Meas.	Meas. − calc.	Salt	Λ_0	Meas.	Meas. − calc.
LiCl	98.93	57.35	6.0	LiCl	90.90	158	− 15
LiIO$_3$	67.35	48.33	1.9	NaCl	96.95	163	− 14
LiNO$_3$	95.24	56.27	5.4	KCl	105.05	185	2
NaCl	108.89	54.69	1.7	RbCl	108.65	199	14
NaIO$_3$	77.42	51.39	3.4	CsCl	113.60	200	12
NaNO$_3$	105.34	58.27	5.8	KF	94.0	167	8
KCl	129.93	59.94	3.5	KBr	109.35	185	− 1
KBr	132.04	62.17	5.5	KI	114.85	184	5
KI	130.52	51.53	− 5.0	NH$_4$Cl	111.00	187	0
KIO$_3$	98.41	54.18	2.8	HCl	193.5	260	17
KClO$_3$	119.47	58.16	3.5	LiNO$_3$	100.25	177	− 3
KNO$_3$	126.41	65.67	9.9	NaNO$_3$	106.45	204	20
KCNS	121.04	54.10	− 0.9	KNO$_3$	114.55	244	55
CsCl	133.08	53.75	− 3.2	RbNO$_3$	118.15	251	59
AgNO$_3$	115.82	62.35	8.3	CsNO$_3$	122.95	268	73
TlNO$_3$	127.55	63.40	7.4	AgNO$_3$	112.95	319	131
Ba(NO$_3$)$_2$	116.95	92.8	5.9	NaBr	101.5	170	− 10
Sr(NO$_3$)$_2$	113.42	97.8	12.3	NaOCH$_3$	98.40	157	− 22
K$_2$SO$_4$	132.23	81.0	− 11.1	NaClO$_4$	115.40	198	8
CaCl$_2$	116.69	88.0	1.9				
Ca(NO$_3$)$_2$	113.56	97.4	11.8				
MgCl$_2$	110.88	83.2	− 0.9				
MgSO$_4$	114.70	242.7	144.4				
CdSO$_4$	115.81	270.7	175.6				

[a]P. Debye and E. Hückel, *Phys. Z. 24*, 305 (1923).
[b]J. E. Frazer and H. Hartley, *Proc. Roy. Soc. A109*, 351 (1925)

dicating an effect similar to the "higher-term effect" in the chemical potentials. The measurements in dilute solutions are fitted by

$$\text{HCl:} \quad \Lambda = 427.1 - 111.0 \; \Gamma^{1/2},$$
$$\text{HIO}_3\text{:} \quad \Lambda = 396.3 - 105.3 \; \Gamma^{1/2} - 1100 \; \Gamma.$$

It is difficult to tell how much of the discrepancies in Table 7–2 should be attributed to experimental error and to the error in determining the limiting slope, and how much to failure of the theory. The worst discrepancies are for the 2–2 salts in aqueous solutions. $ZnSO_4$ and $CuSO_4$ are intermediate between the two given, which agrees with the behavior of the chemical potentials. It is possible to extrapolate to a different limiting conductance with the theoretical limiting slope in extremely dilute solutions, but with an inflection that gives a nearly straight line with a much steeper slope over a wide concentration range. This is probably an effect of the "higher terms" which will be discussed in Chapter 8.

It should be noted from Table 7–2 that the limiting conductance is about the same in alcohol as in water, although the change in solubility is several hundred or a thousand fold. This means that the "salting-out" term, which will be discussed in Sec. 8–4, does not affect the conductance and that the only effect of changing dielectric constant with changing salt concentration should be to change the ratio of κ to the square root of the concentration. So the conductance ratio does not pass through a minimum, as the excess chemical potential or the activity coefficient often does, except in the case of solvents of very low dielectric constant (see Sec. 8–7). The nonionic forces, like the higher terms, should show an effect on the conductance similar to their effect on the activity coefficients.

7–8. Transport Numbers

The terms "transference number" and "transport number" are usually used interchangeably, but we shall distinguish between them. We define the transference number t_i as the number of moles of species i that go with 1 equivalent of positive electricity. The transference number is positive for a cation, negative for an anion, and may be either positive or negative for a neutral molecule. We define the transport number T_i [$= t_i z_i$] as the fraction of the current carried by species i. It is positive for any ion but zero for a neutral molecule. We have also defined Λ_i to be positive for any ion, and we may define the transport number as $T_i = \Lambda_i C_i / \Sigma_j \Lambda_j C_j$. We define the mobility u_i, however, so that it is negative for an anion and need not be zero for a neutral molecule. Then $\Lambda_i = u_i z_i \mathfrak{F}$, $t_i = u_i C_i / \Sigma_j u_j z_j C_j$, and

$T_i = u_i C_i z_i / \Sigma_j u_j z_j C_j$. For t_i and T_i the concentrations occur in ratios, so they may be moles per kilogram of solvent as well as moles per liter of solution. One component may be considered motionless. Hittorf transference numbers are transference numbers relative to water as the motionless component; "true transference numbers" are numbers relative to some neutral solute. In this chapter it is somewhat simpler to talk of transport numbers.

Dividing Eq. (7–47) by Eq. (7–48), simplifying and dropping all terms higher than the first power in κ, we obtain as limiting law for the transport number T_i in terms of its value at zero concentration T_{i0},

$$T_i = T_{i0} + \frac{41.0[(z_+ - z_-)T_{i0} - |z_i|](\Gamma/2)^{1/2}}{(D_0 T)^{1/2} \eta \Lambda_0} . \qquad (7\text{–}54)$$

The transference number also varies as a linear function of the square root of the concentration in very dilute solutions. The amount of change depends not only upon the conductance of the solution, but also upon the magnitude of the absolute value of the transference number. For a symmetric salt $(z_+ = - z_-)$, the change is zero if the value of a transport number is exactly $\frac{1}{2}$ at zero concentration. For other values, the change is in the direction to increase the difference between the transference numbers, and is larger the larger the value of this difference. It is not worth while considering the theory for more than the limiting law since the terms in κ higher than the first are neglected in the original expression. The addition of a term linear in the concentration is sufficient to represent the measurements on the alkali halides up to about 0.1M, but the nitrates require a more complicated function. Table 7–3 contains some of the accurate measurements of MacInnes and Longsworth.

Table 7-3. Transport numbers T_+ of cations in water at 25 C.[a]

Conc.	Ion				
	KCl	NaCl	LiCl	HCl	AgNO₃
0.001	0.4903	—	—	—	—
.002	.4903	—	—	—	—
.005	.4905	—	—	—	—
.01	.4902	0.3919	0.3289	0.8252	0.4648
.02	.4901	.3900	.3261	.8264	.4652
.05	.4900	.3878	.3211	.8292	.4664
.1	.4898	.3853	.3168	.8314	.4682
.2	.4892	.3814	—	—	—
.5	.4887	—	—	—	—
1.0	.4888	At $M = 0.1$ NH₄Cl 0.4900, KNO₃ 0.5128, KBr 0.4834			

[a]D. A. MacInnes and L. G. Longsworth, *Chem. Rev. II*, 171 (1932).

7-9. Conductance Dispersion

When an ion is in an oscillating field, there is a short interval at
each reversal of the field during which its motion is aided by the lag
of the ion atmosphere at the end of the last tack, rather than being
hindered. For the frequencies of ordinary conductance measurements
(time about 10^{-3} sec), the contribution to the conductance during
this time interval is negligible. As the frequency of the oscillation is
increased, however, this contribution increases, until, for times small
relative to the time of relaxation of the ion atmosphere, the ion oscil-
lates about its equilibrium position in the ion atmosphere, aided just
as much as it is hindered, and the time-of-relaxation effect on the conductance vanishes. Since the ion atmosphere is always present, the
ion will always be in a medium bearing the opposite charge, and the
electrophoretic effect should be independent of the frequency. When
the frequency is of the same order of magnitude as the reciprocal of
the time-of-relaxation of the ion atmosphere, there should be a rapid
change in the time-of-relaxation effect with changing frequency,
which may be called conductance dispersion.

At low frequencies, moreover, the lag of the ion atmosphere will
cause the current to lag behind the electromotive force, an effect that
is interpreted by macroscopic measuring instruments as an increase
in the dielectric constant due to the interaction of an ion with its ion
atmosphere. At zero frequency this effect is proportional to the square
root of the concentration. It vanishes completely for very high frequencies, and in the range of conductance dispersion there is also a
dispersion of the macroscopic dielectric constant. Throughout this
range the initial change is proportional to the square of the concentration, but in the concentration range susceptible to measurement it is
approximately proportional to the concentration. Both the conductance and the dielectric-constant effects are complicated by the fact
that the time of relaxation changes with the concentration. In both
cases the effect increases rapidly with increasing valence of the ions.

Wien has measured both effects simultaneously by comparing
magnesium sulfate and barium ferricyanide with alkali chlorides and
calculating the small change for the latter from the theory. The
method consists in changing the concentration of one solution until
both have the same specific conductance at high frequencies and then
comparing the conductances at low frequencies. At the same time
the dielectric constants are measured by changing the concentration
of an alcohol-water mixture in a condenser in parallel with one solution until the capacitances are equalized. Table 7–4 contains the results
for barium ferricyanide, for which the time of relaxation is 0.113

Table 7-4. Percentage increase of conductance and dielectric constant of barium ferricyanide at high frequency.[a]

C	$C^{1/2}$	Wavelength[b]						Infinity,	δ_∞
		10 m		20 m		40 m			
		Meas.	Calc.	Meas.	Calc.	Meas.	Calc.	Calc.	100C
Conductance									
0.00032	0.018	1.8	2.6	1.4	1.5	1.0	1.2	—	
.00073	.027	1.6	3.2	—	1.6	—	1.0	—	
.00160	.040	2.8	3.3	1.2	1.4	0.6	0.7	—	
.00348	.059	2.7	2.6	—	1.0	0.4	0.3	—	
.00672	.082	1.6	1.5	0.7	0.4	0.2	0.1	—	
.0190	.138	1.4	0.6	0.35	0.1	—	0.04	—	
.0428	.207	1.1	0.2	—	—	—	0.00	—	
Dielectric constant									
0.00032	0.018	0.0	0.0	0.11	0.08	0.0	0.2	0.8	25.0
.00073	.027	.15	.13	—	.27	—	.5	1.2	18.4
.00160	.040	.5	.45	.45	.77	1.0	1.3	1.8	11.2
.00348	.059	1.1	1.3	—	1.9	1.5	2.2	2.5	7.2
.00672	.082	2.3	2.9	2.9	3.4	2.8	3.6	3.7	5.5
.0190	.138	4.6	5.9	5.1	—	—	6.2	6.2	3.3
.0428	.207	8.4	9.2	7.6	9.3	—	9.3	9.3	2.2

[a]M. Wien, *Ann. Phys.* (5) *11*, 429 (1931).
[b]Ratio of speed to frequency.

$\times\ 10^{-10}/C$ sec, which corresponds to a wavelength of $3.39 \times 10^{-3}C$ meters. For the conductance measurements the percentage increase is that above the conductance at the same concentration for low frequencies; for the dielectric constant it is that above water itself.

For the conductances the change with the frequency predicted by the theory is found also in the experiments. At the highest frequency the measurements go to dilute enough solutions to demonstrate clearly the maximum predicted by the theory. The error in the dielectric-constant measurements is apparently greater, and the effect of changing frequency is observable only in the dilute solutions, in which the effect itself is small. The measurements are accurate enough, however, to indicate that the conductance theory of Debye and Falkenhagen[26] is sufficient to account for the major part of the change in the dielectric constant with salt concentration, leaving no room for any other effect of the same magnitude. The fact that the dielectric constant and the conductance were measured simultaneously increases the confidence in both measurements. There have been a large number of other measurements of the effect of changing fre-

[26]P. Debye and H. Falkenhagen, *Phys. Z.* 29, 401 (1928).

quency on the conductance that confirm the results given here. Radar and related studies have made possible the use of higher frequencies. Probably this is the strongest evidence for the fundamental correctness of the Debye picture of the ion atmosphere, for it confirms the calculations of effects depending upon the time of relaxation as well as upon the thickness of the ion atmosphere. Since the theory is not given here in detail, it is well to point out that all these conductance effects are calculated from the valences of the ions, their mobilities at zero concentration and low frequencies, the dielectric constant and viscosity of the solvent, the temperature, and universal constants.

Hasted and his collaborators[27] have measured the dielectric constant of several aqueous salt solutions at wavelengths of 10, 3, and 1.25 cm, which are in the dispersion range of liquid water. They find a linear decrease of the low-frequency dielectric constant in moderately dilute solutions. For NaCl at 21°C the effect is linear up to 2-normal at $D = D_0 - 11C$, and then decreases less rapidly. This they attribute largely to the hindrance of rotation of water molecules in the first shell around cations. Above 2-normal the spheres of influence of different ions overlap. They say that the number of nonrotating water molecules per cation is 4 for Na^+, K^+, and Rb^+, 6 for Li^+, 10 for H^+, 14 for Mg^{++} and Ba^{++}, and 22 for La^{+++}.

7-10. Conductance and Field Strength

When an ion moves very rapidly, it should escape entirely from its ion atmosphere, again because of the finite time necessary for the formation of the atmosphere. In this case, since there is no ion atmosphere, the electrocataphoretic decrease of the conductance should disappear as well as the time-of-relaxation effect. The conductance of the ions should approach the conductance at zero concentration as the field strength is increased. The experimental measurements of this deviation from Ohm's law are difficult because of the large heat effects from the large current. Wien[28] has measured the effect by very short pulses of high voltage. The effect appears to be at first proportional to the square of the field strength, then to increase linearly with the field strength through most of the range, and finally to approach asymptotically the equivalent conductance at zero concentration. From the ion-atmosphere theory it is possible to calculate quantitatively that part of the curve which is proportional to the square of the field strength.

[27]J. B. Hasted, D. M. Ritson, and C. H. Collie, *J. Chem. Phys. 16*, 1 (1948); G. H. Haggis, J. B. Hasted, and T. J. Buchanan, *ibid. 20*, 1452 (1952).

[28]M. Wien, *Phys. Z. 32*, 545 (1932).

The initial slope is nearly proportional to the square of the valence product of the ions, and nearly inversely proportional to the square root of the concentration. For small concentrations the curvature is much greater than for large concentrations, for the asymptote lies much nearer to the conductance at low field strengths. For strong electrolytes the effect at 200,000 V/cm varies from a few tenths percent to about 2 percent. Wien once hoped to use these measurements to determine "the true degree of ionization." His work with weak electrolytes shows that this is impossible, for the percentage increase of conductance is about ten times as great for weak as for strong acids, indicating that the magnitude of the increase depends essentially upon the difference between the equivalent conductance at low field strengths and the limiting conductance. The increase with weak acids may be attributed either to increased ionization in the strong electric field, in other words a distribution in space determined in part by the field strength, or to ionization of some of the molecules by impacts of the rapidly moving ions. The first effect appears the more probable, but both may be operative.

7–11. Viscosity

The relative motion of an ion and its ion atmosphere should also determine the first effect, with increasing concentration, of ions on the viscosity. We picture viscous flow as the movements upon each other of parallel plates of very small thickness. Those ions next to the fixed surface are considered to be fixed to the surface, whereas those far away have the velocity of the bulk flow. If the central ion is in an intermediate layer, it will be moving faster than that part of its atmosphere nearer the fixed surface, but slower than that part farther from the surface. The difference in relative velocities of the ion atmosphere and the central ion will be greater the greater the thickness of the ion atmosphere. The calculations are even more complicated than for the conductance. They result in an effect on the viscosity proportional to the square root of the concentration. The effect of displacing the solvent by another medium is proportional to the concentration, so that the total effect up to a few tenths molal may be represented by the equation

$$\eta/\eta_0 = 1 + AC^{1/2} + BC, \tag{7–55}$$

in which η is the viscosity of the solution and η_0 that of the solvent. For uniunivalent electrolytes, A may be represented by the equation

$$A = \frac{1.45}{\eta_0(2D_0T)^{1/2}} \left[\frac{\Lambda_+ + \Lambda_-}{4\Lambda_+\Lambda_-} + \frac{(\Lambda_+ - \Lambda_-)^2}{(3 + 2^{1/2})\Lambda_+\Lambda_-(\Lambda_+ + \Lambda_-)} \right].$$
$$\tag{7–56}$$

For unsymmetric electrolytes, the expression is considerably more complicated.

Some very accurate measurements by Jones and his students are given in Table 7-5. The agreement is very good, considering the small

Table 7-5. Viscosities of aqueous salt solutions at 25 C.[a]

Solute	B meas.	A Meas.	A Calc.
Sucrose	0.8786	0.000	0.000
Urea	.3784	.000	.000
KCl	− .0140	.0052	.0050
$KClO_3$	− .0309	.0050	.0055
$KBrO_3$	− .0008	.0058	.0058
KNO_3	− .0531	.0050	.0052
NH_4Cl	− .01439	.0057	.0050
$CsNO_3$	− .092	.0043	.0051

[a]G. Jones and S. K. Talley, *J. Am. Chem. Soc.* **55**, 624 (1933).

contribution of the square-root term. It must not be supposed that B is negative for all salts. This group was chosen to show the existence of a positive square-root term even when the first-power term is negative. The more usual case, for example with lithium and sodium salts and those with higher valence ions, is a positive B coefficient. Many students of viscosity appear much troubled by salts that reduce the viscosity of water. The effect is probably exactly similar to that of alcohol or of any other small molecules that have viscosities less than that of water. For ions, it is possible to generalize that those which have the larger mobilities and smaller charges have the more negative B coefficient in Eq. (7–55).

8. The Free Energy and Chemical Potentials in Electrolyte Solutions

8-1. The Electric Contribution to the Free Energy

To determine the electric contribution to the free energy it is necessary to consider the work of charging the ions reversibly. We need consider only the work of placing the charge on the surface of the sphere b_i, for the work of concentrating the charge within this sphere is, by our assumptions, independent of the medium and of the concentration of the ions. We shall calculate the work of charging the ions reversibly at constant temperature and pressure in such a way that at each instant the charge of each ion is the same fraction λ of its final charge. We shall indicate the instantaneous value of any quantity by a prime. Then

$$e_i' = \lambda \epsilon z_i, \tag{8-1}$$

$$de_i' = \epsilon z_i d\lambda, \tag{8-2}$$

$$\kappa' = \kappa \lambda, \tag{8-3}$$

$$\psi_b' = (\epsilon z_i \lambda / D)[1/b_i - \kappa \lambda / (1 + \kappa \lambda a)], \tag{8-4}$$

$$W_i = \int_0^{\epsilon z_i} \psi_b' de_i' = \int_0^1 (\epsilon^2 z_i^2 / D)[1/b_i - \kappa \lambda / (1 + \kappa \lambda a)] \lambda d\lambda$$

$$= (\epsilon^2 z_i^2 / 2D)\{1/b_i - [\kappa^2 a^2 - 2\kappa a + 2 \ln (1 + \kappa a)]/\kappa^2 a^3\}$$

$$= (\epsilon^2 z_i^2 / 2D)(1/b_i - X/a), \tag{8-5}$$

in which

$$X = [x^2 - 2x + 2 \ln (1 + x)]/x^2, \tag{8-6}$$

$$x = \kappa a. \tag{8-7}$$

We also define two other quantities already used in Sec. 5–4:

$$Y = x/(1 + x), \tag{8-8}$$

$$Z = Y - X = \tfrac{1}{2}xdX/dx = [1 + x - 1/(1 + x) - 2 \ln (1 + x)]/x^2.$$

The total electric work is

$$W^D = \Sigma_i n_i W_i = (\epsilon^2/2D)\Sigma_i n_i z_i^2 (1/b_i - X/a)$$
$$= (\epsilon^2 N_A/2D)\Sigma_i N_i z_i^2 (1/b_i - X/a). \quad (8\text{–}9)$$

We shall take as standard state an infinitely dilute solution in some solvent at temperature T and pressure p. The system we shall study will include one part of infinite volume, filled with this solvent, and another part at the same temperature and pressure, of finite volume, containing the solvent to be studied, which may or may not be the same as that defining the standard state. Let the initial state of the system be that with all the ions in the part of infinite volume, and the final state that with all the ions in the part of finite volume. The work of going from the initial to the final state will be the same for all isothermal reversible processes, and may be designated as $G - G_0$.

Let the first step be the discharge of the ions in the first part, and designate the work by W_1. By our assumptions,

$$W_1 = -W_0^D = -(N_A \epsilon^2/2D_0)\Sigma_i(N_i z_i^2/b_i), \quad (8\text{–}10)$$

and in the general case,

$$W_1 = -W_0^D - \delta_1. \quad (8\text{–}11)$$

Let the second step be the reversible transfer of the discharged ions from the first (infinite) part of the system to the second (finite) part, and let the work be W_2; then

$$W_2 = RT\Sigma_i N_i \ln(\bar{N}_i/\bar{N}_{i0}) + \delta_2. \quad (8\text{–}12)$$

Let the third step be the charging of the ions in the second solution, and let the work be designated by W_3. By our assumptions,

$$W_3 = W^D = (N_A \epsilon^2/2D)\Sigma_i N_i z_i^2 (1/b_i - X/a), \quad (8\text{–}13)$$

and, in general,

$$W_3 = W^D + \delta_3. \quad (8\text{–}14)$$

For the sum of the three steps,

$$G - G_0 = W_1 + W_2 + W_3 = RT\Sigma_i N_i \ln(\bar{N}_i/\bar{N}_{i0})$$
$$+ W^D - W_0^D - \delta_1 + \delta_2 + \delta_3. \quad (8\text{–}15)$$

If all the deviations from ideality are due to the electric forces considered above, each δ is zero, and in any case the electrostatic contribution to the free energy G^D may be defined as

$$G^D = G - G_0 - RT\,\Sigma_i N_i \ln(\bar{N}_i/\bar{N}_{i0}) - (\delta_2 + \delta_3 - \delta_1) = W^D - W_0^D. \quad (8\text{–}16)$$

It is often considered that $W^D - W_0^D$ is a work-content change at constant volume rather than a free-energy change at constant pressure. It might be taken as either under the proper conditions. The choice of work content would avoid the necessity of the assumption that the volume is unchanged during the charging process at constant pressure and would correspond more closely to the results obtained by more orthodox statistical mechanics. It would add difficulties in the transfer of substance between liquids while the volume of each phase is kept constant. The fact that the measurements of the dielectric constants of liquids correspond to an electric work at constant pressure requires the use of a free energy if these dielectric constants are to be used.

The electric contribution to the chemical potential of any component, electrolyte or nonelectrolyte, to the heat content, heat capacity, volume, or compressibility may be obtained by purely thermodynamic methods by differentiation of W^D with respect to the proper variable. Since we shall always use as independent variables temperature, pressure, and the number of moles of each component, we shall vary only one of these quantities at a time and shall therefore omit the subscripts to denote which variables are being held constant.

8-2. The Chemical Potentials

At constant temperature, pressure, and quantity of the other components, κ varies with N_k directly, and also because the volume and the dielectric constant too are functions of N_k; so x, X, and W_e also depend upon these variables. Then

$$\frac{\partial \kappa}{\partial N_k} = \frac{\kappa}{2}\left(\frac{z_k^2}{\Sigma_i N_i z_i^2} - \frac{\partial \ln D}{\partial N_k} - \frac{\partial \ln V}{\partial N_k}\right), \tag{8-17}$$

$$\frac{\partial x}{\partial N_k} = \frac{x}{2}\left(\frac{z_k^2}{\Sigma_i N_i z_i^2} - \frac{\partial \ln D}{\partial N_k} - \frac{\partial \ln V}{\partial N_k}\right), \tag{8-18}$$

$$\frac{\partial X}{\partial N_k} = \frac{\partial X}{\partial x}\frac{\partial x}{\partial N_k} = Z\left(\frac{z_k^2}{\Sigma_i N_i z_i^2} - \frac{\partial \ln D}{\partial N_k} - \frac{\partial \ln V}{\partial N_k}\right). \tag{8-19}$$

So

$$\frac{\partial W^D}{\partial N_k} = \frac{N_A \epsilon^2}{2D}\left[z_k^2\left(\frac{1}{b_k} - \frac{Y}{a}\right) - \frac{\partial \ln D}{\partial N_k}\Sigma_i N_i z_i^2\left(\frac{1}{b_i} - \frac{Y}{a}\right)\right.$$
$$\left. + \frac{Z}{a}\frac{\partial V}{\partial N_k}\Sigma_i C_i z_i^2\right], \tag{8-20}$$

$$\frac{\partial W_0^D}{\partial N_k} = \frac{N_A \epsilon^2}{2D_0}\frac{z_k^2}{b_k}. \tag{8-21}$$

The first term of Eq. (8–20) depends upon the charge of the molecule of the kth species and is zero for nonelectrolytes. In very dilute solutions the difference between Eqs. (8–20) and (8–21) is proportional to Y and therefore to the square root of the ionic strength. The second term depends upon the effect of species k on the dielectric constant. Since $(\partial \ln D/\partial N_k) \Sigma_i N_i z_i^2$ is also $(\partial \ln D/\partial C_k) \Sigma_i C_i z_i^2$ and $\partial \ln D/\partial C_k$ is approximately independent of the concentrations, this term is proportional to the ionic strength in dilute solutions. It is the most important term in the salting out of nonelectrolytes, although it cannot be distinguished from the nonelectrostatic effects of the δ term since they also are approximately proportional to the concentration. The third term depends upon the molal volume of species k. It is initially proportional to the three-halves power of the ionic strength and is so small that it is usually important only for the solvent.

The chemical potential of an individual ion has only conventional significance, because it is impossible to add or subtract a single ion species without doing a huge amount of electric work whose magnitude depends upon the shape of the phase and its position relative to other bodies. However, for an electrolyte component q each molecule of which gives ν ions, ν_+ of valence z_+ and ν_- of valence z_-, we may write

$$\frac{\partial W^D}{\partial N_q} = \frac{\nu_+ \partial W^D}{\partial N_+} + \frac{\nu_- \partial W^D}{\partial N_-}. \tag{8–22}$$

Since

$$\nu_+ z_+ = -\nu_- z_-, \tag{8–23}$$

it follows that

$$\nu_+ z_+^2 + \nu_- z_-^2 = -\nu z_+ z_-. \tag{8–24}$$

Similarly we may write, as a definition of b_q,

$$-\frac{\nu z_+ z_-}{b_q} = \frac{\nu_+ z_+^2}{b_+} + \frac{\nu_- z_-^2}{b_-}. \tag{8–25}$$

Then

$$\frac{\partial W^D}{\partial N_q} = \frac{N_A \epsilon^2}{2D}\left[-\nu z_+ z_- \left(\frac{1}{b_q} - \frac{Y}{a} \right) - \frac{\partial \ln D}{\partial N_q} \Sigma_i N_i z_i^2 \left(\frac{1}{b_q} - \frac{Y}{a} \right) \right.$$
$$\left. + \frac{Z}{a} \frac{\partial V}{\partial N_q} \Sigma_i C_i z_i^2 \right]. \tag{8–26}$$

8-3. Approximate Expressions

To use these equations we must know something of the variation of the dielectric constant with changing composition of the medium. The measured dielectric constant of an electrolyte solution differs from that of the solvent mostly by the effect of the time of relaxation of the ion atmosphere on the phase lag of the current discussed in Sec. 7–9, which is so large that it masks other effects. Yet this effect must be subtracted to obtain the dielectric constant, which affects the interaction of the ionic charges and should be inserted in these equations. Probably the next largest effect is the polarization of the solvent molecules that are very close to the ions, also discussed in Sec. 7–9. We may assume that the effect of a salt on the dielectric constant is about the same as that of a nonelectrolyte of the same volume and is proportional to the concentration of the salt. If the polarized water molecules are taken into account, they may be considered as water of hydration; see Sec. 8–5.

If the ions do not change the volume or the dielectric constant at all, the concentrations in moles per liter of solution are the same as those in moles per liter of solvent, or ρ times the concentrations in moles per kilogram of solvent. Then Eq. (8–20) for an ion becomes

$$\frac{\partial W^D}{\partial n_k} = (N_A\epsilon^2/2D_0)z_k^2(1/b_k - Y/a), \tag{8-27}$$

and from Eqs. (8–21) and (8–16),

$$\bar{G}_k^D = -(N_A\epsilon^2/2D_0)z_k^2Y/a; \tag{8-28}$$

for the solvent,

$$\bar{G}_s^D = \frac{N_A\epsilon^2\bar{W}_sZ}{2000\ D_0a}\ \Sigma_i m_i z_i^2. \tag{8-29}$$

In Eqs. (8–28) and (8–29), Y and Z are determined with x expressed in moles per kilogram of solvent. These are the forms used in Chapter 5, usually with $x = 1.5/(I/2)^{1/2}$.

If the dielectric constant is proportional to the concentration of water, $D = D_0/(1 + \Sigma_i \bar{V}_i N_i/\bar{V}_s N_s)$, Eq. (8–20) for the solvent leads to the relation (A, α, and M are defined on pages 63 and 64):

$$\phi^D - 1 = (A/\alpha)[-ZI/M + \rho_0\Sigma_i m_i z_i^2\left(\frac{a}{b_i} - Y\right)\Sigma_j m_j \bar{V}_j/M] \tag{8-30}$$

instead of

$$\phi - 1 = -AZI/\alpha M + \Sigma_{ij}\beta_{ij}m_i m_j/M + \cdots. \tag{5-51}$$

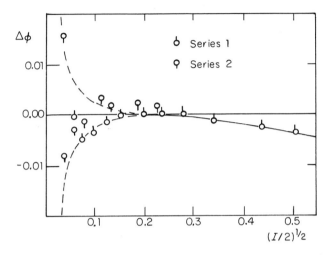

Figure 8-1. Freezing Points of aqueous lanthanum chloride.

The second term of Eq. (8–30) has the form of the second term of Eq. (5–51) with $\beta_{ij} = \rho_0 a \bar{V}_j z_i^2 / b_i$. It should be developed in a power series in $M^{1/2}$, but might be represented to a good approximation by a series in M.

The experimental confimation of the Debye theory was discussed by Scatchard and Prentiss.[1] We may say that the greater the apparent accuracy of the experimental work, the closer is the agreement with theory, with a single exception. Shedlovsky and MacInnes[2] measured the electromotive force of lanthanum chloride cells with transference and calculated from their measurements and the measurements of transference numbers of Longsworth and MacInnes[3] values of the limiting slope only about two-thirds that of theory. Beaumont[4] measured the freezing-point depression of lanthanum chloride solutions up to $I/2 = 0.25$, however, and found excellent agreement with theory. Shedlovsky[5] corrected the conclusions of Shedlovsky and MacInnes,[2] and attributed them to errors in integration. Fig. 8–1 shows the deviations of Beaumont's results from Eq. (8–29) with $\alpha = 2$, expressed as $\Delta\phi$. The curve is for $\beta = -0.0015 \ (I/2)$, and the differences of the broken curves from the full curve show the effect on $\Delta\phi$ of a change of 2×10^{-5} in the freezing-point depression.

[1]G. Scatchard and S. S. Prentiss, *Chem. Rev. 13*, 139 (1933).

[2]T. Shedlovsky and D. A. MacInnes, *J. Am. Chem. Soc. 61*, 200 (1939).

[3]L. G. Longsworth and D. A. MacInnes, *J. Am. Chem. Soc. 60*, 3070 (1938).

[4]G. Scatchard, B. Vonnegut, and D. W. Beaumont, *J. Chem. Phys. 33*, 1292 (1960).

[5]T. Shedlovsky, *J. Am. Chem. Soc. 72*, 3680 (1950).

Fig. 8–2 shows the deviation in the measured electromotive force and that calculated from the equation

$$- \log \gamma_\pm = 1.5254 \, (I/2)^{1/2}/[1 + 2 \, (I/2)^{1/2}] + 0.04 \, (I/2),$$

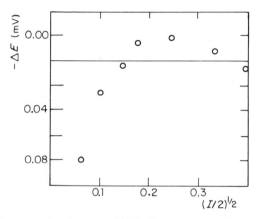

Figure 8-2. Electromotive force at 25°C of aqueous lanthanum chloride cells with transference, expressed as deviation from calculated values derived from equation in text.

assuming that the interpolated transference numbers have no error, plotted as $-\Delta E = E_{calc} - E_{meas}$ in millivolts. The two lowest concentrations are smaller than any for which the transference numbers are measured. An error of 0.001 to 0.002 in the transference numbers is sufficient to explain the difference.

8–4. Electrolytes and Nonelectrolytes. Salting Out

For a nonelectrolyte solute, Eq. (8–20) becomes

$$\frac{\partial W^D}{\partial N_n} = \frac{N_A \epsilon^2}{2D} \left[- \frac{\partial \ln D}{\partial C_n} \Sigma_i C_i z_i^2 \left(\frac{1}{b_i} - \frac{Y}{a} \right) + \frac{Z}{a} \frac{\partial V}{\partial N_n} \Sigma_i C_i z_i^2 \right]. \tag{8-31}$$

Since Y and Z are each proportional to $(I/2)^{1/2}$, and $\partial \ln D/\partial C_n$ and $\partial V/\partial N_n$ are each approximately independent of the concentration, the limiting law for the effect of an electrolyte on the potential of a nonelectrolyte is

$$\frac{\partial W^D}{\partial N_n} = - \frac{N_A \epsilon^2}{2D} \frac{\partial \ln D}{\partial C_n} \Sigma_i C_i z_i^2/b = \frac{N_A \epsilon^2}{2} \frac{\partial (1/D)}{\partial C_n} \Sigma_i C_i z_i^2/b_i. \tag{8-32}$$

We should expect the effect per mole of salt to decrease with increasing salt concentration if $- \partial \ln D/\partial C_n$ is greater than one-third of $\partial \ln V/\partial C_n$, as is usual for nonpolar solutes in water. Usually the effect remains remarkably constant to very high salt concentrations. It is proportional to one specific property of the nonelectrolyte, $\partial(1/D)/\partial C_n$, and to one specific property of each ion other than its valence, $1/b_i$. So this limiting law does not have the sharpness of the limiting law for the effect of an electrolyte on the potential of another electrolyte, both because it starts out with the same power of the concentration as nonelectrostatic effects and because of the specificity. Mole for mole, the electrostatic effect is inversely proportional to the radius of the small ion and, for similar nonelectrolytes, proportional to the volume of the nonelectrolyte. The nonelectrostatic effect in aqueous solutions is largely due to the displacement of water and is roughly proportional to the product of the volumes of the ion and of the nonelectrolyte. Therefore, the nonelectrostatic effects will not change greatly the relative effects of two nonelectrolytes. Since they usually have the opposite sign to that of the electrostatic effects, they will reduce the effect of large ions below that calculated from Eq. (8–32). In fact, an electrolyte with two large ions may salt in rather than salt out. For very small ions the effect is also less than that calculated from Eq. (8–32) because the nonelectrolyte medium is not homogeneous and the maximum effect obtainable is the effect of repelling all the solute from the regions around the ions.[6,7]

Eq. (8–32), in the absence of nonelectrostatic effects, leads to the expression for the salting out,[8]

$$\ln \frac{S_0}{S} = \frac{1}{RT} \frac{\partial W_e}{\partial N_n} = \frac{N_A \epsilon^2}{2RT} \frac{\partial(1/D)}{\partial C_n} \Sigma_i C_i z_i^2 / b_i. \qquad (8\text{–}33)$$

It corresponds to the rule discovered empirically by Setschenow[9] that the logarithm of the solubility of a nonelectrolyte is proportional to the salt concentration. Fig. 8–3 shows the solubility of oxygen in various salt solutions as $\log S_0/S$, and Fig. 8–4 shows the solubilities of two proteins in ammonium sulfate solutions as $\log S$. The slopes of the solubilities of the proteins are about ten times that for oxygen. In a logarithmic equation this makes an enormous difference. A change in salt concentration that salts out half the oxygen will salt out all but one-thousandth of the protein. However, this salting-out effect corresponds to a much smaller change of dielectric constant per unit

[6]P. Debye, *Z. Phys. Chem.* *130*, 56 (1927).
[7]G. Scatchard, *J. Chem. Phys.* *9*, 34 (1941).
[8]P. Debye and J. McAulay, *Phys. Z.* *26*, 22 (1925).
[9]J. Setschenow, *Z. Phys. Chem.* *4*, 117 (1889).

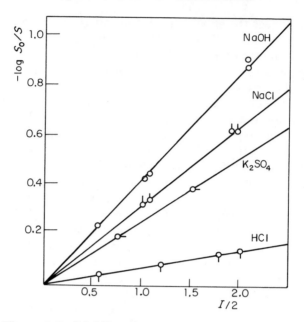

Figure 8–3. Solubility of oxygen in electrolyte solutions.

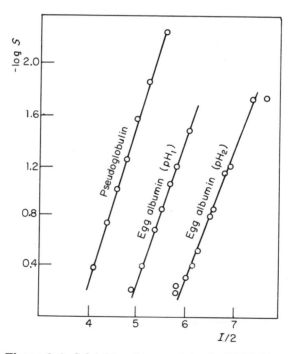

Figure 8-4. Solubility of two proteins in $(NH_4)_2SO_4$.

Table 8-1. The effect of some nonelectrolytes on the dielectric constant of water.

Substance	$-d \ln D/dC_n$	$\dfrac{-d \ln D/dC_n}{\overline{W}_n}$	Method of determination
Hydrogen	0.05	25	Solubility
Oxygen	.06	1.9	Solubility
Ethyl alcohol	.045	1.0	Direct
Ethyl acetate	.09	1.0	Direct
Ethyl ether	.075	1.0	Direct
Sucrose	.08	0.23	Direct
Egg albumin	.6	.02	Solubility
Pseudoglobulin	.7	.01	Solubility
Urea	$-$ .04	$-$ .75	Direct

volume of protein than per unit volume of oxygen. Table 8–1 shows the values of $-d \ln D/dC_n$ for various nonelectrolytes, some measured directly and some inferred from Eq. (8–33) by comparison of the effect of a given salt on the solubility of this nonelectrolyte and on that of another whose effect on the dielectric constant is known. The table also shows the effect per unit weight, which, except for hydrogen and perhaps for oxygen, corresponds roughly to the effect per unit volume. The effect of the proteins per unit mass is very small, and intermediate between that of sucrose and urea, as might be expected from their composition.

The change in excess free energy of the electrolyte may well be divided into two parts: first, the change in electric work as the solvent is changed by the addition of the nonelectrolyte, and then the electric work of concentrating the electrolyte in the solvent of changed dielectric constant. The first term may be written

$$\bar{G}^{D}_{k0} = -\nu(N_A \epsilon^2 z_+ z_-/2b_k)(1/D - 1/D_0). \qquad (8\text{–}34)$$

This is often called the Born equation.[10]

Fig. 8–5 shows the quantity $\ln\gamma_{\pm} = \bar{G}^{D}_{k0}/2RT$ for NaCl and KCl in water–alcohol mixtures as a function of the mole fraction of alcohol.[11] The full lines represent Eq. (8–34), with the values of b_k determined from the points for pure alcohol.

Table 8–2 shows the values of b and a calculated from the crystal radii of Pauling[12] as $b = 2/(1/r_+ + 1/r_-)$ and $a = r_+ + r_-$ for some

[10]M. Born, Z. Phys. 1, 45 (1920).
[11]G. Scatchard, J. Am. Chem. Soc. 47, 2098 (1925); Chem. Rev. 3, 383 (1927); Trans. Faraday Soc. 23, 454 (1927).
[12]L. C. Pauling, J. Am. Chem. Soc. 50, 1036 (1928).

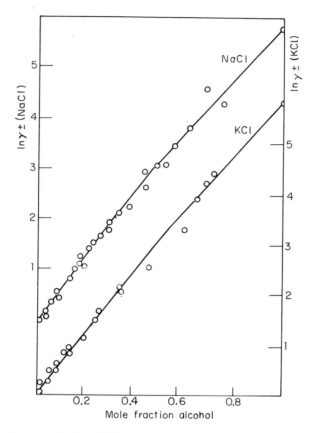

Figure 8-5. Solubilities of salts in water-ethanol mixtures.

alkali halides and the values of b determined from Eq. (8–33) or (8–34), and a and B determined by Hückel's equation,[13]

$$\ln \gamma_{\pm} = - A \kappa/(1 + \kappa a) + BC,$$

by Harned and his coworkers.[14] The values of b for the alkali chlorides show the effects of heterogeneity,[6] and those for potassium iodide and nitrate show the effects of nonelectrostatic interactions. One of the unsolved problems of solution chemistry is the reason for the small salting out of a large number of nonelectrolytes.

Series of salts arranged in decreasing order of magnitude of some effect, such as salting out, are commonly called Hofmeister series,

[13]E. Hückel, *Phys. Z. 26*, 93 (1925).

[14]H. S. Harned and B. B. Owen, *Physical Chemistry of Electrolytic Solutions*, (Reinhold, New York, 3rd ed., 1958), Table (12-5-2), p. 510.

Table 8-2. Sizes of some ions.[a]

Ion	Calculated (Å)		Measured (Å)		
	b	a	b	a	B
LiCl	0.84	2.16	1.35	4.25	0.121
NaCl	1.13	2.46	1.31	4.00	.0521
KCl	1.35	2.76	1.35	3.8	.0202
KBr	1.39	2.87	1.5	3.84	.0282
KI	1.44	3.04	2.3	3.94	.0462
KNO₃	—	—	1.7	—	—
KOH	—	—	1.0	3.7	.1294
HCl	—	—	6.4	4.3	.133

[a]From references 12–14.

after one of the first investigators to publish such a series. Many have been published since, particularly by colloid chemists, and they all place the ions in essentially the same order. The differences are often attributed to hydration, adsorption, or some more mystical cause. The present theory attributes them to the ionic charge—ionic radius ratio.

The potential of the solvent in a mixed solution of an electrolyte q and a nonelectrolyte n should be given, in accordance with Eq. (5–51), by the relation

$$(\phi - 1)M = -2A[1 + x - 1/(1 + x) - 2\ln(1 + x)]/\alpha^3 + \beta_{qq}m_q^2$$
$$+ 2\beta_{qn}m_qm_n + \beta_{nn}m_n^2 + \cdots.$$

$$(8-35)$$

The first terms omitted are those due to the change in dielectric constant on the addition of nonelectrolyte, and they are proportional to the three-halves power of the electrolyte concentration and the first power of the nonelectrolyte concentration. We might expect that they should resemble the difference between Eqs. (8–31) and (8–32) for $\partial W_e/\partial N_n$ and become quite small.

To the above approximation, the similar relation for the non-electrolyte alone at the same concentration is

$$(\phi - 1)_n\, m_n = \beta_{nn}m_n^2$$

and that for the electrolyte alone is

$$(\phi - 1)_q\, \nu_q m_q = -2A[1 + x - 1/(1 + x) - 2\ln(1 + x)]/\alpha^3 + \beta_{qq}m_q^2.$$

Therefore

$$(\phi - 1)M - (\phi - 1)_n\, m_n - (\phi - 1)_q\, \nu_q m_q = 2\beta_{qn}m_qm_n$$
$$= -\nu_q m_q\, m_n(N_A\epsilon^2 z_+z_-/2RTb_q)\partial(1/D)/\partial C_n.$$

$$(8-36)$$

This difference is shown in Fig. 8–6 for several salts and ethyl alcohol in water solution, determined from the freezing-point measurements of Sachs.[15] Prentiss and I[16] measured the freezing points of mixtures of sodium chloride with glycine, ethyl alcohol, and mixtures of the two, and Benedict and I[17] measured mixtures of dioxane with lithium, sodium, and potassium chlorides and compared the measurements with theory.

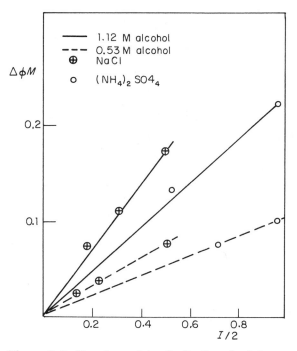

Figure 8–6. Freezing points of salt-ethanol mixtures.

8–5. Hydration

Although it is sometimes possible to determine that there is chemical combination between solute and solvent, it is very seldom possible to show that there is none, and the amount of solvation depends very greatly upon the definition of solvation, that is, upon the definition of normal behavior without solvation. It is possible to ascribe to solvation all negative deviations from Raoult's law, that is, all in-

[15]H. N. Sachs, M.S. thesis, M.I.T. (1925).
[16]G. Scatchard and S. S. Prentiss, *J. Am. Chem. Soc. 56*, 1486, 2314 (1934).
[17]G. Scatchard and M. A. Benedict, *J. Am. Chem. Soc. 58*, 837 (1936).

creases of the osmotic coefficient or of the activity coefficient of the solute with increasing solute concentration. This is done more often for ions than for nonelectrolytes, and salting out was attributed to hydration of the ions long before it was to electrostatic effects. The principles are more easily shown however for simple nonelectrolyte solutions.

We shall assume that a solute molecule I always combines with h_i molecules of water, so that the only species present in appreciable quantities are hydrated solute molecules and free water S. We also assume that the activity a_k of species k (that is, hydrated molecules in which the solute K is combined with h_k molecules of water) is given by

$$a_k = N'_k/N'_s = N_K/(N_S - \Sigma_i N_i h_i)$$
$$= r_k/(1 - \Sigma_i r_i h_i), \qquad (8\text{--}37)$$

in which N'_k and N'_s are moles of species, N_i and N_S are moles of components, and $r_i = N_i/N_S$, so that $r_i = m_i \bar{W}_S$. A Gibbs-Duhem integration shows that

$$\ln a_S = - \Sigma_j r_j/(1 - \Sigma_i r_i h_i). \qquad (8\text{--}38)$$

However, the activity of component K is then given by

$$\ln a_K = \ln a_k - h_k \ln a_S$$
$$= \ln r_k - \ln(1 - \Sigma_i r_i h_i) + \frac{h_k \Sigma_j r_j}{1 - \Sigma_i r_i h_i} \qquad (8\text{--}39)$$

The potential of component K is affected by the hydration of any other solute molecule, and by the concentration of any other molecule if it is itself hydrated.

8-6.   Mixed Electrolytes. Specific Ion Interaction

Much of the interaction between electrolytes is described by Eq. (8–28), which is a sufficient description for very dilute solutions. The behavior of mixtures can be determined from that of solutions of a single electrolyte in much more concentrated solutions by means of Brönsted's "Theory of Specific Ion Interaction."[18] The theory may be stated very simply, and it is reminiscent of the famous chapter in a natural history of Ireland, "*The Snakes of Ireland.* There are no snakes in Ireland." The theory is that the electrostatic forces are so strong that ions that have charges of the same sign never approach

[18]J. N. Brönsted, *J. Am. Chem. Soc. 44*, 877 (1922); *45*, 2898 (1923).

close enough for their short-range nonelectrostatic interactions to become appreciable. Brönsted was considering the two-by-two interactions that we have included in the β terms, but his relation can be extended to the higher terms without rapid loss in accuracy. For example, a cluster of two anions with one cation will have a strong tendency to be linear, with the cation in the middle so that the two anions will still be far apart. Of course, there will be even less tendency for the formation of a cluster of three ions all of the same sign than for the formation of a pair. Breckenridge and I have discussed the basis of this theory.[19]

Let us use odd numbers to designate cations and even numbers to designate anions, and consider first the single salt q, with ions 3 and 4. The theory of specific ion interaction says that $\beta_{33} = \beta_{44} = 0$, and the general thermodynamic theory says that $\beta_{34} = \beta_{43}$. Therefore,

$$2\beta_{qq} = \Sigma_{ij}\nu_i\nu_j\beta_{ij} = \nu_3^2\beta_{33} + \nu_3\nu_4(\beta_{34} + \beta_{43}) + \nu_4^2\beta_{44} = 2\nu_3\nu_4\beta_{34}.$$

$$(8\text{--}40)$$

So β_{34} may be determined from measurements on the single salt q. Similarly, β_{36} may be determined from the salt with ions 3 and 6, β_{54} from the salt with ions 5 and 4, and so on.

Consider, for simplicity, the case of all univalent ions. The direct application of Eq. (5–48) gives

$$\ln \gamma_3 = - AY/\alpha + 2 \sum_{j \text{ even}} \beta_{3j}m_j,$$

$$\ln \gamma_4 = - AY/\alpha + 2 \sum_{i \text{ odd}} \beta_{4i}m_i,$$

$$(8\text{--}41)$$

In a mixture of salts with ions 3,4 and 5,4,

$$\ln \gamma_3 = - AY/\alpha + 2\beta_{34}m_4,$$
$$\ln \gamma_5 = - AY/\alpha + 2\beta_{54}m_4,$$
$$\ln \gamma_4 = - AY/\alpha + 2\beta_{34}m_3 + 2\beta_{54}m_5.$$

$$(8\text{--}42)$$

At constant ionic strength, $m_4 = m_3 + m_5$ is constant, so $\ln \gamma_3$ and $\ln \gamma_5$ are independent of the composition, whereas $\ln \gamma_4$ is a linear function of the composition. If we define γ_{34} as the mean activity coefficient of the electrolyte with ions 3 and 4, and so on,

$$\ln \gamma_{34} = - AY/\alpha + 2\beta_{34}m_3 + (\beta_{34} + \beta_{54})m_5,$$
$$\ln \gamma_{54} = - AY/\alpha + (\beta_{34} + \beta_{54})m_3 + 2\beta_{54} m_5,$$
$$\ln \gamma_{34} - \ln \gamma_{54} = (\beta_{34} - \beta_{54})(m_3 + m_5) = (\beta_{34} - \beta_{54})m_4.$$

$$(8\text{--}43)$$

The difference between the logarithms of the two mean activity

---

[19]G. Scatchard and R. G. Breckenridge, *J. Phys. Chem.* 58, 596 (1954).

coefficients is independent of the composition. Each is a linear function of the composition; the slopes are the same and equal to half the difference between the values for the pure electrolytes. If we use superscripts to denote the medium, at constant total concentration,

$$\ln \gamma_{34}^{54} = (\ln \gamma_{34}^{34} + \ln \gamma_{54}^{54})/2 = \ln \gamma_{54}^{34}.$$

In a mixture of two salts without a common ion, $m_3 = m_4 = m_{34}$, $m_5 = m_6 = m_{56}$, and

$$
\begin{aligned}
\ln \gamma_3 &= - AY/\alpha + 2\beta_{34}m_4 + 2\beta_{36}m_6, \\
\ln \gamma_4 &= - AY/\alpha + 2\beta_{34}m_3 + 2\beta_{54}m_5, \\
\ln \gamma_{34} &= - AY/\alpha + 2\beta_{34}m_{34} + (\beta_{36} + \beta_{54})m_{56}, \\
\ln \gamma_{56} &= - AY/\alpha + (\beta_{36} + \beta_{54})m_{34} + 2\beta_{56}m_{56}.
\end{aligned}
\tag{8-44}
$$

The logarithms of the mean activity coefficients are linear in the composition at constant total ionic strength, so

$$\ln \gamma_{34}^{56} = (\ln \gamma_{54}^{54} + \ln \gamma_{36}^{36})/2 = \ln \gamma_{56}^{34}. \tag{8-45}$$

The activity coefficient of 56 in 34 is the same as that of 34 in 56, but they are the mean of the values for the reciprocal-salt pair, 54 and 36, and not the mean for the two salts themselves.

Hydration of a cation should remove solvent water from another cation as well as from an anion, and vice versa, according to the discussion of the last section. This does not alter the calculation of the mean activity coefficient in any mixture of salts of the same valence type from the mean activity coefficients of single salts, but it does introduce a small error in the calculation in mixtures of different valence types, and a larger one in the calculation of individual ion activities, except as they are used conventionally to be combined as mean activity coefficients. As a useful approximation we may still write

$$\ln \gamma_k = \frac{-Az_k^2(I/2)^{1/2}}{1 + 1.5(I/2)^{1/2}} + \Sigma_j f_{kj}(I)m_j \tag{8-46}$$

if

$$f_{kj}(I) = 0 \tag{8-47}$$

when $z_k z_j$ is positive and

$$f_{kj}(I) = \left\{ \frac{\ln \gamma_{kj}^{kj} - \dfrac{z_k z_j A(I/2)^{1/2}}{1 + 1.5(I/2)^{1/2}}}{} \right\} \frac{(z_k - z_j)^2}{2I} \tag{8-48}$$

when $z_k z_j$ is negative.

The most thorough test of theory comes from its application[20] to the results of the freezing-point measurements of mixtures of potassium and lithium chlorides and nitrates,[21] chosen because both the lithium salts have large positive and nearly equal values of β, potassium nitrate has a large negative value, and for potassium chloride β is small if $\alpha = 1$. Fig. 8–7 shows the values of $\Delta \ln \gamma_{\pm} \; [= \ln \gamma_{\pm} + AY]$

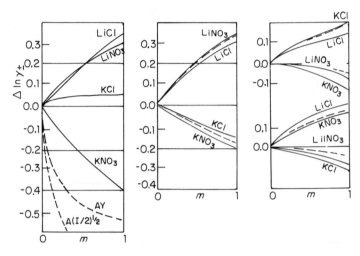

Figure 8–7. Deviations of logarithms of activity coefficients at constant composition of solute.

as a function of the ionic strength $I/2$. On the left are the pure salts. The broken lines are the Debye limiting law, $- A(I/2)^{1/2}$ and $- AY$. The other two figures have the curves grouped in pairs, and each curve represents the limit of $\ln \gamma_{\pm}$ for the salt with which it is tagged in the pure salt with which the other curve is tagged. Thus the lowest curve in the middle figure is $\ln \gamma_{\mathrm{KNO_3}}^{\mathrm{KCl}}$ and the next full curve is $\ln \gamma_{\mathrm{KCl}}^{\mathrm{KNO_3}}$. The broken curve is $(\ln \gamma_{\mathrm{KCl}}^{\mathrm{KCl}} + \ln \gamma_{\mathrm{KCl}}^{\mathrm{KNO_3}})/2$, and all three curves should be the same if the theory of specific ion interaction were strictly valid. In the lower part of the right-hand figure, the upper broken curve is $\Delta(\ln \gamma_{\mathrm{KCl}}^{\mathrm{KCl}} + \ln \gamma_{\mathrm{LiNO_3}}^{\mathrm{LiNO_3}})/2$ and the lower is $\Delta(\ln \gamma_{\mathrm{LiCl}}^{\mathrm{LiCl}} + \ln \gamma_{\mathrm{KNO_3}}^{\mathrm{KNO_3}})/2$. The agreement is very much better for the reciprocal-salt pair than for the average of the salts themselves.

Fig. 8–8 shows the variation with changing composition of the binary systems at $I/2 = 1$, which was the limit of the measurements. The theory of specific ion interaction says that the curves should all be linear and that in the middle figure the pairs KCl–KNO$_3$ and LiCl–LiNO$_3$ should be parallel in the first and third sections, that the

[20]G. Scatchard, *Chem. Rev.* **19,** 309 (1936).

[21]G. Scatchard and S. S. Prentiss, *J. Am. Chem. Soc.* **56,** 2320 (1934).

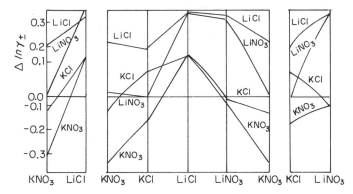

Figure 8-8. Deviations of activity coefficients at constant ionic strength of unity.

slope in the third section should be the negative of that in the first, and that the slope for KCl–KNO$_3$ should be half the difference between the values of $\Delta \ln \gamma_{KNO_3}^{KNO_3}$ and $\Delta \ln \gamma_{KCl}^{KCl}$, and so on. The same should hold for the pairs KCl–LiCl and KNO$_3$–LiNO$_3$ in the second and fourth sections. The approximation is certainly good enough to be very useful.

Harned and his coworkers[14] measured the potential of electric cells without transference, which give the chemical potential of hydrochloric acid in mixtures with the alkali halides, and found that $\Delta \ln \gamma_{HCl}$ is nearly proportional to m and agrees well with the specific ion interaction theory, up to $I/2 = 3$. For the alkali hydroxides in the corresponding halides, they found that the β terms are not adequate, and the agreement with the specific-ion interaction theory is not good at all.

The applications to the effect of one salt upon the solubility of another are very interesting. For example, the solubility of silver chloride in potassium nitrate and potassium chloride solutions is given by

$$\tfrac{1}{2}\ln \left(\frac{(Ag^+)(Cl^-)-}{(Ag^+)_0(Cl^-)_0} \right) = AY/\alpha - (\beta_{AgNO3} + \beta_{KCl})m_{KNO_3} \\ - (\beta_{AgCl} + \beta_{KCl})m_{KCl} - 2\beta_{AgCl}m_{AgCl}.$$

Here β_{AgNO_3} and β_{KCl} can be determined from measurements on soluble salts, but it is not possible to obtain a high enough concentration of silver chloride to determine β_{AgCl} directly. However, determination of the solubilities in one silver salt or in one chloride, as in potassium chloride above, is sufficient to determine this coefficient, so that the solubility in any other salt with a common ion may be calculated as that in mixtures without a common ion.

The specific-ion interaction theory has received surprisingly little recognition. There are a few measurements of the electromotive force of hydrogen chloride cells in mixtures with salts of different valence types. There are many such measurements of solubilities in such mixtures. The discussion of Scatchard and Breckenridge[19] deals particularly with changing valence type.

8-7. Higher-Term Correction. Electrostatic Association

The error involved in Debye's expansion of the Boltzmann exponential and dropping all terms after the second was investigated very soon after the theory was announced. Mathematical extensions have been made by Debye himself at the request of LaMer,[22] by Müller,[23] by Gronwall, LaMer, and their collaborators,[24] by Kirkwood,[25] and by Mayer.[26]

We shall be more interested in the physicochemical approach of Bjerrum,[27] who calculated the number, $n_{ik}dr$, of ions of charge z_i in a spherical shell of thickness dr at a distance r from an ion with valence k in a solution so dilute that the potential at r due to ions other than k may be neglected:

$$n_{ik}dr = c_{i0}4\pi r^2 \exp(-\epsilon^2 z_k z_i/kTDr)dr, \qquad (8\text{--}49)$$

in which c_{i0} is the concentration of ions of species i where the electrostatic potential is zero. Differentiation of n_{ik} with respect to r yields

$$dn_{ik}/dr = c_{i0}4\pi(2r + \epsilon^2 z_k z_i/kTD)\exp(-\epsilon^2 z_k z_i/kTDr), \qquad (8\text{--}50)$$

which is zero when

$$r = -\epsilon^2 z_k z_i/2kTD = a'. \qquad (8\text{--}51)$$

If z_k and z_i are of opposite signs, the minimum occurs at a positive value of r, the distance at which the mutual energy of the two ions is $2kT$. Fig. 8–9 shows n_{ik} for $z_i = 0$, $z_i = -z_k$, and $z_i = +z_k$. The broken lines are the values of the latter two that correspond to the Debye approximation.

$$n_{ik} = c_{i0}4\pi r^2(1 - \epsilon^2 z_k z_i/kTDr). \qquad (8\text{--}52)$$

---

[22]V. K. LaMer and C. A. Mason, *J. Am. Chem. Soc. 49*, 410 (1927).

[23]H. Müller, *Phys. Z. 28*, 324 (1927).

[24]T. H. Gronwall, V. K. LaMer, and K. Sandved, *Phys. Z. 29*, 358 (1928); V. K. LaMer, T. H. Gronwall, and L. Greiff, *J. Phys. Chem. 35*, 2245 (1931).

[25]J. G. Kirkwood, *Chem. Rev. 19*, 275 (1936).

[26]J. E. Mayer, *J. Chem. Phys. 18*, 1426 (1950); J. C. Poirier, *ibid.*, *21*, 965, 972 (1953).

[27]N. Bjerrum, *Kgl. Dansk. Vid. Selsk. 7*, 9 (1926).

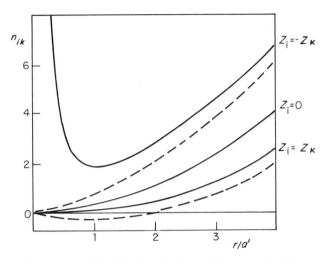

Figure 8-9. Distribution of charge close to central ion.

We note that Eq. (8–52) makes the total number in a shell the same as though all the ions were uncharged, and we note also that this approximation makes the concentration of ions with valence $+ z_k$ negative for all values of r less than $2a'$.

Bjerrum notes that there are very few ions near the minimum, and in very dilute solutions the probability of there being more than one ion closer than a' is very small. If we designate the second ion as j, all the other ions will be so far away that $1/r_{ik}$ and $1/r_{ij}$ are very nearly equal. Therefore the mutual energy of ion i with the pair j and k will be very nearly zero and the contribution of this ion pair to the interionic free energy and the ionic strength practically disappears. For higher concentrations, the number of such pairs may be calculated from the law of mass action corrected for electrostatic interaction. Bjerrum makes this assumption for symmetric salts by treating such pairs as associated molecules, and assuming that all ions that are not members of such pairs may be treated by the Debye-Hückel approximation with $a = a'$. If a is equal to or greater than a', the association constant is zero. If a is less than a', the constant is $1/c_{i0}$ times the number of ions of type i in the shell between $r = a$ and $r = a'$. If

$$t = - \epsilon^2 z_k z_i / kTDr,$$

$$n_{ik}dr = n_{i0}4\pi\left(\frac{\epsilon^2 z_k z_i}{kTD}\right)^3 \frac{e^t}{t^4}\, dt$$

and

$$\int_a^{a'} n_{ik}dr = - n_{i0}4\pi\left(\frac{\epsilon^2 z_k z_i}{kTD}\right)^3 \int_2^{2a'/a} \frac{e^t}{t^4}dt = K_{ki}c_{i0},$$

in which c_{i0} is the concentration in moles per liter, so $n_{i0}/c_{i0} = N_A/1000$. Then

$$K_{ki} = -\frac{4\pi N_A}{1000} \left(\frac{\epsilon^2 z_k z_i}{DkT} \right)^3 \int_2^{2a'/a} \frac{e^t}{t^4}\, dt. \tag{8-53}$$

Fig. 8–10 shows (solid curve) the values of the integral, usually

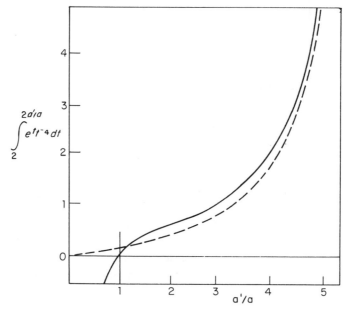

Figure 8–10. Bjerrum integral, Eq. (8–53) (solid curve) and Kirkwood integral, Eq. (8–55) (dashed curve).

designated by Q, as a function of a'/a. The negative values for a'/a less than 1 have no real significance.

Fig. 8–11 shows log $\gamma_{\pm}$ calculated by Eq. (8–53) vs. the Debye-Hückel limiting law for various values of a'/a. When a'/a is greater than unity, $-$ log $\gamma_{\pm}$ is greater than the Debye-Hückel approximation for the same a, but the difference is unimportant until a'/a becomes greater than 2. For very large values of this ratio, $-$ log $\gamma_{\pm}$ becomes even larger than the value for the limiting law, with an inflection in dilute solutions. The result is a nearly linear relation to the square root of the ionic strength over a fairly large concentration range, with a slope much greater than that of the limiting law. For a uni-univalent electrolyte in water at 18°C, a' is 3.52Å, and it varies but little with the temperature. Here the corrections are negligible, for

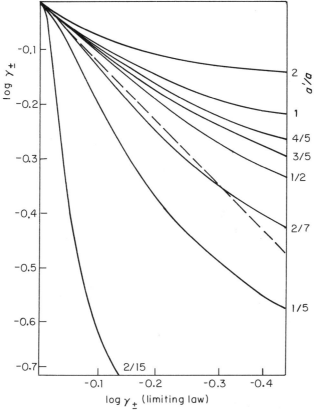

Figure 8–11. Effect of higher terms (Bjerrum).

it is doubtful whether any of the atomic diameters are less than 2Å, but the correction should become important for salts of higher valence types, and for 1–1 salts in solvents of lower dielectric constants.

Müller[23] treated the same problem by graphical integration of the complete Boltzmann exponential. Gronwall, LaMer, and Sandved[24] give an analytic treatment of the expanded Boltzmann integral. For symmetric electrolytes, all the even-powered terms are zero. They have tabulated values for the third and fifth powers as functions of κa. They also give an infinite series expressing the complete expression to terms in $(\kappa a)^2$. Mayer[26] obtains the same series by a very different method. Poirier[26] has tabulated values of the first 16 of Mayer's series. LaMer, Gronwall, and Greiff[24] have tabulated values for the second and third powers for unsymmetric electrolytes. Kirkwood[25] has obtained a closed expression for the series of terms in $(\kappa a)^2$ from the third term on, which is therefore complete for symmetric electrolytes.

Table 8-3. Some values of Q_{ik} and B_{ik}.

b_{ik}	Q_{ik}	B_{ik}	$-b_{ik}$	$-B_{ik}$
1	—	0.046	1	0.037
2	0.000	.105	2	.069
3	.325	.181	3	.097
4	.550	.285	4	.120
5	.755	.414	5	.141
6	1.041	.645	6	.159
7	1.417	.979	7	.176
8	1.996	1.525	8	.192
9	2.950	2.450	9	.206
10	4.547	4.023	10	.219
12	13.51	12.84	12	.242
15	101.8	101.2	15	.272

If

$$b_{ik} = 2a'/a_{ik} = -\epsilon^2 z_i z_k / DkT a_{ik}, \tag{8-54}$$

$$
\begin{aligned}
B_{ik} &= \int_0^{b_{ik}} (1/t^4)(e^t - 1 - t - t^2/2 - t^3/6)\, dt \\
&= \{b_{ik}^3[E_i(b_{ik}) - \ln|b_{ik}| - C_0] - e^{b_{ik}}(2 + b_{ik} + b_{ik}^2) \\
&\quad + (2 + 3b_{ik} + 3b_{ik}^2 + 11b_{ik}^3/6)\}/6b_{ik}^3,
\end{aligned}
\tag{8-55}
$$

in which $E_i(b_{ik})$ is the exponential integral of b_{ik} and C_0 is Euler's constant, 0.5772. The corresponding contribution to $\ln \gamma_k$ is

$$\ln \gamma_k''' = -\frac{\epsilon^4 z_k^3}{(DkT)^2}\frac{\kappa^2}{\Sigma n_i z_i^2}\Sigma_i n_i z_i^3 B_{ik}. \tag{8-56}$$

Table 8–3 contains some values of Bjerrum's integral Q_{ik} and of B_{ik}. The dashed curve in Fig. 8–10 is $(B_{ik} - B_{-ik})$. For symmetric electrolytes, Q_{ik} should equal $(B_{ik} - B_{-ik} - \frac{1}{2}b_{ik} + 1/4)$. Bjerrum's method gives values that are too small for small values of b_{ik} and too large by 0.145 in the limit of large values of b_{ik}. Although both errors are small, Epstein and I[28] suggested that Bjerrum's value of the association constant, Eq. (8–53), be replaced by

$$K_{ki} = -\frac{4\pi N_A}{1000}\left(\frac{z_k z_i \epsilon^2}{DkT}\right)^3\left[B_{ki} - \frac{(z_k/z_i)^2 B_{kk} + (z_i/z_k)^2 B_{ii}}{2}\right]. \tag{8-57}$$

This agrees with theory to terms proportional to the concentration and has all the other assets of the Bjerrum method.

Eq. (8–57) is written so as to include the binary compound for unsymmetric electrolytes such as Na_2SO_4 or $LaCl_3$, and the agreement

---

[28] G. Scatchard and L. F. Epstein, *Chem. Rev.* **30**, 211 (1942).

with Bjerrum's method is about the same as for symmetric electrolytes. The logical extension of Bjerrum's method is to take into account secondary and higher associations until the neutral molecule is formed. For example, in $LaCl_3$, we should consider three reactions:

$$La^{+++} + Cl^- = LaCl^{++},$$
$$LaCl^{++} + Cl^- = LaCl_2{}^+,$$
$$LaCl_2{}^+ + Cl^- = LaCl_3.$$

There is no way of checking the higher association with physical theory. Epstein and I recommended using the same size for the higher association.

Unsymmetric electrolytes introduce another complication. The third term in the expansion of the Boltzmann exponential leads to a term proportional to $\kappa^2 \ln \kappa$. When κ approaches zero, $d(\kappa^2 \ln \kappa)/d(\kappa^2)$ approaches infinity although $d(\kappa^2 \ln \kappa)/d\kappa$ approaches zero. By the Debye charging process, which we have used, of determining the electrostatic free energy by charging all the ions together and then determining the potentials by differentiation, the contribution of this term to $\ln \gamma_k$ is given by the relations[24]

$$\ln \gamma_k'' = \left(\frac{\epsilon^2}{DkTa} \right)^2 z_k^3 \frac{\Sigma_i n_i z_i^3}{\Sigma_i n_i z_i^2} \, [\tfrac{1}{2} X_2 - Y_2], \tag{8–58}$$

$$X_2(\kappa a) = -\tfrac{1}{2} \frac{(\kappa a)^2}{(1 + \kappa a)^3} \, e^{3\kappa a} \, E_i(3\kappa a), \tag{8–59}$$

$$Y_2(\kappa a) = \frac{1}{(\kappa a)^4} \int_0^{\kappa a} x^3 X_2(x) dx. \tag{8–60}$$

The Güntelberg-Müller charging process[23, 29] ignores the change of volume and of dielectric constant on addition of an ion, and determines the chemical potential directly by charging a single ion at constant κ. Strictly speaking, the process should be carried out for enough ions that $\Sigma_i \nu_i z_i = 0$, but the solution may be of infinite extent and the result is the same. For the Debye-Hückel approximation the result is the same as for the Debye process. From Eq. (7–31), when the charge on ion i is e,

$$\psi_b = \frac{e}{D} \left(\frac{1}{b_i} - \frac{\kappa}{1 + \kappa a} \right), \tag{8–61}$$

and

$$\bar{G}_{ie}^e = \int_0^{z_i \epsilon} \psi_b \, de = \frac{\epsilon^2 z_i^2}{2D} \left(\frac{1}{b_i} - \frac{\kappa}{1 + \kappa a} \right). \tag{8–62}$$

[29]N. Bjerrum, Z. Phys. Chem. 119, 145 (1926).

For the higher terms, however, the two charging processes give different answers. For example, by the Güntelberg-Müller process,

$$\ln \gamma_k'' = \left(\frac{\epsilon^2}{DkT}\right)^2 z_k{}^3 \frac{\Sigma_i n_i z_i{}^3}{\Sigma_i n_i z_i{}^2} \; (\tfrac{1}{3} \, X_2). \qquad (8\text{-}63)$$

The reason for the difference is that the expanded equations violate the condition of integrability, and $\partial\psi_i/\partial e_k \neq \partial\psi_k/\partial e_i$, or the work of charging the ions depends upon the order in which they are charged. This condition is not violated for the Debye-Hückel approximation. We shall assume later that it is not violated for other modifications that stop with the second term in the expansion of the Boltzmann exponential. The condition of integrability is not violated for the terms proportional to $(\kappa a)^2$, as in the Kirkwood integral, and no integration is involved in the Bjerrum treatment, so these two methods are free from this uncertainty. We may hope that in other terms the error of the other charging process is not greater than the difference between the two processes. Probably the Debye process gives the better answer.

Values of X_2 and Y_2 are tabulated by LaMer, Gronwall, and Greiff[24] up to $\kappa a = 0.4$ and by Breckenridge and me[19] from $\kappa a = 0.4$ to 3.2. In most cases this second-term contribution is very small compared with those of the higher terms at all concentrations at which measurements are possible. However, it is well to investigate before neglecting it in any case.

8-8. Other Models

We have assumed that the mean collision diameter a is the same for all the ions in a solution, even though the radii from salting-out measurements appear quite different. When there are only two ions, one cation and one anion, we have assumed that $a = b_+ + b_-$. Let us assume that $b_+ < b_-$. Then around each cation there is a spherical shell from $2b_+$ to $b_+ + b_-$ into which other cations can penetrate, although we have assumed that they cannot. Around each anion there is a spherical shell from $b_+ + b_-$ to $2b_-$ into which we have assumed that other anions can penetrate, although they cannot. The electrostatic forces prevent many cations from entering the shell from which we have incorrectly assumed them absent, and the recognition of electrostatic forces has prevented us from assuming that many anions have entered the shell in which we have incorrectly assumed them present. Moreover, the two errors tend to compensate. However, the shell from $b_+ + b_-$ to $2b_-$ has a larger volume and the electrostatic forces are smaller, so the compensation is not exact.

It is important to recognize that, if all the ions do not have the same size, the Debye-Hückel equations do not apply. If the diameter a is not the same for all the ions, there is no a for any ion.

It is possible to extend the Debye-Hückel treatment to any number of exclusion diameters that prevent ions of different types from nearer approach to the central ion. I have carried it out for two diameters,[30] and Mayer[26] treats any number of different sizes. The results confirm our simplifying assumptions, but become important when ions of one type are very much larger than the others, as in protein solutions.[31] Kirkwood and I[32] gave an approximate treatment for a dumbbell, a molecule consisting of two spheres, with charges at the center of each, equal in magnitude and of the same or opposite signs, kept apart at a constant distance R. Perhaps the chief interest was the proof that even for this extreme model the Debye limiting law is valid.

Kirkwood[33] has determined the potential of a spherical molecule containing any distribution of charges, and Westheimer and Kirkwood[34] have determined the potential of an ellipsoid of revolution with any distribution of charges, or of dipoles of infinitesimal length and any orientation, along the axis of revolution. Both of these treatments include the effects of changing ionic strength and changing dielectric constant, but in the limit of zero concentration of the molecule whose potential is being determined. All the treatments of this section use the Güntelberg-Müller charging process.

The most complete theoretical treatment of any solutions is a study of the osmotic coefficients of the alkali halides.[35] The crystal-lattice radii of Pauling were used to determine $b_+ = r_+$, $b_- = r_-$, $a = r_+ + r_-$, $V_+ = kr_+^3$, and $V_- = kr_-^3$. The equation used was

$$\phi = \frac{\ln(1 + \nu_s N_s/N_0)}{\nu_s N_s/N_0} + \frac{\epsilon^2 N_A}{2RTD_0} \left[z_+ z_- \frac{Z}{\alpha} + \frac{z_+ z_- V_s m}{1 + \kappa a} \right.$$

$$\left. + \frac{2\nu_1 \nu_2}{\nu_s} \left(\frac{V_- z_+^2}{b_+} + \frac{V_+ z_-^2}{b_-} \right)m \right]$$

$$+ \frac{2\nu_+ \nu_- V_+ V_- (2a_{12} - a_{10} - a_{20})m}{\nu_s RT(1 + V_s m)^2} . \tag{8-64}$$

[30] G. Scatchard, *Phys. Z. 33*, 22 (1932).

[31] G. Scatchard, A. C. Batchelder, and A. Brown, *J. Am. Chem. Soc. 68*, 2320 (1946).

[32] G. Scatchard and J. G. Kirkwood, *Phys. Z. 33*, 297 (1932).

[33] J. G. Kirkwood, *J. Chem. Phys. 2*, 351 (1934).

[34] J. G. Kirkwood and F. H. Westheimer, *J. Chem. Phys. 6*, 506 (1938); F. H. Westheimer and J. G. Kirkwood, *ibid.*, 513 (1938).

[35] G. Scatchard, *Chem. Rev. 19*, 309 (1936).

This equation assumes that the electrostatic forces prevent ions of the same charge from approaching close enough to have any influence on a, on the ion-molecule effect in $V_i z_j^2/b_j$, or on the molecule-molecule term $(2a_{ij} - a_{i0} - a_{j0})V_iV_j$. The molecule-molecule term is calculated from simple theory for nonpolar molecules. The first term corrects to mole fractions. I assumed that $(2a_{12} - a_{10} - a_{20})$ is the same for all the alkali halides, and determined its value and the ratio of V to r^3 from the measurements on molal lithium bromide and cesium bromide. This leads to $k = 0.008$, or to a volume in solution 3.15 times the volume of the spheres. A volume somewhat less than 4 is to be expected. This leads to a value of $(2a_{12} - a_{10} - a_{20})$, about one-third of that for an aliphatic hydrocarbon in water.

Fig. 8–12 shows the values for the alkali chlorides, bromides, and

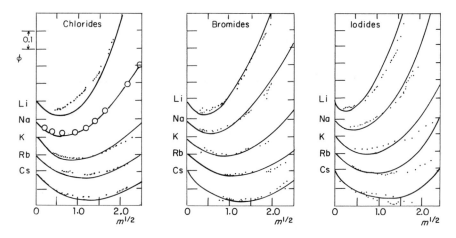

Figure 8–12. Osmotic coefficients of alkali halides in aqueous solution at 25°C. Dotted curves: experimental results; solid curves: predictions from Eq. (8–64). The points for NaCl are from measurements in the author's laboratory; the other points come from Robinson and Sinclair, *J. Amer. Chem. Soc.* **56**, 1830 (1934); **57**, 1161 (1935).

iodides. The calculated results are too low in dilute solutions and too high in concentrated solutions, particularly with the iodides. They do show that the sizes and valences of the ions are sufficient to characterize the alkali halides. In concentrated solutions the differences in the last two terms are the only ones that are important. The first is positive, and is large when one ion is small, the other large; the second is negative, and is large when both ions are large. The calculated results agree with the experimental ones in making the osmotic coefficient decrease with increasing size of the cations for the iodides, bromides, and chlorides but change in the opposite order for the fluorides. Theory and experiment also agree in making the osmotic coefficients of the lithium, sodium, and potassium halides increase with increasing size of the anion and to change in the opposite order for cesium. Theory puts rubidium with the smaller cations, the measurements group it with cesium, but neither makes it change much. It would be legitimate to tinker with the radii to give a closer fit with experiment.

8-9.   Electromotive Force and Chemical Potentials. Liquid-Junction Potentials

We shall consider a simple cell with two electrodes and one liquid junction, in which the composition of the solution is constant throughout any plane perpendicular to the line through the two electrodes. The electrode at the left is surrounded by solution A and that at the right by solution B. The change in free energy through the part of the cell in which the composition is changing from that of A to that of B per faraday of electricity, $\Delta G_\ell/N$, is

$$\Delta G_\ell/N = \int_A^B \Sigma_i t_i d\bar{G}_i, \tag{8-65}$$

in which t_i $[= m_i u_i/\Sigma_j m_j z_j u_j]$ is the number of moles of the ith species that are transported in the direction of the positive current per faraday at the point x (see Sec. 7–8) and $\bar{G}_i$ is the chemical potential of that species at the same point. Nonelectrolytes may not be neglected in this summation, but it is usually more convenient to fix the point x, not so that its distance from either electrode is unchanged when electricity passes through the cell, but so that the amount of one component, usually the solvent, on either side of a plane through x is not changed during the passage of the current or, more accurately, so that the amount of that component on either side is increased by the amount produced at the electrode on that side. Thus the transfer of that component is zero. It is convenient to split this change of free energy into four parts:

$$\Delta G/N = RT\int \Sigma_i t_{io} d\ln m_i + RT\int \Sigma_i t_i^* d\ln m_i + RT\int \Sigma_i t_{io} d\ln \gamma_i$$
$$+ RT\int \Sigma_i t_i^* d\ln \gamma_i, \tag{8-66}$$

in which

$$t_{io} = m_i u_{io}/\Sigma_j m_j z_j u_{jo}, \tag{8-67}$$
$$t_i^* = t_i - t_{io}, \tag{8-68}$$

and u_i is the mobility of the ith species, which is reckoned negative for an anion, as are its transference number and its valence. For perfect solutes with constant mobilities the last three terms are each zero.

The first term of Eq. (8–66) shows that the change in free energy must depend upon the way in which the concentrations of the different species vary in going from solution A to solution B unless there are only two kinds of ion in the boundary. For only two kinds of ion the law of electric neutrality fixes their relative concentration, so that each t_i and each γ_i may be expressed as a function of the concentration of either ion.

Planck[36] determined the first term for a boundary at which the composition is fixed as A at the left of one plane and as B at the right of another plane and the composition between the two planes is fixed by diffusion. Such a junction has been closely approximated by having one solution flow down each side of a vertical mica plate with a small hole bored through it for the junction.[37] Taylor[38] and Guggenheim[39] attempted to calculate this first term for the junction made by placing A and B in contact along a true plane and allowing them to mix by diffusion. Such a junction is approximated by a "flowing junction" in which the flow is stopped. Henderson[40] calculated the first terms for a junction formed by mechanical mixing. Any part of this "mixture boundary" may be made up by mixing x parts of B and $(1 - x)$ parts of A. This junction is closely approximated by the two types of junction that have been found experimentally to give steady and reproducible electromotive forces. In the "flowing junction" a tube shaped between a T and a Y is placed on its side with the three arms in the same vertical plane.[41] The heavier solution flows up through one arm, the lighter down through the other at the same rate and the two flow out together through the horizontal arm. Most of the electric current goes through the region of first meeting where diffusion has had little effect even though the junction is very thin. The Clark[42] junction, which is usually used in practice, is made by sucking the lighter liquid into the top of an inclined tube holding the heavier and gives a mixture boundary about a centimeter through,

[36]M. Planck, *Wied. Ann.* **39**, 161, 561 (1890).

[37]E. J. Roberts and F. Fenwick, *J. Am. Chem. Soc.* **49**, 2787 (1927).

[38]P. B. Taylor, *J. Phys. Chem.* **31**, 1478 (1927).

[39]E. A. Guggenheim, *J. Am. Chem. Soc.* **52**, 1315 (1930).

[40]P. Henderson, *Z. Phys. Chem.* **59**, 118 (1907); **63**, 325 (1908).

[41]A. B. Lamb and A. T. Larson, *J. Am. Chem. Soc.* **42**, 229 (1920); D. A. MacInnes and Y. L. Yeh, *ibid.*, **43**, 2563 (1921); G. Scatchard, *ibid.* **47**, 696 (1925).

[42]W. M. Clark, *The Determination of Hydrogen Ions* (Williams and Wilkins, Baltimore, 1923, 1928).

so that any change due to diffusion is slow. Henderson's expression is therefore the most useful as well as the simplest. We use x as the measure of change from A to B:

$$m_j = m_{jA} + (m_{jB} - m_{jA})x,$$
$$dm_i = (m_{iB} - m_{iA})dx.$$

Then the first term of Eq. (8–66) gives

$$\frac{\Delta G_t}{NRT} = \int_A^B \frac{\Sigma_i m_i u_{io} d\ln m_i}{\Sigma_j m_j u_{jo} z_j} = \int_A^B \frac{\Sigma_i u_{io} d\,m_i}{\Sigma_j m_j u_{jo} z_j}$$

$$= \int_0^1 \frac{\Sigma_i(m_{iB} - m_{iA})u_i dx}{\Sigma_j[m_{jA} + (m_{jB} - m_{jA})x]u_{jo} z_j}$$

$$= \frac{\Sigma_i(m_{iB} - m_{iA})u_{io}}{\Sigma_i(m_{iB} - m_{iA})u_{io} z_i} \ln \frac{\Sigma_i m_{iB} u_{io} z_i}{\Sigma_i m_{iA} u_{io} z_i} \,. \qquad (8\text{–}69)$$

Eq. (8–69) reduces to simple forms if there are only two ions or if the solutions A and B each contain a single electrolyte with an ion in common and of the same valence type and the same concentration.

The last three terms of Eq. (8–66) can be integrated if the mobilities and the activity coefficients are known as functions of the composition. In the few cases for which they have been tested, the last two terms tend to cancel the second, so that the first term alone is a very fair approximation of the whole expression. The integration of the third term appears to require a knowledge of the individual ion activities, but it does not. If the two electrodes are each reversible to the kth ion, the free-energy change for the electrode reactions is

$$(\bar G_{kB} - \bar G_{kA})/z_k = (RT/z_k) \ln (m_{kB}/m_{kA}) - RT \int (1/z_k)d\ln \gamma_k. \qquad (8\text{–}70)$$

Since $\Sigma_i t_i z_i = 1$, we may write the last term of Eq. (8–70) as $-RT \int \Sigma_i(t_i z_i/z_k)d\ln \gamma_k$ and combine it with the last two terms of Eq. (8–66), to give

$$RT \int \Sigma_i t_i d\,[\ln \gamma_i - (z_i/z_k) \ln \gamma_k]. \qquad (8\text{–}71)$$

If z_i and z_k have opposite signs, the expression in brackets is $(1 - z_i/z_k)$ times the logarithm of their mean activity coefficient. If z_i and z_k have the same sign, this expression may be related to their two mean activity coefficients with any ion of the opposite sign. If the two electrodes are not reversible to the same ion, the summation can be carried out for the ion that reacts at either electrode and the electromotive force of the cell can be calculated by adding to this sum the electromotive force of the cell without transference involving the two electrodes and the solution in which the one not used is present in the real cell. In either case this procedure gives the total free-energy change of the cell process without any individual ion activity coefficients. More complicated cells can always be treated as a series of simple cells provided that there is a single reaction at each electrode.

It should be noted that cells which we describe as "without transference" are not as simple as they are usually written. The second description is necessary for the cell:

$$Pt(s), H_2(g), HCl(m), AgCl(s), Ag(s),$$
$$Pt(s), H_2(g), HCl(m) + H_2(sat), HCl(m) + AgCl(sat), AgCl(s),$$
$$Ag(s), Pt(s),$$

if the solubility of the platinum and the silver may be neglected. The difference between the two cells becomes important in two cases: first, when the HCl concentration is so small that the Cl$^-$ from the AgCl is not negligible relative to that from HCl; and second, when the pressure and resultant concentration of hydrogen are so large that the free energy of transfer of HCl from one solution to another is not negligible. A difference in nonelectrolyte concentration at the two sides of a liquid junction may have as much effect as a difference in electrolyte concentration.

A study of Eq. (8–69) will show why saturated potassium chloride solutions are used to eliminate liquid-junction potentials, and how effective they are in particular cases as far as the first term is concerned. Let the B solution be saturated potassium chloride for which the transference numbers are nearly $\frac{1}{2}$. If they were exactly $\frac{1}{2}$, $\Sigma_i m_{iB} u_{io}$ would be zero; $\Sigma_i m_{iA} u_{io}$ is always less than $\Sigma_i m_{iA} u_{io} z_i$; the ratio in the logarithm is the ratio of the conductances. If it is very small, and we represent it by x, the whole expression will be less than $RT\,x\,\ln x$, which approaches zero as x approaches zero. If the mobilities of the anions and cations in A are also nearly equal, G_1/N will be much less than $RT\,x\,\ln x$.

8–10.   Electrode Potentials and Single-Ion Activities

We have just seen that the liquid-junction potentials cannot be determined without a knowledge of single-ion activities, and the inverse is also true. Neither of these quantities is capable of exact thermodynamic definition. The electric potential difference between two media of different dielectric constants is not definable because the work of moving unit charge from one medium to the other will depend upon the radius of the cavity in which it is placed, just as it does for an ion according to Eq. (8–34).

The standard-electrode potential is obtained from experimental measurements and is obviously not the difference in potential between the electrode and the solution when the activity of each dissolved substance is unity. It may be thought of as $1/N\mathfrak{F}$ times the decrease of free energy that accompanies the reaction when 1 faraday of electricity passes from the electrode to the solution. The chemical potentials necessary for the calculation of this free-energy decrease are usually determined from electromotive-force measurements. The best measurements involve the extrapolation to zero concentration of the excess chemical potentials of neutral bodies, often to be treated as the mean for certain ions. Some of the older measurements do involve computations of liquid-junction potentials and therefore an indeterminate error.

The numerous values for single-ion activities given in the literature, although not thermodynamically exact, are not complete nonsense, and it is important to see how much the more important ones of the hydrogen-ion activity may be trusted. The more careful workers define pH as a constant plus $(N\mathfrak{F}/2.3RT)$ times the electromotive force of a certain cell. The cell involves an electrode reversible to the hydrogen ion in the solution in question, a salt bridge, and a standard electrode, and the constant will depend upon the nature of each electrode and of the salt bridge. This pH is usually interpreted as $-\log C_{\mathrm{H}^+}$ or as $-\log a_{\mathrm{H}^+}$. The second expression would be correct if the electromotive force of the cell Pt(s), H$_2(g)$, solution A, salt bridge, solution B, H$_2(g)$, Pt(s) were a measure of the free energy of transfer of hydrogen ion from solution B to solution A, which would be true only if the free energy of the processes occurring at each liquid junction had the same absolute value but opposite signs.

The pH is also measured by indicators. We may write the general indicator reaction as for an acid of dissociation constant K as

$$\mathrm{H}^+ + \mathrm{I}^z = \mathrm{HI}^{z+1}.$$

In the use of indicators it is $C_{\mathrm{HI}^{z+1}}/C_{\mathrm{I}^z}$ that is measured and pH is defined as

$$
\begin{aligned}
p\mathrm{H} = {}&- \log(C_{\mathrm{HI}^{z+1}}/C_{\mathrm{I}^z}) - \log K = -\log C_{\mathrm{H}^+} - \log \gamma_{\mathrm{H}} + \\
&\log (\gamma_{\mathrm{I}^z}/\gamma_{\mathrm{HI}^{z+1}}) = -\log a_{\mathrm{H}^+} - \log (\gamma_{\mathrm{I}^z}/\gamma_{\mathrm{HI}^{z+1}}).
\end{aligned}
\tag{8-72}
$$

The pH so defined is an experimental quantity, and $\log a_{\mathrm{H}^+}$ must have meaning within the limits with which the last term may be determined. There is no question but that it has meaning within the approximation to which I^z and HI^{z+1} may be treated as perfect solutes. Of course, it is impossible to determine rigorously what these limits are, but any use of the hydrogen-ion activity should be in an equation similar to that above, say to determine the ratio $C_{Bz'}/C_{HBz'+1}$:

$$\log C_{Bz'}/C_{HBz'+1} = \log K_b - \log a_H + \log \gamma_{Iz}\, \gamma_{HBz'+1}/\gamma_{HIz+1}\, \gamma_{Bz'},$$

and the last term again may be determined from mean activity coefficients alone, so that the expression for the activity coefficient of the hydrogen ion is useful to the approximation to which mean activities may be determined from the assumptions used.

Hammett[43] defines the acidity function H_0 by the relation

$$H_0 = - \log a_H \gamma_B^0/\gamma_{AB}^1 = \log C_{HB^1}/C_{B0} - \log k, \qquad (8\text{–}73)$$

which is the pH measured by an indicator with a neutral base. He has found the relation useful with a large variety of solvents and a wide range of acidity.

I have suggested[44] that the single-ion activities in aqueous solutions be defined by assuming that the liquid-junction potential with saturated potassium chloride is the same for every solution, which is equivalent to defining $\log a_{H^+}$ as $- p$H, and seems the most convenient arbitrary definition of single-ion activities. If followed consistently it cannot contradict any experimental measurements, and it is very useful in many cases where measurements are difficult.

In the derivation of the Debye theory of electrostatic interaction we have used the electric potential difference between two parts of the same solution which differ in the concentration of ions and which are, therefore, not strictly speaking the same medium. The success of this theory and of the Born equation (8–34) gives us confidence that there is no great error in this assumption for dilute solutions. Within this error we may also assume that the electric potential difference is the same in the solution as in the standard state and give more than formal significance to the chemical potential and the activity of a single ion. We can also determine the difference of electrostatic potential within the same error by measuring the equilibrium distribution of two ions of different charges, say across a semipermeable membrane. If we let the concentrations of sodium ion and of chloride ion on the two sides be C_{Na^+}, C'_{Na^+}, C_{Cl^-}, and C'_{Cl^-} and the electrostatic potentials be ψ and ψ', Boltzmann's distribution gives

$$C_{Na^+} = C'_{Na^+} e^{-(\psi-\psi')\epsilon/kT},$$
$$C_{Cl^-} = C'_{Cl^-} e^{(\psi-\psi')\epsilon/kT},$$
$$C_{Na^+}C'_{Cl^-}/C'_{Na^+}C_{Cl^-} = e^{-2(\psi-\psi')\epsilon/kT},$$

[43]L. P. Hammett, *Physical Organic Chemistry* (McGraw-Hill, NY, 1940).
[44]G. Scatchard, *Science* 95, 27 (1942).

and

$$(\psi - \psi') = - (kT/2\epsilon) \ln C_{Na^+} C'_{Cl^-}/C'_{Na^+} C_{Cl^-}. \qquad (8\text{-}74)$$

A necessary, but not sufficient, condition for the identity of the medium on the two sides is that $C_{Na^+} C_{Cl^-} = C'_{Na^+} C'_{Cl^-}$. The assumption that the pH is $-\log a_{H^+}$ is equivalent to the assumption that this potential $\psi - \psi'$ is the electromotive force of the cell electrode: E, salt bridge, solution A, membrane, solution B, salt bridge, electrode E if ψ is the potential in A and ψ' that in B.

8-11. Other Thermodynamic Functions

The complete differentiation with respect to temperature or pressure of Eq. (8–13) or (8–20) involves the variation of D, V, κ, a, and b. Epstein and I[28] proposed that $\kappa a = x$ be considered independent of both temperature and pressure. This makes X, Y, and Z also independent of temperature and pressure. Then it is convenient to write the expressions for the chemical potentials and the free energy of electrolyte solutions as

$$\bar{G}_k^{es}/RT = - (AY/x)z_k^2(I/2)^{1/2} + 2\Sigma_j \beta_{jk} m_j, \qquad (8\text{-}75)$$

$$\bar{G}_S^E/RT = [2(AZ/x)(I/2)^{3/2} - \Sigma_{ij}\beta_{ij} m_i m_j]\bar{W}_S/1000, \qquad (8\text{-}76)$$

$$G^E/RT = [- 2(AX/x)(I/2)^{3/2} + \Sigma_{ij}\beta_{ij} m_i m_j]N_S \bar{W}_S/1000$$
$$= [- (AX/x)\Sigma_i N_i z_i^2(I/2)^{1/2} + \Sigma_i \beta_{ij} N_i m_j]; \qquad (8\text{-}77)$$

A is defined in Eqs. (5–45) and (5–46), x in Eqs. (5–53) and (5–47), X in Eq. (8–6), Y in Eqs. (5–50) and (8–8), and Z in Eqs. (5–52) and (8–8). Then any other thermodynamic function may be obtained by replacing A and the β's by the appropriate function obtained by differentiation. The electrostatic contribution will have the same form for each of these functions.

By Eqs. (5–45) and (5–46),

$$A = N_A^2 \epsilon^3 (\pi \rho_0/500)^{1/2}/(RTD_0)^{3/2}.$$

For water at 1 atm,

$$A = 1.124[1 + 0.15471 \, (t/100) + 0.03569 \, (t/100)^2 + 0.02389 \, (t/100)^3],$$
$$dA/dT = 1.124 \, [0.0015471 + .0007138 \, (t/100) + .0007167 \, (t/100)^2]. \qquad (8\text{-}78)$$

For the enthalpy,

$$H^{ms} = - T^2 \partial(G/T)/\partial T$$

$$= RT^2 \left[\frac{dA}{dT} \frac{X}{x} \Sigma_i N_i z_i^2(I/2)^{1/2} + \Sigma_{ij} \frac{\partial \beta_{ij}}{\partial T} N_i m_j \right]. \qquad (8\text{-}79)$$

At 25°C in water, the enthalpy in calories is

$$H^{ms} = 352 \frac{X}{x} \Sigma_i N_i z_i^2 (I/2)^{1/2} + 1.767 \times 10^5 \Sigma_{ij} \frac{\partial \beta_{ij}}{\partial T} \Sigma_i N_i m_j.$$

(8–80)

The limiting value of X/x is $2/3$.

Lange and his coworkers made very accurate measurements of the heat of dilution. Their results were carefully reviewed by Young and Seligmann,[45] who found excellent agreement with the theory for many uniunivalent salts and for the alkali sulfates. The agreement for the alkaline-earth chlorides is less satisfactory, and the bibivalent salts show little or no tendency to reach the limiting slope. Part of this difficulty may come from hydrolysis, but that cannot explain all of it. The effect of the higher terms should be much more evident in the enthalpy than in the free energy. If a_{ik} is independent of the temperature, b_{ik} of Eq. (8–54) varies as $1/D_0 T$. If a_{ik} decreases slightly, as we have assumed, the variation is even more rapid; B_{ik} of Eq. (8–55) varies almost exponentially with b_{ik}, and $\ln \gamma_k'''$ of Eq. (8–56) or K_{ki} of Eq. (8–57) varies as $B_{ik}/(D_0 T)^3$.

The β's that correspond to salting out are almost independent of the temperature, and make but a slight contribution to the enthalpy. The β's that correspond to nonelectrolyte interaction, on the other hand, have almost the same value in the enthalpy as in the free energy. Since the electrostatic terms have the opposite sign, this means that salts of two large ions, such as KNO_3, will show very large deviations from the limiting law instead of almost none at all. This difference between the heats and free energies has been taken as evidence against the electrostatic theory. It has also been suggested[46] that the nature of any effect, electrostatic, ion-molecule or molecule-molecule, can be determined by comparing the free energy and the enthalpy.

A differentiation of the enthalpy with the temperature gives the heat capacity. The contribution corresponding to Eqs. (8–79) and (8–80) is

$$C_p^{ms} = RT \left\{ \left[2 \left(\frac{\partial A}{\partial T} \right)^2 + T \frac{\partial^2 A}{\partial T^2} \right] \frac{X}{x} \Sigma_i N_i z_i^2 (I/2)^{1/2} \right.$$
$$\left. + \Sigma_{ij} \left[2 \left(\frac{\partial \beta_{ij}}{\partial T} \right)^2 + T \frac{\partial^2 \beta_{ij}}{\partial T^2} \right] N_i m_j \right\}.$$

(8–81)

At 25°C in water, the heat capacity, in calories per degree, is

$$C_p^{ms} = 3.85 \frac{X}{x} \Sigma_i N_i z_i^2 (I/2)^{1/2} + 686 \Sigma_{ij} \left[2 \left(\frac{\partial \beta_{ij}}{\partial T} \right)^2 + \frac{\partial^2 \beta_{ij}}{\partial T^2} \right].$$

(8–82)

---

[45] T. F. Young and P. Seligmann, *J. Am. Chem. Soc.* 60, 2379 (1938).
[46] G. Scatchard, *Chem. Rev.* 13, 7 (1933).

The theoretical value of the coefficient 3.85 cannot be considered as exactly determined because the variation of the dielectric constant with the temperature is not known accurately enough. However, the coefficient should be the same for all electrolytes. It is rather surprising that $C_p^{ms}/\Sigma_i N_i z_i^2$ is very often a linear function of $(I/2)^{1/2}$ over large ranges in fairly concentrated solutions, say from $I/2 = 0.2$ to 3 molal. The slopes differ for different salts and are usually much larger than the limiting law of Eq. (8–76) would indicate. The measurements in very dilute solutions are not sufficiently accurate to know whether the limiting slope agrees with theory or not.

For properties like the heat capacity or the volume, absolute values of which are measured directly, it is customary to define another function, the apparent molal property, often given the symbol ϕ or Φ. Since it is closely related to the partial molal properties, we shall use the symbol for the property with a double bar over it, as $\bar{\bar{C}}_{p2}$, or for a general property, $\bar{\bar{F}}_2$. The definit on is

$$F = N_1 \bar{F}_{11} + N_2 \bar{\bar{F}}_2. \tag{8–83}$$

Since

$$F = N_1 \bar{F}_1 + N_2 \bar{F}_2,$$
$$\bar{\bar{F}}_2 = \bar{F}_2 + (N_1/N_2)\,(\bar{F}_1 - \bar{F}_{11}). \tag{8–84}$$

In the limit of zero concentration of 2 in 1, $\bar{\bar{F}}_2 = \bar{F}_2$. Therefore,

$$F^{m1} = N_2(\bar{\bar{F}}_2 - \bar{\bar{F}}_{21}) = N_2(\bar{F}_2 - \bar{F}_{21}).$$

There is an electrostatic contribution to $\bar{C}_{p2}^1$ which arises from the work of charging the ion at infinite dilution.

$$\bar{G}_k^{1D} = \frac{N_A \epsilon^2 z_k^2}{2 D_o b_k}. \tag{8–85}$$

$$\bar{H}_k^{1D} = \left(\frac{N_A \epsilon^2 z_k^2}{2 D_o b_k}\right)\left(1 + T\,\frac{\partial \ln D_o b}{\partial T}\right), \tag{8–86}$$

$$\bar{C}p_k^{1D} = \left(\frac{N_A \epsilon^2 z_k^2 T}{2 D_o b_k}\right)\left[\frac{\partial^2 \ln D_o b}{\partial T^2} - \left(\frac{\partial \ln D_o b}{\partial T}\right)^2\right]. \tag{8–87}$$

Wyman's equation for the dielectric constant of water gives, at 25°C,

$$\frac{\partial^2 \ln D_o}{\partial T^2} = -\,(0.00460)^2 + 0.0000176,$$

$$\left(\frac{\partial \ln D_o}{\partial T}\right)^2 = (0.00460)^2,$$

and $N_A \epsilon^2/2 D_o R T$ at 25°C in water is 3.57. So

$$\bar{C}p_k^{1D} = \left\{-0.111 + 6308\left[\frac{\partial^2 \ln b_k}{\partial T^2} - \left(\frac{\partial \ln b_k}{\partial T}\right)^2\right]\right\}\frac{z_k^2}{b_k}\ \text{cal/deg}$$

in water at 25°C if b_k is measured in angstroms. The experimental values of $\bar{C}^i_{pk}$ for the uniunivalent salts are -10 to -15 cal, indicating a relatively large negative value for the derivatives in brackets, which probably depends upon the fact that the solvent is not a continuous medium.

The volume change is determined by differentiation with respect to the pressure. Eq. (8–77) yields

$$\bar{V}^E = RT\left[-\frac{AX}{x}\Sigma_iN_iz_i{}^2\,(I/2)^{1/2}\frac{\partial\ln A}{\partial p} + \Sigma_{ij}\frac{\partial\beta_{ij}}{\partial p}\,N_im_j\right]$$

$$= RT\left[\left(\frac{AX}{x}\right)\Sigma_iN_iz_i{}^2\,(I/2)^{1/2}\left(1.5\frac{\partial\ln D_o}{\partial p} - 0.5\frac{\partial\ln V_o}{\partial p}\right)\right.$$

$$\left. + \Sigma_{ij}\frac{\partial\beta_{ij}}{\partial p}\,N_im_j\right]. \tag{8–88}$$

The assumption that $\partial\ln D_o/\partial p = -\partial\ln V_o/\partial p$ is probably more accurate than the experimental measurements. With this assumption,[47] for water at 25°C,

$$\bar{V}^E = 1.29\,\frac{X}{x}\,\Sigma_iN_iz_i{}^2\,(I/2)^{1/2} + 2\,RT\Sigma_{ij}\frac{\partial\beta_{ij}}{\partial p}\,N_im_j. \tag{8–89}$$

There have been many very careful measurements of the volumes of salt solutions. Redlich[48] found from such measurements on sodium chloride a limiting slope 7 percent greater than that given by Eq. (8–84). As for the heat capacities, the apparent molal volumes appear to be linear functions of the square root of the concentration up to high concentrations, but the slope usually differs from the theoretical limiting slope.

Eq. (8–80) yields

$$\bar{V}^{1D}_k = \frac{N_A{}^2z_k{}^2}{2D_ob_k}\left(-\frac{\partial\ln D_o}{\partial p} - \frac{\partial\ln b_k}{\partial p}\right).$$

The first term gives the electrostriction effect, and is $-4.5z_k{}^2/b_k$ ml/mole for water at 25°C if b_k is measured in angstroms. For polyvalent ions this term often leads to a negative partial volume.

The apparent molal coefficients of compressibilities and of thermal expansion can be calculated by further differentiation. The measurements are not accurate enough to determine the coefficients with any precision, but in each case the limiting law should be a linear function of the square root of the ionic strength, with the limiting slope a common factor times the square of the valence.

---

[47]E. J. Cohn and J. T. Edsall, *Proteins, Amino Acids and Peptides* (Reinhold, New York, 1943), p. 62.
[48]O. Redlich, *J. Phys. Chem.* 44, 619 (1940).

APPENDICES
A. Symbols

Each symbol is given with a brief definition, followed by the number of the section or appendix in which it is first used with that definition. Symbols are grouped as follows: (1) English capital letters; (2) English lower-case letters; (3) Greek letters. The more usual definitions are listed first.

A	work content or Helmholtz free energy	2–1, C–1
	parameter in Debye-Hückel equation	5–4
	chemical formula of component	1–3, C–2
	parameter in general equation	B–1
	parameter in viscosity equation	7–11
A_o	parameter in Beattie-Bridgeman equation	3–6
Å	angstrom unit	5–4
B	parameter in Van der Waals vapor-pressure equation	D–9
	chemical formula of component	1–3
	function in Kirkwood equation	8–7
	parameter in general equation	B–1
	parameter in Hückel equation	8–5
	parameter in viscosity equation	7–11
	availability	C-7
$T_B^\wedge$	Boyle point	3–3
B_o	parameter in Beattie-Bridgeman equation	3–6
C	concentration (mole/lit)	1–6, B–2
	parameter in general equation	B–1
	heat capacity	2–1
C_V	heat capacity at constant volume	2–1
C_p	heat capacity at constant pressure	2–1
°C	degree Celsius	B–3

167

D	dielectric constant	5–4, D–2
	parameter in general equation	B–1
$\underset{\vee}{D}$	Debye-Hückel	
$\hat{D}$	of dipole	D–5
E	energy	2–1, C–2
	electromotive force	6–2
$\underset{\vee}{E}$	formula of component	1–3
	excess quantity of mixing	B–2
F	general function	8–11
	force	7–6
	chemical formula of component	1–3
^{FH}	Flory-Huggins	5–2
$\mathfrak{F}$	Faraday constant	6–8, B–3
	field strength	D–2
G	Gibbs free energy, free enthalpy	2–1, B–2
$\bar{G}_i$	chemical potential	1–3, C–1
G_i	standard $\bar{G}_i$ (temperature function)	2–1
G_i^0	G_i at temperature T^0	2–1
H	enthalpy	2–1, B–2
	Hammett function	8–10
	factor defined in Sec. 6–39	6–6
I	double ionic strength $\Sigma_i m_i z_i^2$	5–4
$\underset{\vee}{I}$	ideal	5–2, B–2
$\hat{\imath}$	ionic	D–5
J	joules	6–4
K	equilibrium constant	2–3, B–2
	K_c in terms of concentrations	2–3
	K_p in terms of partial pressures	2–3
K	degree Kelvin	B–3
L	specific conductance	7–1
	Langevin function	D–2
$\hat{L}$	Debye-Hückel limiting law	5–4
	of London dispersion forces	D–5
M	total molality $\Sigma_i m_i$	6–1
$\underset{\vee}{M}$	of mixing	4–3, B–2
N	number of moles	1–5
	number of electrochemical equivalents	6–8
N_A	Avogadro's number	1–5
$\bar{N}$	mole fraction	1–5, B–2
$\hat{N}$	number average	6–6
P	polarization	D–2
	standard pressure	6–1
	osmotic pressure	6–6

	parachor	D–9
$\hat{P}$	of polarization	D–5
Q	heat	C–1
	function in Bjerrum equation	8–7
	function in Fig. 6–4	6–1
R	gas constant	2–1, B–2
	parameter in square-well potential	3–3
$\hat{R}$	reversible	C–1
S	entropy	2–1, B–2
	solubility	6–7
T	absolute temperature (K)	1–6, B–2
	transport number $T_k = u_k z_k m_k / \Sigma_i u_i z_i m_i$	7–8
T_B	Boyle temperature	3–3
V	volume	1–2, B–2
W	weight	1–2, B–2
	work	6–1, C–1
$\hat{w}$	weight average	6–6
$\tilde{W}$	molecular weight	5–2, C–6
$\tilde{W}_i$	weight fraction	1–2, B–2
X	Debye-Hückel function $Y - Z$	5–4, 8–1
	external electrical force	7–6
Y	Debye-Hückel function $x/(1 + x)$	5–4
Z	Debye-Hückel function $[1 + x - 1/(1 + x)$	
	$- 2 \ln(1 + x)]/x^2$	5–4
$\hat{z}$	Z average	6–6
a	activity	1–5, C–7
	parameter in van der Waals equation	
	of state	3–2
	mean ion diameter	5–4
	parameter in Beattie-Bridgeman equation	
	of state	3–6
	parameter in exponential repulsion	D–6
	coefficient in Eq. (D–27)	D–9
	number of moles of component A	1–3
	index of absorption	D–2
b	parameter in van der Waals equation	
	of state	3–2
	parameter in Beattie-Bridgeman equation	
	of state	3–6
	parameter in Kirkwood equation	8–7
	radius of ion in Debye equation	7–3
	parameter in exponential repulsion	D–4
	number of moles of component B	1–3

μ	dipole moment	D–2
ν	number of moles, particularly of ions, in one formula weight	7–1
	number of moles in change of state	1–3, C–2
	index in Eq. (5–5)	5–1
π	geometric ratio	5–4, D–2
	square of weight of measurement	B–1
$\hat{\pi}$	of osmotic pressure	C–6
ρ	density	3–9, D–5
	electric density	7–4
σ	closest distance of approach (billiard ball, etc.) distance of zero energy	3–3, D–4
τ	$1/T$	2–6
	turbidity	6–6
ϕ	volume fraction	1–5, 5–1, 5–2, 5–5
	osmotic coefficient, from $\bar{G}_S$ if without subscript	5–4, C–6, D–2
ψ	electrostatic potential	7–4, D–1
	refractive increment, $\partial n/\partial m$	6–6
ω	angular velocity	6–6
$\check{\omega}$	last part of system	C–2
$\hat{\omega}$	last point	3–9
$\check{0}$	standard state of reference infinite dilution	B–2
$1,2,i,$	component 1, 2, i	1–2, B–2
$'$	to indicate species as opposed to component	2–3
	to indicate vapor as opposed to liquid	6–1
	first part of system	C–2
	first derivative at standard conditions	2–2
	second approximation in virial coefficient	B–2
$''$	second derivative at standard conditions	2–2
$||$	determinant	6–6
$||$	absolute value	7–8
$!$	factorial	
$\bar{F}$	molal quantity, $F/\Sigma N$	1–2, B–2
$\bar{F}_k$	partial molal quantity, $(\partial F/\partial N_k)_{T,p,N}$ except	1–2, B–2
$\bar{N}_k$	mole fraction, $N_k/\Sigma N_i$	1–2, B–2
$\bar{\bar{F}}$	average	D–2
$\tilde{F}$	specific quantity, F/W	1–2, B–2
$\tilde{F}_k$	partial specific quantity, $(\partial F/\partial W_k)_{T,p,W}$ except	1–2, B–2
$\tilde{W}_k$	weight fraction, $W_k/\Sigma W_i$	1–2, B–2
$\bar{\bar{F}}$	apparent molal quantity	8–11

B. The Analytic Expression of Experimental Results

B-1. Least Squares

The least-squares method is objective in the sense that judgment is used only to set up rules before computation is started. The computation itself is entirely automatic. The least-squares method minimizes the sum of the squares of the deviations of the individual measurements from the analytic expression for a given assignment of error to the different variables, a given weighting of the individual measurements, a given rule for eliminating points with large deviations, and a given form of the equation, or if it is a power series, a given rule for the number of terms to be used. It is often said that any two workers will obtain the same expression from the same data, but this is true only if they agree on each of the foregoing *a priori* rules.

One should not attempt to use this method without first studying a good treatise on the subject,[1] but we shall discuss one very simple case. If $\Phi(x, y, z, \ldots)$ may be expressed as

$$\Phi_i = A + Bx_i + Cy_i + Dz_i + \cdots,$$

and if all the error is attributed to Φ, with each x, y, z, ... absolutely accurate, then the method is very simple. Each of the variables y, z, ... may be a function of x, such as x^2, e^x, or $\ln x$. Each equation is multiplied by a factor π_i, where π_i^2 is the weight of the ith measurement, and the resulting equations are added, to give

$$\Sigma_1 \pi_i \Phi_i = \Sigma_i \pi_i A + \Sigma_i \pi_i Bx_i + \Sigma_i \pi_i Cy_i + \Sigma_i \pi_i Dz_i + \cdots.$$

Then each equation is multiplied by x_i and the new equations are added to give

$$\Sigma_i \pi_i \Phi_i x_i = \Sigma_i \pi_i Ax_i + \Sigma_i \pi_i Bx_i^2 + \Sigma_i \pi_i Cx_i y_i + \Sigma_i \pi_i Dx_i z_i + \cdots.$$

---

[1]For example, W. J. Youden, *Statistical Methods for Chemists* (Wiley, New York, 1951).

The same procedure is followed with y_i, z_i, ..., to give

$$\Sigma_i \pi_i \Phi_i y_i = \Sigma_i \pi_i A y_i + \Sigma_i \pi_i B x_i y_i + \Sigma_i \pi_i C y_i^2 + \Sigma_i \pi_i D y_i z_i + \cdots,$$
$$\Sigma_i \pi_i \Phi_i z_i = \Sigma_i \pi_i A z_i + \Sigma_i \pi_i B x_i z_i + \Sigma_i \pi_i C y_i z_i + \Sigma_i \pi_i D z_i^2 + \cdots,$$

and so on. This gives as many simultaneous equations as there are parameters A, B, C, D, ..., and solution of this set of equations gives the proper values of these parameters. The method also includes rules for determining the probable error of a single measurement and of each of the parameters.

B-2. Graphic Deviation Curve

In the graphic deviation-curve method, judgment may be used at each stage of the calculation and the agreement is always directly visible. The function Φ must be chosen so that all the measurements have nearly equal weight, but some adjustment may be made by eye. If the parameters are determined one or two at a time, however, the weighting may be changed. For example, the absolute error of measurement of the pressure is usually independent of the pressure for small pressures, but proportional to the pressure for larger pressures. So, for low pressures the initial function should be proportional to the pressure, and for high pressures it should be proportional to the ratio p/C. If the molecular weight is known, it is convenient to choose p/RT and p/RTC. If the virial expansion is used, we have

$$p/RT = C + \beta C^2 + \gamma C^3 + \delta C^4 + \cdots,$$
$$p/RTC = 1 + \beta C + \gamma C^2 + \delta C^3 + \cdots.$$

We may well work with both equations and see how they compare. We may determine all the parameters together, or we may determine first the second virial coefficient β, and perhaps smooth that for temperature before proceeding to the third virial coefficient γ, and so on. In either case it will probably be worth while to work with deviation functions determined by fitting roughly the highest concentration and a second concentration about half as large. Then

$$p/RT - C - \beta' C^2 - \gamma' C^3 = (\beta - \beta')C^2 + (\gamma - \gamma')C^3 + \delta C^4 + \cdots,$$
$$p/RTC - 1 - \beta' C - \gamma' C^2 = (\beta - \beta')C + (\gamma - \gamma')C^2 + \delta C^3 + \cdots.$$

To determine the parameters all at once we plot the deviation function against some convenient function of the concentration, draw a smooth curve, and read from it, at approximately equal intervals, as many points as there are parameters to be determined. Deviations from the resulting equation should then be plotted to determine if better values are possible.

If the parameters are to be determined one at a time, the pressure deviation is plotted against the square of the concentration, or the virial deviation against the concentration itself, and the limiting slope at zero concentration is plotted. Since a small slope may be determined more accurately than a large one, it is well to replot the new deviation curve against the same function of the concentration before proceeding further. Then the new deviation function is plotted against C^3 for the pressure or against C^2 for the virial. After the coefficient γ is satisfactorily determined, the resulting deviation function is plotted against C^4 for the pressure or against C^3 for the virial, and this procedure is repeated until the parameters are all determined. It will often be convenient to determine β from the pressure plot and the higher coefficients from the virial plot, but the proper place for transition should be determined from the deviations themselves. (Sometimes the coefficients are determined one at a time by dividing the ordinate by the concentration for each successive stage instead of multiplying the abscissa; this is good once, but repetition gives too little weight to the concentrated solutions.) At any stage it is easy to recheck on any coefficient by plotting the deviation against the appropriate power of the concentration.

For the specific volumes discussed in Sec. 1–2, we would determine the specific volume of each component directly for two reasons: the measurements are usually more accurate than those for mixtures, and it would be intolerable to have $\tilde{V}_{11}$ depend on component 2. We define the specific volume of mixing, $\tilde{V}^M$, by the relation

$$\tilde{V}^M = \tilde{V} - \tilde{W}_1\tilde{V}_{11} - \tilde{W}_2\tilde{V}_{22}.$$

For a two-component system it is convenient to express $\tilde{V}^M$ symmetrically,

$$\tilde{V}^M = \tilde{W}_1\tilde{W}_2[\tilde{V}_{12}^0 + \tilde{V}_{12}'(\tilde{W}_1 - \tilde{W}_2) + \tilde{V}_{12}''(\tilde{W}_1 - \tilde{W}_2)^2 + \cdots].$$

The volume of mixing is zero for a pure component. The value for the 0.5 mixture depends only on $\tilde{V}_{12}^0$. It is convenient to determine the higher parameters in pairs, one antisymmetric and one symmetric, using the 0.25–0.75 combination for $\tilde{V}_{12}'$ and $\tilde{V}_{12}''$, and combining with the 0.125–0.875 combination for the next pair.

For nonelectrolyte mixtures the enthalpy of mixing, and the corresponding heat capacity, compressibility, and so forth, behave like the volume of mixing. For any of these quantities we may use moles as units rather than weights. The entropy and free energy, however, are conveniently split into two terms, ideal and excess:

$$\bar{G}^M = \bar{G}^I + \bar{G}^E, \qquad \bar{G}^I = RT \, \Sigma_i \bar{N}_i \ln \bar{N}_i,$$
$$\bar{S}^M = \bar{S}^I + \bar{S}^E, \qquad \bar{S}^I = -R \, \Sigma_i \bar{N}_i \ln \bar{N}_i,$$

and $\bar{G}^E$ and $\bar{S}^E$ behave like $\tilde{V}^M$ and $\bar{H}^M$.

With functions of the temperature, the measurements at the extremes of the range are seldom more accurate than those near the middle, and they are often much less accurate. Moreover, the greatest interest is often in the value of the function itself and its temperature derivatives at a standard temperature T^0 near the middle of the range. So it is useful to use the function $(T - T^0)$. There is room for the use of judgment in spacing the points at which the parameters are determined, but usually the higher coefficients are so small that the problem is not serious. So we use the function

$$\Phi = A + B(T - T^0) + C(T - T^0)^2 + \cdots.$$

Often the temperature dependence of the pressure, equilibrium constant, or rate constant is determined by plotting log p, log K, ... vs. $1/T$. If this function is linear, T log p, or T log K, vs. T is also linear. If the functions are not linear, the T log K power series gives properly the value of the change in heat capacity and its temperature coefficients.

B-3. Values of Physical Constants

Avogadro number[a] N_A	6.02217×10^{23} molecules/mole
$(pV)_{0°C}^{p=0} = RT_{0°C}{}^a$	(22.4136 ± 0.0004) lit atm/mole
$T_{0°C}$	273.150K
Gas constant[a] R	8.3143 J/mole K
1 calorie[b]	4.1840 J
1 atmosphere[b]	1,013,250 dyne/cm^2
1 liter[b]	$(1,000.028 \pm 0.004)$ cm^3
Acceleration of gravity[b] g	980.665 cm/sec^2
Boltzmann constant[a] k	1.38062×10^{-23} J/K
Electron charge[a] ϵ	1.602192×10^{-20} e.m.u.
	4.80325×10^{-10} e.s.u.
Faraday constant[a] $\mathfrak{F}$	9.64867×10^4 coulombs/mole

[a]B. N. Taylor, W. H. Parker, and D. N. Langenberg, *Rev. Mod. Phys.*, *41*, 375 (1969).
[b]E. R. Cohen and J. W. M. DuMond, *Rev. Mod. Phys.*, *37*, 537 (1965).

B-4. Temperature Scales

The thermodynamic absolute or Kelvin temperature T (units K) is defined by the relation

$$T = 273.16(\alpha_T/\alpha_{tp}),$$

where α is the extrapolated value of p/C for a gas at zero pressure and the subscript tp refers to the triple point of water. On this scale the absolute temperature of the "ice point," which is the equilibrium

temperature of ice and air-saturated water at 1 atm, falls at 273.15 K.
A corresponding thermodynamic Celsius temperature t_{th} can be obtained as $(T - 273.15)°C$.

The International Practical Temperature Scale is based, in the
temperature range $0°C$ to $630°C$, on the assertion that the resistance
of a properly constructed platinum-resistance thermometer is a
quadratic function of temperature in this range, and on fixed Celsius
temperatures of $0°C$, $100°C$, and $444.6°C$ for the ice point, the normal
boiling point of water, and the normal boiling point of sulfur, respectively. Celsius temperatures t_{int} on this practical scale differ from
thermodynamic Celsius temperatures t_{th} by small amounts, according
to the formula[1].

$$t_{th} - t_{int} = (0.01t)[- 0.0060 + (0.01t - 1)(0.04106 - 7.363 \times 10^{-5}t)],$$

where $t = t_{int}$ on the right-hand side.

A more complete account of temperature scales is given in a chapter
by Swindells,[2] where, however, the formula for the difference $(t_{th} - t_{int})$
is misprinted.

B-5.   Series Expansions

We very often have occasion to expand various functions in a power
series of the independent variable, say x, particularly when x is small;
the ones most often used, with the limits within which they converge,
are:

$$(1 + x)^{-1} = 1 - x + x^2 - x^3 + \cdots \qquad (x^2 < 1)$$
$$(1 - x)^{-1} = 1 + x + x^2 + x^3 + \cdots \qquad (x^2 < 1)$$
$$(1 + x)^{1/2} = 1 + x/2 - x^2/8 + x^3/16 - 5x^4/128 + \cdots \qquad (x^2 < 1)$$
$$(1 - x)^{1/2} = 1 - x/2 - x^2/8 - x^3/16 - 5x^4/128 - \cdots \qquad (x^2 < 1)$$
$$e^x = 1 + x + x^2/2 + x^3/6 + x^4/24 + \cdots \qquad (x^2 < \infty)$$
$$e^{-x} = 1 - x + x^2/2 - x^3/6 + \cdots \qquad (x^2 < \infty)$$
$$\ln (1 + x) = x - x^2/2 + x^3/3 - x^4/4 + \cdots \qquad (x^2 < 1)$$
$$\ln (1 - x) = - x - x^2/2 - x^3/3 - x^4/4 - \cdots \qquad (x^2 < 1)$$

[1]J. A. Beattie, M. Benedict, B. E. Blaisdell, and J. Kaye, *J. Chem. Phys.*
42, 2274 (1965).
[2]J. F. Swindells, in ed. E. U. Condon and H. Odishaw, eds., *Handbook of
Physics* (McGraw-Hill, New York, 2nd ed., 1967), pp. 5–34 to 5–44.

C. Thermodynamic Review[1]

C-1. Definitions

The first big advance in thermodynamics is the replacement of functions of the extent of a process, such as work or heat, by functions of the state of the system. This is accomplished by the invention of the functions energy and entropy, which are defined for a closed system by the equations

$$dE = dQ - dW, \qquad \text{(C-1)}$$

$$TdS = dQ_R, \qquad \text{(C-2)}$$

in which Q is the heat added to the system, W is the work done by the system on its surroundings, E is the energy, S is the entropy, T is the absolute temperature, and the subscript R indicates that the process is reversible. It follows that

$$dE = TdS - dW_R. \qquad \text{(C-3)}$$

Much of chemical thermodynamics can be derived with the additional restriction that the only work is that of expansion, $W_{R\ \text{exp}} = pdV$. Then

$$dE = TdS - pdV. \qquad \text{(C-4)}$$

Eq. (C-4) contains only functions of the state of the system, as do the rest of the equations of this appendix.

For open systems we must include the potential, intrinsic potential, or chemical potential of each component to give

$$dE = TdS - pdV + \Sigma_i W_i \tilde{G}_i \qquad \text{(C-5)}$$

---

[1]The thermodynamics in this book is almost all based on J. W. Gibbs, *Trans. Connecticut Acad. 3*, 108–248 (1875–76), 343–524 (1877–78); or *The Collected Works of J. Willard Gibbs* (Longmans, Green, New York, Vol. I, 55–371, 1906, 1928).

or

$$dE = TdS - pdV + \Sigma_i N_i \bar{G}_i. \qquad \text{(C-6)}$$

It is convenient to define three related functions: the enthalpy H, the work content, or Helmholtz free energy, A, and the free energy, or Gibbs free energy, G, by the relations

$$H = E + pV, \qquad \text{(C-7)}$$

$$A = E - TS, \qquad \text{(C-8)}$$

$$G = E - TS + pV = H - TS = A + pV. \qquad \text{(C-9)}$$

These equations of definition are not restricted to the case of no work other than expansion work, but the following differential relations are so restricted:

$$dH = dE + pdV + Vdp$$
$$= TdS + Vdp + \Sigma_i N_i \bar{G}_i, \qquad \text{(C-10)}$$

$$dA = dE - TdS - SdT$$
$$= - SdT - pdV + \Sigma_i N_i \bar{G}_i, \qquad \text{(C-11)}$$

$$dG = dE - TdS - SdT + pdV + Vdp$$
$$= - SdT + Vdp + \Sigma_i N_i \bar{G}_i. \qquad \text{(C-12)}$$

Eqs. (C-6), (C-10), (C-11), and (C-12) define the chemical potential per mole as

$$\bar{G}_k = (\partial E/\partial N_k)_{S,V,N} = (\partial H/\partial N_k)_{S,p,N}$$
$$= (\partial A/\partial N_k)_{T,V,N} = (\partial G/\partial N_k)_{T,p,N}, \qquad \text{(C-13)}$$

in which the subscripts indicate the quantities that are held constant during the differentiation, and $_N$ means the quantities of all components other than $_k$.

Some students find a logical difficulty in Gibbs's definition of the chemical potential, the first statement of Eq. (C-13), because no matter can be added to a system without adding entropy at the same time. It is possible, however, to decrease the entropy by an auxiliary process in which no matter is added or removed, so that the conditions of this definition are fulfilled by the two-stage process. Examination of either of the last two statements of Eq. (C-13) will show that there is no difficulty in adding matter while keeping constant the temperature, the quantities of all other components, and either the volume or the pressure.

C-2. Criteria of Equilibrium

Gibbs has shown that the following equivalent criteria of equilibrium may be derived from the relations discussed above as the two laws of thermodynamics:

I. For the equilibrium of any isolated system it is necessary and sufficient that in all possible variations of the state of the system which do not alter its energy, the variation of its entropy shall either vanish or be negative.

II. For the equilibrium of any isolated system it is necessary and sufficient that in all possible variations of the state of the system which do not alter its entropy, the variation of its energy shall either vanish or be positive.

From these criteria Gibbs derives another, which is much more useful for our purposes even though its applicability is limited to cases in which the work is restricted to work of expansion. Our statement of it will differ considerably from his, for he states the criterion for a special case, discusses the limitations, and then gives more general conditions for cases in which each of the limitations is removed in turn. We shall state the general criterion and then amplify it for the special cases.

Consider a system in which the variation of energy due to motion in gravitational, electric, or magnetic fields, to change in the area of interfaces, or to change in stresses on rigid bodies is negligible, and let the system be divided into several parts in such a way that each part is homogeneous. Thus any phase may be divided into several parts but no part may include portions of more than one phase. We shall designate a property of the whole system by the appropriate symbol without a superscript, and those of the various parts by symbols such as E', E'', ..., E^ω. The system is isolated, and we shall apply the second criterion, keeping the volume and the entropy constant. For each part,

$$dE' = T'dS' - p'dV' + \bar{G}'_1 dN'_1 + \cdots + \bar{G}'_z dN'_z,$$

$$\cdot \qquad \cdot \qquad \cdot \qquad \cdot \qquad \qquad \cdot$$

$$\cdot \qquad \cdot \qquad \cdot \qquad \cdot \qquad \qquad \cdot$$

$$dE^\omega = T^\omega dS^\omega - p^\omega dV^\omega + \bar{G}^\omega_1 dN^\omega_1 + \cdots + \bar{G}^\omega_z dN^\omega_z.$$

If each of the variations represented in this set of conditions is possible and independent, subject to the conditions

$$dS = dS' + dS'' + \cdots + dS^\omega = 0,$$
$$dV = dV' + dV'' + \cdots + dV^\omega = 0,$$
$$dN_1 = dN'_1 + dN''_1 + \cdots + dN^\omega_1 = 0,$$

$$\cdot \qquad \cdot \qquad \cdot \qquad \qquad \cdot$$

$$\cdot \qquad \cdot \qquad \cdot \qquad \qquad \cdot$$

$$\cdot \qquad \cdot \qquad \cdot \qquad \qquad \cdot$$

$$dN_z = dN'_z + dN''_z + \cdots + dN^\omega_z = 0,$$

the necessary and sufficient conditions for the second criterion of equilibrium are

$$T' = T'' = \cdots = T^{\omega},$$
$$p' = p'' = \cdots = p^{\omega},$$
$$\bar{G}_1' = \bar{G}_1'' = \cdots = \bar{G}_1^{\omega},$$
$$\cdot \qquad \cdot \qquad \cdot$$
$$\cdot \qquad \cdot \qquad \cdot$$
$$\cdot \qquad \cdot \qquad \cdot$$
$$\bar{G}_z' = \bar{G}_z'' = \cdots = \bar{G}_z^{\omega}.$$

The variations need not all be independently possible, however. If two groups of parts are isolated from each other so that they cannot exchange entropy, either with matter or as heat, they need not be at the same temperature. If two groups are separated by a rigid membrane so that neither can gain volume at the expense of the other, they need not be at the same pressure. If a component cannot be transferred alone from any part to any other, its chemical potential need not be the same in those two parts. We shall therefore express the criterion of equilibrium in such a system as follows: *all parts of the system that can exchange heat must be at the same temperature; all parts of the system that are not separated by a rigid diaphragm must be at the same pressure; for any process that can occur in the system in either direction the variation in free energy must vanish; and for any process that can occur in one direction only the variation in free energy must vanish or be positive.*

The free energy of transfer of dN_i moles of the ith component from a part of the system where its chemical potential is $\bar{G}_i'$ to another part where its potential is $\bar{G}_i''$ is $(\bar{G}_i'' - \bar{G}_i')dN_i$. If the transfer may be made in either direction that component is an *actual component* of each part, and at equilibrium

$$\bar{G}_i'' = \bar{G}_i'.$$

$$(\text{C--14})$$

If there is none of that component in the second part at equilibrium, it is only a *possible component* of that part and the transfer cannot be made from the second part to the first, so at equilibrium

$$\bar{G}_i'' \geqq \bar{G}_i'.$$
$$(\text{C--15})$$

Gibbs notes in a later part of the paper that the $>$ may be dropped in those cases for which $\bar{G}_i''$ approaches minus infinity as C_i approaches zero, and he considered it probable that this includes all cases. Later evidence has overwhelmingly justified this opinion. We shall retain the inequality, however, since the justification for its omission does not come from the two laws of thermodynamics.

Sometimes the transfer of two or more components in definite ratios from one phase to another or to two or more different phases is possible when the transfer of any one of the components alone is not possible. Examples are the melting of a hydrate $(')$ to a solution of the salt in water $('')$ and, for three phases, the decomposition of the solid hydrate $(')$ to anhydrous salt $('')$ and water vapor $(''')$. If the transfer is in the ratio $\nu_j/\nu_i, \ldots$, the free energy of the transfer of dN_i moles of the ith component with the corresponding quantities of the other components is

$$[\nu_i(\bar{G}_i'' - \bar{G}_i') + \nu_j(\bar{G}_j'' - \bar{G}_j') + \cdots] \, dN_i/\nu_i$$

or

$$[\nu_i(\bar{G}_i'' - \bar{G}_i') + \nu_j(\bar{G}_j''' - \bar{G}') + \cdots] \, dN_i/\nu_i,$$

and at equilibrium

$$\nu_i\bar{G}_i'' + \nu_j\bar{G}_j'' + \cdots = \nu_i\bar{G}_i' + \nu_j\bar{G}_j' + \cdots \tag{C–16}$$

or

$$\nu_i\bar{G}_i'' + \nu_j\bar{G}_j''' + \cdots = \nu_i\bar{G}_i' + \nu_j\bar{G}_j' + \cdots. \tag{C–17}$$

For a chemical reaction, in a single phase or in several, we may write the general equation $\Sigma_i\nu_iA_i = 0$, in which A_i represents the formula of a component and ν_i the number of formula weights of that component produced in the reaction and is therefore negative for a reactant and positive for a product. The change in free energy when dN_k moles of the kth component are produced is $(\Sigma_i\nu_i\bar{G}_i)dN_k/\nu_k$, and at equilibrium

$$\Sigma_i\nu_i\bar{G}_i = 0. \tag{C–18}$$

Eq. (C–18) is the thermodynamic basis of the law of mass action.

This third criterion of equilibrium has the advantage over the others that, although the conditions of constant entropy or of constant energy are very difficult to realize experimentally, the conditions of constant temperature and constant pressure are relatively easy. Moreover, there is no need for the system to be isolated, provided: that the other systems in contact with it have the same temperature, that they have the same pressure or the surfaces of contact are rigid so that the pressure on the system considered is kept constant, and that each component is at the same chemical potential in the surrounding systems as in the system under consideration or cannot pass through the surface of contact. These simplifications come partly from the restriction that the energy associated with the interfaces is negligible. Otherwise it would not be possible to separate the energy, for example, sharply into one part to be assigned to the system under consideration and another part to be assigned to the surrounding systems.

The two laws of thermodynamics tell us one important fact not covered by these criteria of equilibrium, namely, that the free energy of a system that is not in equilibrium is never less than that of a system in equilibrium composed of the same quantities of the same components at the same temperature and pressure. Therefore, if there is any reaction in a system it will be in the direction that diminishes the free energy. The negative of the free-energy change of a process is known as the *affinity* of that process whenever the term affinity is given a precise meaning. Since the affinity of a process is generally measurable only by determining the potentials in the initial and final states, we shall have little use for the term.

The chief operations of chemical thermodynamics are the evaluation of differences of chemical potential by Eqs. (C–14) to (C–18), the determination of their changes with changing temperature and pressure, and the use of them in these same equations. The importance of chemical thermodynamics lies in the fact that the potential of a component in any state of a phase is the same for all processes, physical or chemical, by which that component may be added to the phase or removed from it.

C–3.   The Phase Rule

We are now in a position to reconsider the question of the state of a system. We are interested in the following functions: E, S, H, A, G, V, p, T, and W_i and $\tilde{G}_i$ for each of the n components, or $2n + 8$ quantities altogether. Each of them must be the same at two different times if the system is to be in the same state at those two times. These quantities are not, however, all independent, and H, A, and G may be determined from the others by their equations of definition, leaving $2n + 5$ quantities. Gibbs has shown that these are all determinable from equations relating $n + 3$ of them to one another, which he calls *fundamental equations*. The variables that must be related to give a fundamental equation are in one of the following sets:

$$E,\ S,\ V,\ W_1,\ \ldots,\ W_n;$$
$$H,\ S,\ p,\ W_1,\ \ldots,\ W_n;$$
$$A,\ T,\ V,\ W_1,\ \ldots,\ W_n;$$
$$G,\ T,\ p,\ W_1,\ \ldots,\ W_n;$$
$$T,\ p,\ \tilde{G}_1,\ \ldots,\ \tilde{G}_n.$$

The last set, which relates only $n + 2$ quantities, fails to give one item of information that is given by the others, namely, the total quantity of matter in the system.

Suppose that we have a system of n components in r phases. By the third criterion of equilibrium the temperature, pressure, and chemical potential of each component will be the same throughout, or, if Eq. (C–15) for any two components must be replaced by Eq. (C–17) or (C–18), there will be a corresponding equation of condition among the relative quantities of these components that can be transferred. If we are not interested in the quantity of matter in any phase or in the whole system, the $2n + 8$ variables will be reduced by the three equations of definition mentioned above and the relation $W_1/\Sigma_j W_j = 1 - (W_2 + \cdots + W_n)/\Sigma W_j$ to $2n + 4$, and these may be related by an equation for each of the r phases, from which $n + 2$ of these variables may be determined from the other $n + 2$. There are then r equations between $n + 2$ variables, and the *variance* v is given by the equation

$$v = n + 2 - r. \tag{C–19}$$

This equation is known as the phase rule.

Gibbs assumed that systems with negative variance, that is, with the number of phases exceeding the number of components by more than 2, do not exist in nature. A vast amount of experimental observation has shown no example of negative variance. A system with two more phases than components is *invariant* and can exist at only one temperature, pressure, and composition of each phase. The composition of the whole system may vary, of course, between the extremes of the various phases. If the number of phases exceeds the number of components by only 1, the system is *univariant* and may persist within limits when either the temperature, the pressure, or the composition of its phases is changed. But fixing either the temperature or the pressure will determine the other and the composition of each phase; fixing the composition of any phase will determine the temperature, pressure, and composition of the other phases only if that phase is one whose composition may vary. If the number of phases is equal to the number of components, the system is *bivariant* and two variables must be fixed to determine the others, with the same reservation that fixing the composition of a phase may not be a sufficient condition. Systems of higher variance are obtained when the number of phases is less than the number of components. It is obvious, since no material system can exist in less than one phase, that the maximum variance is 1 greater than the number of components.

In the use of the phase rule it is important that the components be independent and that their number be sufficient to represent the composition and change in composition of every phase. For the case in which there are e equations of condition relating the variables, the phase rule may be extended to

$$v = n + 2 - r - e. \qquad (C–20)$$

C–4. The Clapeyron Equation

We may derive an important relation for univariant systems in which no change can occur to alter the composition of any phase. The most important example is a one-component, two-phase system, but the relation also applies to two-component, three-phase system if two of the phases are the two components and the third contains the two in constant proportions, such as a salt hydrate in equilibrium with the anhydrous salt and water, and to azeotropes, which are two phases in equilibrium with two or more components. It should apply to more complicated systems if any exist to which the condition applies. Let the symbols $\Delta G \cdots$ represent the change in the corresponding functions when any arbitrary amount of matter changes from the first state to the second. By the third criterion of equilibrium, the condition that the system remain in equilibrium is equivalent to the condition that the change in ΔG be zero. We have, however,

$$d\Delta G = - \Delta S dT + \Delta V dp,$$

so

$$(dp/dT)_{\Delta G} = \Delta S/\Delta V = \Delta H/T\Delta V. \qquad (C–21)$$

The second equality depends upon the fact that the change of state is reversible, isothermal and at constant pressure. Eq. (C–21) is known as the *Clapeyron equation* or the *Clapeyron-Clausius equation*.

C–5. The Gibbs-Duhem Equation

A very important proposition regarding extensive properties was first obtained by Euler purely mathematically. Its first use with thermodynamic functions was Gibbs's application to the chemical potential, which was exploited by Duhem and is therefore often called the *Gibbs-Duhem equation*. The rather physical proof that follows is due to Gibbs. If Φ is any extensive property of a system,

$$(d\Phi)_{T,p} = \Sigma_i (d\Phi/dN_i)_{T,p,N} \, dN_i. \qquad (C–22)$$

We may integrate Eq. (C–22) at constant ratios of $N_2/N_1, N_3/N_1, \ldots,$ so that each $(d\Phi/dN_i)$ is constant by the definition of an extensive property, to give

$$\Phi = \Sigma_i N_i (d\Phi/dN_i)_{T,p,N}. \qquad (C–23)$$

Differentiating Eq. (C–23) completely at constant T and p gives

$$(d\Phi)_{T,p} = \Sigma_i (d\Phi/dN_i)_{T,p,N}\, dN_i + \Sigma_i N_i [d(d\Phi/dN_i)_{T,p,N}]_{T,p}.$$
$$\text{(C–24)}$$

Subtracting Eq. (C–22) from Eq. (C–24) leaves the *Gibbs-Duhem equation*,

$$\Sigma_i N_i [d(d\Phi/dN_i)_{T,p,N}]_{T,p} = 0. \qquad \text{(C–25)}$$

The conditions of constant temperature and pressure restrict the possible changes to increase or decrease in the quantity of matter. For any such change at constant composition, that is, at constant ratios N_2/N_1, N_3/N_1, ..., each $d\Phi/dN_i$ is constant; so the important changes are variations in composition. If each $d\Phi/dN_i$ except one is known as a function of the composition, that one may be determined by integration. The commonest application is to the determination of the potential of one component of a binary mixture from that of the other without going through the free energy.

C–6. Standard States and Reference States

In Sec. 2–2 the standard state is defined as 1 atm pressure and a standard temperature, usually 25°C. The standard free energy, $\bar{G}_k^0$, and standard enthalpy, $\bar{H}_k^0$, are defined as the values of the free energy and enthalpy under the standard conditions, with the convention that $\bar{G}_k$ and $\bar{H}_k$ are zero for an element in the state of aggregation that is most stable under these standard conditions. For gases, the standards are often chosen as G_k^0 and H_k^0, the reference state of 25°C and zero pressure, where $G_k^0 = G_k - RT \ln RTC_k$ and C_k is so small that decreasing it further does not alter the value of G_k^0. It would be advantageous to use G_k^0 and H_k^0 always, and the difference is often not negligible.

For substances in aqueous solution, we usually use the reference state of zero molality of all solutes:

$$\bar{G}_k^0 = G_k - RT \ln m_k,$$

and for ions we use the additional conventions that $\bar{G}_{H+}^0$ and $\bar{H}_{H+}^0$, the standard free energy and enthalpy of the hydrogen ion, are zero. There have not been enough measurements in other solvents for an extensive standard system to have been developed.

In the study of solutions we often know changes in free energy, and so on, more precisely than we know these standard values, so we define the change on mixing of any function as the value of that function for the mixture minus the sum of the values for the unmixed components at the same temperature and pressure, $\bar{\Phi}_i^i$:

$$\Phi^M = \Phi - \Sigma_i N_i \bar{\Phi}_i^i, \tag{C–26}$$

or

$$\bar{\Phi}^M = \bar{\Phi} - \Sigma_i \bar{N}_i \bar{\Phi}_i^i \tag{C–27}$$

and

$$\bar{\Phi}_k^M = \bar{\Phi}_k - \bar{\Phi}_k^k \tag{C–28}$$

for partial quantities. We note that there are four types of functions: for the volume and heat capacity, Φ and Φ_i^i are measured independently; for the enthalpy, Φ^M is measured directly; for the free energy, the individual values of $\bar{\Phi}_i - \bar{\Phi}_i^i$ are measured and

$$\bar{\Phi}^M = \Sigma_i \bar{N}_i(\bar{\Phi}_i - \bar{\Phi}_i^i), \ \bar{\Phi} = \Sigma_i \bar{N}_i \bar{\Phi}_i^i + \bar{\Phi}^M; \tag{C–29}$$

and the entropy of mixing is calculated from the temperature coefficient of the free energy, or from the free energy and the enthalpy.

If the standard state of reference is an infinitely dilute solution in the solvent S and the concentration unit is the mole fraction, we use

$$\Phi^{MS} = \Phi - \Sigma_i N_i \bar{\Phi}_i^S, \tag{C–30}$$

$$\bar{\Phi}^{MS} = \bar{\Phi} - \Sigma_i \bar{N}_i \bar{\Phi}_i^S, \tag{C–31}$$

$$\bar{\Phi}_k^{MS} = \bar{\Phi}_k - \bar{\Phi}_k^S. \tag{C–32}$$

For the solvent, $\bar{\Phi}_S^{MS} = \bar{\Phi}_S^M$.

If the standard state of reference is an infinitely dilute solution in the solvent S and the concentration unit is the molality, we use lower-case superscripts:

$$\Phi^{ms} = \Phi - \Sigma_i N_i \bar{\Phi}_i^s, \tag{C–33}$$

$$\bar{\Phi}_k^{ms} = \bar{\Phi}_k - \bar{\Phi}_k^s. \tag{C–34}$$

For the free energy or the entropy it is very convenient to subtract the value for the ideal solution, so we make the following definitions:

$$\bar{\Phi}_k^E = \bar{\Phi}_k^M - \bar{\Phi}_k^I, \tag{C–35}$$

$$\bar{\Phi}_k^{ES} = \bar{\Phi}_k^{MS} - \bar{\Phi}_k^{IS}, \tag{C–36}$$

$$\bar{\Phi}_k^{es} = \bar{\Phi}_k^{ms} - \bar{\Phi}_k^{Is}. \tag{C–37}$$

It follows that

$$\Phi^E = \Phi^M - \Phi^I, \tag{C–38}$$

$$\bar{\Phi}^E = \bar{\Phi}^M - \bar{\Phi}^I, \tag{C–39}$$

and so on. Since $\bar{\Phi}_k^I$ and $\bar{\Phi}_k^{Is}$ are zero for the enthalpy, the volume, and their temperature and pressure derivatives, there is no advantage in the use of excess enthalpy, volume, and so on. For the free energy and entropy,

$$\bar{G}_k^I = - T\bar{S}_k^I = RT \ln \bar{N}_k, \qquad (C\text{--}40)$$

$$\bar{G}_k^{Is} = - T\bar{S}_k^{Is} = RT \ln m_k, \qquad (C\text{--}41)$$

except for the solvent, for which

$$G_S^{Is} = - T\bar{S}_S^{Is} = - RT\ \bar{W}_S\Sigma_i' m_i/1000, \qquad (C\text{--}42)$$

in which the Σ_i' is the sum over all species except the solvent. When the concentrations are expressed as molalities, the excess potential of the solvent is not as useful a function as the osmotic coefficient, defined as

$$\phi = - 1000\ \bar{G}_S^{ms}/RT\ \bar{W}_S\Sigma_i' m_i = 1 - 1000\ \bar{G}_S^{es}/RT\ \bar{N}_S\bar{W}_S\Sigma_i' m_i. \qquad (C\text{--}43)$$

Obviously $\phi^{Is} = 1$. This value of ϕ is sometimes called the practical osmotic coefficient, and $- \bar{G}_S^{MS}/RT \ln \bar{N}_S$ is defined as the theoretical osmotic coefficient. Since this function gives no advantage over $\bar{G}_S^{MS}$, we shall not include it. Often the term osmotic coefficient is used for approximate values calculated from other quantities such as the osmotic pressure. We shall use identifying subscripts for these, such as ϕ_π from the osmotic pressure, ϕ_f from the freezing-point depression, ϕ_b from the boiling-point elevation, and ϕ_p from the logarithm of the partial pressure of the solvent. We shall reserve ϕ without subscript for the value from Eq. (C–43).

C-7. Other Thermodynamic Functions

The form of functions used by Gibbs and given in the foregoing summary is, of course, not the only form that could be used. For example, the work content, free energy, and chemical potential are as closely related to the entropy as to the energy and could have been replaced throughout by the functions $-A/T$, $-G/T$, and $-\bar{G}/T$, with some simplification in the temperature differentiation. Such a method was devised by Massieu[2] for closed systems, and developed by Planck[3] for open systems.

We shall consider especially the system developed by Lewis,[4] which utilizes the difference in chemical potential divided by RT so

[2] F. Massieu, *Comptes rendus 69*, 858, 1057 (1869).

[3] M. Planck, *Treatise on Thermodynamics* (Longmans, Green, 1903).

[4] G. N. Lewis, *Proc. Am. Acad. 43*, 259 (1907); *Z. Phys. Chem. 61*, 129 (1907); G. N. Lewis and Merle Randall, *Thermodynamics* (McGraw-Hill, New York, 1923).

that the result is dimensionless, and expresses it as an exponential so that it is proportional to the concentration rather than to its logarithm. We shall follow Lewis in using this system in parallel with that of Gibbs.

The *fugacity* f for a component of a gas may be defined by the equation

$$\ln f_k = \bar{G}_k - \bar{G}_k^0. \tag{C-44}$$

For a perfect gas the fugacity of each component is equal to its partial pressure pN_k.

The *activity* a may be defined by the equation

$$a_k = kf_k, \tag{C-45}$$

in which k is a function of the temperature but may be given any arbitrary value at each temperature. The values usually chosen are such that, at the same temperature and pressure as the system in which the activity is measured, the activity is unity for the pure substance in the same state of aggregation, or is equal to the mole fraction, or the weight molality in extremely dilute solutions. These three definitions would make $RT \ln a_k$ equal to $\bar{G}_k^M$, $\bar{G}_k^{MS}$, $\bar{G}_k^{ms}$, respectively. Sometimes other definitions are chosen, and it is usually left to the context to show which is meant.

The *activity coefficient* γ is defined as the activity divided by the mole fraction or by the weight molality. The first is used for the first two definitions of the activity, the other for the third definition. These four definitions make $RT \ln \gamma_k$ equal to $\bar{G}_k^E$, $\bar{G}_k^{ES}$, $\bar{G}_k^{es}$, respectively. Just as for the activities, the distinction between these definitions is usually left to the context.

The great advantage of the use of activities and activity coefficients is that most of the equations expressing equilibria, in particular the mass-action equation, which are familiar to chemists in terms of concentrations or of mole fractions, but which are only approximately true in these terms, are rigorously true in terms of activities or if each concentration or mole fraction is multiplied by the appropriate activity coefficient. Important exceptions are those equations for ions that involve the law of electric neutrality expressed in terms of each ion concentration multiplied by the appropriate charge. The system of Lewis has been very helpful in popularizing exact thermodynamics among chemists.

In problems involving the transfer of heat, the temperature T_s of the sink is as important as the temperature of the system, and there is a great advantage in using a new function, the *availability B*, defined by the relation

$$B = H - T_s S. \tag{C-46}$$

D. The Classification of Molecules

D-1. Ions

A molecule consists of one or more atomic nuclei and, except for a very few such as the neutron, the proton, or the α-particle, a certain number of electrons. That is, molecules are made up of positive and negative electric charges. The behavior of any molecule is determined largely by the number and distribution of these charges. We shall consider first the behavior of a few simple cases in an external field.

An *ion* is a molecule that is not electrically neutral. If it has more positive charges than negative it is a cation, and if it has an excess of negative charges it is an anion. In an electric field, the electrostatic energy of an ion with valence z_i is $\psi \epsilon z_i$, if ψ is the electrostatic potential and ϵ the charge on a proton. A cation will migrate to a position of the lowest possible (least positive) potential and an anion will go the highest possible potential. Since the ions have electric charges, this migration carries electricity, or ionic solutions conduct electricity. This characteristic makes it very easy to determine the presence of ions and to measure approximately their concentration.

D-2. Dipoles[1]

If a molecule is neutral but the centroid of the negative charges does not coincide with the centroid of the positive charges, it is called a *dipole*. The *dipole moment* is the product of the distance l between the centroids and the absolute value of either charge. The unit is the debye, which is 10^{-18} e.s.u. Since the charge on the proton is 4.8×10^{-10} e.s.u., the dipole moment in debyes is about five times the separation of the centroids measured in angstroms (10^{-8} cm) multiplied by the

---

[1]This treatment is all taken from P. Debye, *Polar Molecules* (Chemical Catalogue Co., New York, 1929).

charge (in protonic units) of the positive centroid. If the time-average positions of the centroids do not coincide when there is no external electric field, the dipole is called a *permanent dipole*. If there is an average moment when the molecule is in an external field that disappears when the field is removed, it is called an *induced dipole*. The ease with which an induced dipole is formed is called the polarizability of the molecule. Many substances have no permanent dipoles, but every molecule containing both positive nuclei and electrons is polarizable. However, we shall consider first an imaginary neutral molecule with a permanent dipole, but with zero polarizability.

The energy of a dipole in a field is $-\mathfrak{F}\mu \cos\theta$, if $\mathfrak{F}$ is the field strength, μ the dipole moment, and θ the angle between the dipole and the field. This energy may be calculated from the energy of the two poles. The energy of the positive pole is $e\psi_+$ and that of the negative pole is $-e\psi_-$, so the total energy u is

$$u = e(\psi_+ - \psi_-) = e\,\frac{d\psi}{dx}\,l\cos\theta$$
$$= -\mu\mathfrak{F}\cos\theta. \tag{D–1}$$

In a homogeneous field, a neutral dipole will not migrate, but it will tend to orient itself in the position of lowest energy. This tendency will be opposed by the thermal motion. The average energy may be calculated by the use of Boltzmann's principle:

$$n_i/n_{i0} = e^{-u_i/kT}, \tag{D–2}$$

in which n_i is the number of molecules in the ith state, u_i is the energy of a single molecule in that state, and n_{i0} is the number there would be if the energy were the same for every state. The *a priori* probability that the relative orientation of two vectors is between θ and $\theta + d\theta$ is proportional to the area $\sin\theta\,d\theta = d\cos\theta$.

Then the average energy is given by

$$\bar{u} = \int u_i\,dn_i \Big/ \int dn_i. \tag{D–3}$$

We abbreviate $\mu\mathfrak{F}/kT$ to x. Then

$$\bar{u} = -\,\frac{\int_{-1}^{+1} kTx\cos\theta e^{x\cos\theta}\,d\cos\theta}{\int_{-1}^{+1} e^{x\cos\theta}\,d\cos\theta} = -\,kTx\left(\frac{e^x + e^{-x}}{e^x - e^{-x}} - \frac{1}{x}\right)$$
$$= -\,kTx\,L(x). \tag{D–4}$$

The function $L(x)$, given in parentheses, is called the Langevin of x because it was first developed by Langevin for the study of magnetic dipoles. For large values of x, it approaches unity, which means physically that all the dipoles are completely oriented and that the energy of each is $- kTx = - \mu\mathfrak{F}$. This state is called saturation and is approached for magnetism. For electric polarization, however, the value of x is so small that we can often be content with the first approximation, that $L(x) \cong x/3$, or

$$\bar{u} \cong - kTx^2/3 = - \mu^2\mathfrak{F}^2/3kT. \tag{D–5}$$

This state corresponds to the dipoles being nearly randomly oriented. The Langevin of x is shown in Fig. D–1. This treatment is obviously

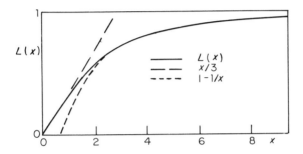

Figure D–1. Langevin function.

classical, but the use of quantum mechanics makes very little difference, and none at all in the first approximation.

To treat induced polarizability, we assume that the displacement of negative charge relative to positive is proportional to the field strength $\mathfrak{F}$. This displacement gives a dipole antiparallel to the field, which will have a moment proportional to the field strength and an energy proportional to the moment and to the field strength. Over all, the energy of an induced dipole, like that of a permanent dipole, is proportional to the square of the field strength. The proportionality constant is called the polarizability, α_0. The polarizability is, at least as a good first approximation, independent of the orientation of the molecule. There is, therefore, no variation with temperature. For low field strengths the permanent dipole may be considered independent of the polarization, so the two effects may be taken as additive, and the total average energy as

$$\bar{u} = - \mathfrak{F}^2(\alpha_o + \mu^2/3kT). \tag{D–6}$$

The molar polarization P is defined as

$$P = - 4\pi N_A\bar{u}/3\mathfrak{F}^2 = (4\pi N_A/3)(\alpha_0 + \mu^2/3kT). \tag{D–7}$$

Experimentally, the polarization is defined by the Clausius-Mossotti relation,

$$P = \frac{D-1}{D+2} \frac{1}{C}, \tag{D–8}$$

D being the dielectric constant. If P, or for a dilute gas D itself, is plotted against $1/T$, the slope is proportional to μ^2, and the intercept at $1/T = 0$ is proportional to α_0. So the measurement of dielectric constants of gases at various temperatures gives a means of determining both the dipole moment and the induced polarizability.

The polarization due to permanent dipoles and that due to induced polarizability can also be distinguished by measuring the dielectric constant at different frequencies. We recall that D may be divided into a real part D' and an imaginary part D'', which are related to the indices of refraction n and of absorption a:

$$D' = n^2(1 - a^2), \tag{D–9}$$

$$D'' = 2an^2, \tag{D–10}$$

$$D = D' - iD'' = n^2(1 - ia)^2. \tag{D–11}$$

Each type of motion has a natural frequency, at which the index of absorption is maximum. For frequencies very different from the natural frequency, the absorption is so small that the dielectric constant is practically equal to the square of the index of refraction. For frequencies much lower than the natural frequency the contribution of any type of motion to the dielectric constant is nearly constant, and for frequencies much higher than the natural frequency the contribution is zero. Induced polarities are largely due to the displacement of electrons and have natural frequencies usually in the ultraviolet. The rotations of ordinary molecules, which are associated with the contributions of the permanent dipoles to the dielectric constant, have natural frequencies corresponding to a wavelength of a few millimeters or centimeters. It is therefore possible to separate the two effects approximately by measuring the dielectric constant for wavelengths of a few meters or more and the index of refraction for ordinary light. The first measurement gives the sum of the two effects, and the second gives the induced polarity alone.

It should be noted that another difference between actual molecules and the rigid model is to be expected in addition to the displacement of electrons. Two oppositely charged parts of the molecule, say two atoms, may have a relative rotation or vibration without any motion of the molecule as a whole. If these parts contain atomic nuclei, the natural frequency of the motion will be small and thermal agitation will oppose the contribution to the dielectric constant. It seems that the effect should resemble in every experimentally observable way the rotation of a polar molecule, so that it is possible to measure a dipole moment in a molecule that is on the whole symmetric. In other words, a molecule that has no dipole moment in its equilibrium configuration, but that can assume other configurations that do have a moment, will behave in an electric field as though it possessed a moment. Similar effects should be important with long-chain polymers.

For nonpolar substances we should expect α_o and P to be independent of the concentration. In fact, there is only a very slight change in these quantities in going from a very dilute gas to a liquid, and for liquid mixtures as well as for gaseous mixtures,

$$P = \Sigma_i P_i x_i. \tag{D–12}$$

For polar substances, however, the permanent dipoles interact with each other so that the Clausius-Mossotti relation may be not even approximately valid. However, in a solution of a polar solute in a nonpolar solvent the limiting value of $(P + \partial P/\partial x_2)$, as x_2 approaches zero, should be the polarization of the solute unaffected by interactions between dipoles. It should be independent of the solvent and the same as the limiting value in the gas as the gas concentration approaches zero. For most substances, measurements in dilute solution are so much easier than measurements in the vapor phase that this is probably the commonest method of determining dipole moments.

The dielectric constants of solutions of dipoles at higher concentrations are especially important for the amino acids and proteins, which have extremely small vapor pressures or solubilities in nonpolar solvents. They have been studied particularly by Wyman,[2] Onsager,[3] and Kirkwood.[4]

D-3.   Dipole Moment and Chemical Structure[5]

The models of dipolar molecules are based upon two hypotheses. The first is that the four valences of a tetravalent atom are directed toward the apices of a regular tetrahedron, that is, they make angles of about 110° with each other, and that the valences of atoms of lower valence have essentially the same directions, or that one or two of the apices are removed in such atoms without much change in the angles between

[2]J. Wyman, Jr., *Chem. Rev. 19*, 213 (1936); *J. Am. Chem. Soc. 58*, 1482 (1936).
[3]L. Onsager, *J. Am. Chem. Soc. 58*, 1186 (1936).
[4]J. G. Kirkwood, *J. Chem. Phys. 7*, 911 (1939).
[5]A much more complete description is given in C. P. Smyth, *Dielectric Behavior and Structure* (McGraw-Hill, New York, 1955).

those that are left. The change from the tetrahedral angle to 90° or to 120° is a small change from our present point of view, though it may be very important for quantitative calculations. The second hypothesis is that no dipole is produced by the linkage of two like atoms, but that one is usually produced when two unlike atoms are combined, and that it usually lies entirely between the two atoms. In other words, a bond between two like atoms is purely covalent, a bond between unlike atoms has some ionic character, and the nature of a bond is not influenced by the other linkages of either of the two atoms. A dipole between two atoms of the same kind means some influence of the other linkages of the atoms, so that the dipole of at least one of these linkages is extended to a neighboring linkage.

Elements should yield completely nonpolar molecules regardless of the number of atoms in the molecule or of their arrangement. The contrary hypothesis has been made about the heavier halogens, but the experimental evidence appears to be against it, and to indicate that, if these substances do dissociate into positive and negative ions without combining with other molecules, the dissociation occurs with a quantum jump and without appreciable concentrations of intermediate stages.

The molecules of the hydrogen halides, on the other hand, should have large dipole moments, as do the compounds of hydrogen with the fifth and sixth families, in which the two or three hydrogens are on the same side of the nucleus of the other atom. In all these cases the moment is smaller the larger the negative atom, because the larger negative atoms are more polarized by the proton, or, expressed differently, because the proton is drawn farther inside the larger negative ions.

In methane, the dipole due to one carbon-hydrogen linkage is just compensated by the projection of the other three on its axis, so that methane has no moment. If one hydrogen is replaced by a methyl group, the projections of the three added hydrogens just equal that of the displaced hydrogen and the carbon-carbon linkage introduces no moment, so ethane, like methane, has no moment. The higher saturated hydrocarbons may be built up in the same way and have no moments. Benzene, as we might expect from its symmetry, is also nonpolar; if a hydrogen is replaced by a methyl group, however, a small moment is introduced. Either the moment of hydrogen attached to an aliphatic carbon is not equivalent to that of hydrogen attached to a carbon in the benzene ring, or there is a dipole between the two types of carbon, or both. Certainly the carbon of the benzene ring is in some way not equivalent to that in an aliphatic hydrocarbon.

When a halogen, oxygen, or nitrogen is introduced into an organic molecule, it imparts a dipole moment, unless, as in carbon tetrachloride, the symmetry of the molecules causes cancellation. The dipole moments of water and the alcohols are about 1.8 debyes. The moments of ethers are smaller, about 1.2 debyes. The aldehydes and ketones have moments of about 2.7 debyes. In esters the carbonyl and ether dipoles are so nearly antiparallel that the net moment is reduced to 1.8 debyes.

The polysubstituted derivatives of benzene are very interesting. For example, nitrobenzene has a moment of 4.0, and those of the dinitrobenzenes are: ortho-, 6.0; meta-, 3.8; and para-, 0.0. The reasons for the differences are that the moments add vectorially, and the angle between the two dipoles is different for the different isomers. If the angles increased by 60° for each carbon atom around the ring and there were no interaction between the groups, the moments for the disubstituted derivatives should be 6.9, 4.0, and 0.0. From the moments of the para- derivatives with two different groups it is possible to determine the relative orientations of the dipoles in different groups. For example, the moment of chlorobenzene is 1.6 and that of phenol is 1.7; the corresponding paranitro- derivatives have moments of 2.4 and 5.0, which are 1.6 less and 1.0 greater than that of nitrobenzene itself. Obviously the chlorobenzene dipole is antiparallel to the nitro, and the hydroxy-benzene dipole is (at least approximately) parallel. If the dipole in the nitro- group is arbitrarily considered as pointing in to the benzene ring, that in the chloro- group does also, but that in the hydroxy group points out. Not all the measurements have the precision we should desire, and the reality is sometimes apparently more complicated than our simple picture, but taking both theory and measurements as they stand we find the same orientation for the dipoles of –F, –Cl, –Br, –I, –NO$_2$, –CHO, –CN, –NC, and –SCN and the opposite orientation for –CH$_3$, –OH, –OCH$_3$ and –NH$_2$. We should expect the first series to have the positive end of the dipole toward the benzene ring, and the second to have it toward the substituting radical.

D-4. Mutual Energy of Two Molecules. Repulsion

All molecules repel one another at sufficiently small distances of separation. Otherwise, they would collapse toward zero volume at very high pressures. Two monatomic molecules at low pressures repel each other approximately as rigid spheres or billiard balls, that is, their mutual energy is infinite if the distance of separation is less than σ, and the energy of repulsion is zero for distances greater than σ.

But real molecules, like real billiard balls, are not perfectly rigid. Better approximations for the mutual repulsive energy u_R are

$$u_R = k(\sigma/r)^n \tag{D-13}$$

$$u_R' = k'e^{b(1-r/\sigma)}. \tag{D-14}$$

If $k' = k$ and $b = n$, $u_R' = u_R$ and $du_R'/dr = du_R/dr$ when $r = \sigma$, but for smaller distances of separation Eq. (D-14) gives smaller values of the slope than does Eq. (D-13). That is Eq. (D-14) gives a somewhat softer molecule than Eq. (D-13) for $b = n$. The difference is not great, however, except for very small values of r. For moderate values of r, n or b may be taken as 12. Slater and Kirkwood[6] calculated $b = 12.08$ for helium, and Lennard-Jones[7] has been very successful with $n = 12$.

The mutual repulsive energy of two polyatomic molecules depends upon their relative orientation as well as upon their distance of separation, and Eq. (D-13) or (D-14) should hold for the individual atoms about as well as they do for monatomic molecules. Then the repulsion averaged over all angles will resemble that of much softer molecules, and σ and n will both appear to vary with the temperature.

D-5.　Mutual Attraction of Two Molecules

Except for quantized chemical interaction, the mutual attractive energy of two molecules may be calculated as the electrostatic coulomb energy between their charges provided that account is taken of the motion of the charges and the orientation and polarization of each molecule by the field of the other. For each charge e_i on the first molecule and each charge e_j' on the second, this energy is $e_i e_j'/r_{ij}$, so the total attractive energy is

$$u = \Sigma_i e_i \Sigma_j e_j'/r_{ij}. \tag{D-15}$$

If the molecules are far apart relative to the dimensions of either, we may consider the charges all located at the centers of the molecules, and the energy of two ions is

$$u_{II} = \Sigma_i e_i \Sigma_j e_j'/r = \epsilon^2 zz'/r \tag{D-16}$$

if $\Sigma_i e_i = \epsilon z_i$ and $\Sigma_j e_j' = \epsilon z'$.

This energy is zero unless both of the molecules have a net charge, that is, unless both are ions. It is repulsive if both ions have the same sign, attractive if their signs are different.

[6]J. C. Slater and J. G. Kirkwood, *Phys. Rev. 37*, 682 (1931).
[7]J. E. Lennard-Jones, *Proc. Roy. Soc. 106A*, 463 (1924); *Physica 4*, 941 (1937).

Any molecule is polarized by the field of an ion. For small fields the relative displacement of positive and negative charges is proportional to the strength of the field of one molecule and to the polarizability of the other. The mutual energy is proportional to this displacement and to the field strength, or the energy due to polarization by ionic charges is

$$u_{IP} = - \epsilon^2(z^2\alpha' + z'^2\alpha)/r^4 \tag{D-17}$$

This energy is always attractive.

Similarly, a permanent dipole tends to orient in the field of an ion. If the orientation is nearly random, Eq. (D–5) leads to

$$\bar{u}_{ID} = - (z^2\mu'^2 + z'^2\mu^2)\epsilon^2/3kTr^4. \tag{D-18}$$

This energy is also always attractive.

At large distances the square of the field strength of a permanent dipole is proportional to the inverse sixth power of the distance. The resulting polarization of another molecule, and orientation of another dipole lead to energies

$$\bar{u}_{DP} = - (\mu^2\alpha' + \mu'^2\alpha)/r^6, \tag{D-19}$$

$$\bar{u}_{DD} = - 2\mu^2\mu'^2/3kTr^6. \tag{D-20}$$

Even a monatomic molecule with completely spherical distribution of the average charge has an instantaneous dipole due to the motion of its electrons.[8] The magnitude of this dipole has been calculated from various properties of the atom. Perhaps the easiest to extend to more complex molecules is the two-thirds power of the polarizability. This gives rise to a mutual energy between two molecules that is proportional to the polarizability of the second molecule. London called it the dispersion energy. It is also called the Van der Waals energy or the London energy. We shall use the last term, which avoids duplication of initials. Then

$$\bar{u}_{L} = - K(\alpha^{2/3}\alpha' + \alpha\alpha'^{2/3})/r^6. \tag{D-21}$$

For polyatomic molecules, the best method is probably to sum the interactions of each atom in the first molecule with those of each atom in the second. Some workers have preferred to picture the electrons as centered around the bonds rather than around the nuclei. London[9] has warned against the danger of any use of spherical distributions for polyatomic molecules, and has noted particularly that series of

---

[8]R. Eisenschitz and F. London, Z. Phys. 60, 491 (1930); F. London, ibid. 63, 245 (1930); W. Heitler and F. London, ibid. 44, 455 (1927); Y. Sugiura, ibid. 45, 484 (1927).

[9]F. London, J. Phys. Chem. 46, 305 (1942).

conjugated double bonds should give very asymmetric elliptical oscillators that include several atoms.

If both molecules are ions, the mutual energy is very long range. Even if only one is an ion, the energy falls off only as $1/r^4$. If both molecules are neutral, however, there is no interaction for which the energy decreases less rapidly than $1/r^6$, and the mutual energy of two permanent dipoles, the dipole-polarization energy, and the London energy all vary as the inverse- sixth power of the separation. The dipole-polarization effect is always small relative to the other two, and usually the London energy is large relative to the dipole-dipole energy. Probably it is always relatively large at large distances of separation except for those molecules with abnormally large dipoles, which are called zwitterions, dipolar ions, or internal salts. The simplest example is glycine. It is interesting to see the magnitude of the dipole effect by comparing the isomers:

Glycine   Nitroethane

$$
\begin{array}{cc}
\overset{\text{H}}{\underset{\text{H}}{\overset{|}{\underset{|}{\text{H--N--C--C--}}}}} \quad & \quad \overset{\text{H}}{\underset{\text{H}}{\overset{|}{\underset{|}{\text{H--C--C--N}}}}}
\end{array}
$$

They have the same electronic pattern and differ only in the interchange of the nitrogen with the terminal carbon atom. Nitroethane is a low-boiling liquid that is soluble in nonpolar organic solvents and insoluble in water. Glycine is a nonvolatile solid that cannot be melted without decomposition. It is insoluble in nonpolar solvents but soluble in water. The big difference between these two molecules is in the charged NH_3^+ and CO_2^- groups of glycine, which give it a dipole moment of about 15 debyes.

D–6. Mutual Energy of Simple Molecules

For monatomic molecules and some more complicated nonpolar ones Lennard-Jones has been very successful in combining the inverse-sixth-power attraction with the inverse-twelfth-power repulsion, to give

$$u = 4\epsilon[(\sigma/r)^{12} - (\sigma/r)^6] = \epsilon[(r_m/r)^{12} - 2(r_m/r)^6], \qquad \text{(D–22)}$$

in which σ is the distance at which $\bar{u}$ is zero, $-\epsilon$ is the minimum energy, and r_m, the value of r when $u = -\epsilon$, is $2^{1/6}\,\sigma$. Actually Lennard-Jones used empirical integral values of the two exponents, but most of the applications have been with the 6-12 Lennard-Jones formula.

Others have preferred to use the exponential repulsion to give

$$u = \epsilon[ae^{b(1-r/r_m)} - 2(r_m/r)^6], \tag{D-23}$$

in which a and b are so related that the minimum energy is $-\epsilon$. If ϵ is to be the same in Eq. (D–23) as in Eq. (D–22), and if u is to have the same value at large values of r/r_m, a must be unity and b must be 12, to give

$$u = \epsilon[e^{12(1-r/r_m)} - 2(r_m/r)^6]. \tag{D-24}$$

For values of r greater than r_m, u from Eq. (D–24) is slightly larger than that from Eq. (D–22) with the same values of ϵ and r_m, but is never more than 5 percent larger. For values of r smaller than r_m, the difference is much larger, and u from Eq. (D–24) passes through a maximum at r/r_m about 1/3, and becomes negative for r/r_m less than about 1/4. This behavior does not correspond to a physical reality, and corrections must be made in calculations from the exponential repulsion.

D-7. Hydrogen Bonds

As two molecules approach each other, the higher moments become important. They are small, though not negligible, for monatomic molecules, and may be all-important for more complicated ones. The over-all moment becomes less important than the detailed structure. For example, the two nitro- groups of the dinitrobenzenes react almost independently of each other, so that the para- compound has as large an energy as the ortho- or the meta- compound. The polarization displacement of charge is no longer proportional to the field strength, and the resultant energy may become very large.

This effect is most important when there are dipoles very near the surface that can approach one another very closely when they are parallel. The positive dipole is a proton attached to a halogen–the smaller the better–to oxygen, to nitrogen, or to a carbon that is also attached to several halogens. The negative dipole is an electron pair not associated with a double bond in oxygen, in a halogen, or in nitrogen, and apparently also the π electrons of the benzene ring. In such cases the proton is on a line between the two heavier nuclei. This interaction is so important that it is often called a hydrogen bond.

The carboxylic acids form very effective hydrogen bonds, and the reaction is apparently limited to dimers and trimers by ring formation:

```
    O   H—O                          O              H—O
    ‖    \                           ‖               \
R—C        C—R              R—C                         C—R
    \      ‖                    \                     ‖
    O—H   O                      O                   O
                                  \                 /
                                   H               H
                                    \             /
                                     O           O
                                      \         /
                                         C
                                         |
                                         R
```

The alcohols apparently cannot form small rings. With only one bonding proton per molecule, they do form unstable chains or linear polymers. Water is so abundant that its hydrogen bonding is the most important. The tetrahedral water molecule has two protons and two nonbonding electron pairs. Each oxygen atom in ice is surrounded by four protons, two of which are closer to it than to the oxygen atom on the other side, and two farther away. This same hydrogen bonding to form indefinite highly branched polymers is important in liquid water, and apparently in gaseous water also.

D–8.   Chemical Combination

Far stronger is the mutual energy of two molecules that form a covalent bond by the pairing of an odd electron in the first with an odd electron in the second that have antiparallel spins. The energy minimum is so deep and so narrow that it is not difficult to divide the molecules into two classes: those at a distance of separation so small that the two may be considered as reacting chemically to form a single molecule, and those at a distance so great that they may be regarded as two independent molecules.

In many other cases it is difficult or impossible to define what we mean by chemical action precisely enough to measure it. If there is a difference of color, such as the difference between nitrogen dioxide and nitrogen tetroxide, there is little doubt of compound formation. This difference may lie outside the visible, usually in the infra-red, and it may appear only in the Raman spectrum. There may be a sharp distinction in some other physical property, such as the conductance of ions. From the conductance of salt vapors we know that only a very few of the molecules are electrically charged. In solutions, the coulomb energy of two ions is inversely proportional to the dielectric constant. Therefore, in a solution of high dielectric constant the ions will not be paired because of this interaction. Since the coulomb energy drops off only as the inverse first power of the distance, the conductance will increase less rapidly than the concentration. For some substances, such as acetic acid, there is another type of mutual energy which leads to the formation of many nonconducting pairs when the concentration of ions is still too small to cause much change in the conductance-concentration ratio because of coulomb interaction.

Agreement with the law of mass action is not sufficient evidence for a chemical reaction. If the mutual energy of two atoms were always zero, or even always repulsive, we might still say that atoms closer together than 3 Å make a compound molecule and apply the law of mass action. We might also say that only atoms more than 9.9 but less than 10.0 Å apart make a molecule and apply the law of mass action with equal success. The first definition of a molecule would appear reasonable to many people, the second would not.

If the mutual energy is not zero, we can say that a molecule is formed if the probability of atom B being in a spherical shell of thickness dr around atom A passes through a maximum and a minimum as r increases. Then we can say that all those atoms within the minimum distance form a compound. We shall try this for Lennard-Jones molecules, for which the number of atoms in a spherical shell between r and $r + dr$ is proportional to

$$n \cong (r/r_m)^2 \exp - \{[(r_m/r)^{12} - 2(r_m/r)^6]\epsilon/kT\}. \qquad (D\text{--}25)$$

Fig. D-2 shows n vs. r/r_m for $\epsilon/kT = 0, 2/3, 1,$ and 2. There is a mini-

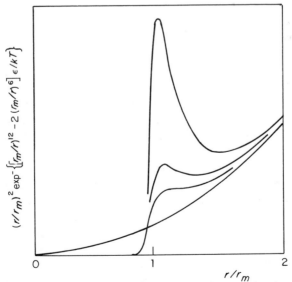

Figure D-2. Distribution of Lennard-Jones molecules.

mum for all values of ϵ/kT greater than 2/3. This corresponds to a temperature $1.5 \times 0.78 = 1.17$ times the critical temperature. We should expect this temperature to be greater than the critical temperature because the presence of one atom close to the central atom makes it more difficult for another to approach closely, so at higher concentrations the probability is decreased for r/r_m near unity. This minimum is probably closely related to the critical state, but the relation is not simple.

A definition that leads to chemical reaction between nonpolar substances above the critical point is not a convenient one. It would be possible to require that the minimum have a certain depth relative to the maximum, say kT. Better still would be to require that the energy in the basin be quantized. This corresponds to the criterion of color change, but the definition is not always easy to use in practice.

It is convenient to apply the law of mass action when an interaction is limited, as by a maximum valence or coordination number. It is convenient to talk in terms of chemical reactions when the law of mass action is used. It should be remembered, however, that sometimes there is no justification for belief that these chemical reactions are real. There is usually no convenience in using the law of mass action or in talking of chemical reaction when an action is unlimited, that is, when the mutual energy of atom A and atom B is unaffected by the presence or absence of atom C. The coulomb interaction of ions is unlimited. In this case indiscriminate use of the law of mass action leads to errors because of the long range of coulomb energy.

There are interactions that are partially limited for which the use of the law of mass action will sometimes be convenient and sometimes not. The types of hydrogen bonding give a good illustration. Two carboxylic acid molecules can interact so that the two hydrogens are both used in bonding and they form dimers and a few trimers, but almost no higher polymers. The use of the law of mass action is most convenient. Alcohol molecules cannot interact in this way, and the energy of formation of a hydrogen bond seems to depend but little on the degree of polymerization of either reacting molecule. So large polymers are formed. Since each alcohol molecule has only one bonding hydrogen, these polymers are linear chains. Water, on the other hand, has two bonding hydrogens per molecule. In ice these form a three-dimensional tetrahedral lattice like that of diamond; in liquid water the bonding is less complete. The use of the law of mass action is less convenient with alcohols than with acids, and still less convenient with water. Limiting the reaction to the formation of dimers, trimers, or other small polymers is usually misleading for either water or the alcohols. The coordination number of nonpolar spherical molecules approaches 12, and the use of the law of mass action is much less convenient. Although the coordination number increases rapidly with the number of atoms in the molecule, we have at present no other way of treating the interaction of complicated molecules with interaction in different parts that are not the same but are of the same order of magnitude.

D-9. Condensed Phases

Statistical mechanics calculates the average behavior of many molecules acting two at a time from the behavior of a pair of molecules. The calculation of the behavior of many molecules actually interacting strongly at the same time is a much more difficult one, and the difficulties are the fundamental ones of the mechanical problem of three or more bodies. There are, however, certain approximations that can be made when the molecules are so close together that the variations in distance can be ignored.

At absolute zero we should expect any substance to have a close-packed structure. The distance between spherical molecules would be just a bit less than r_m, the mutual energy of two neighboring molecules a little bit less than $-\epsilon$, and the energy of next neighbors very much less. Consider a plane between two layers of close-packed spheres at A (Fig. D–3). If the molecules were in completely static equilibrium,

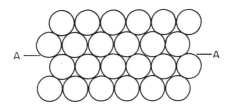

Figure D–3. Close-packed spheres.

the force between the molecules above this plane and those below it must be zero. The attraction decreases less rapidly than the repulsion, so the net interaction between next-nearest neighbors will be attraction; therefore nearest neighbors will be crowded to a position slightly nearer than r_m where the energy is less than ϵ. The total energy does not differ appreciably from ϵ for each nearest-neighbor pair. Since each molecule would have 12 nearest neighbors, the nearest-neighbor energy to be attributed to it would be -6ϵ. The total energy of sublimation is about 8ϵ per molecule. The volume per molecule would be $(4/3)\pi r_0^3/0.79$. Therefore, the intermolecular energy per unit volume would be $-1.51\ \epsilon/r_0^3$. At finite temperatures the molecules will oscillate about the equilibrium distances, with a somewhat smaller mutual energy. Since the curve is much steeper for distances smaller than r_m than for larger distances, the average separation will be larger than r_m, but it will not be very much larger at temperatures well below the critical, and the energy will not differ greatly from $-\epsilon$. We should therefore expect the volume of the liquid and its energy of evaporation to the ideal-gas state to be closely related to the second virial coefficient. I have given the name *cohesive energy* to the energy of the ideal gas at the same temperature minus that of the liquid, and *cohesive energy density* to the ratio of this quantity to the volume of the liquid.[10]

The cohesive energy density is also called the internal pressure[11] because it is often approximately equal to $(\partial E/\partial V)_{T,N}$, which is more properly called the *internal pressure*. The so-called thermodynamic equation of state is derived by taking the partial derivative of the energy with respect to the volume, and replacing $(\partial S/\partial V)_T$ by $(\partial p/\partial T)_V$ by means of a Maxwell relation:

$$p + (\partial E/\partial V)_{T,N} = T(\partial p/\partial T)_{V,N}. \qquad (D\text{-}26)$$

The name *thermal pressure* has been given to $T(\partial p/\partial T)_{V,N}$, and p is, of course, the external pressure. So it was natural to consider the difference as an internal pressure. For a van der Waals fluid the thermal pressure is $RTC/(1 - bC)$ and the internal pressure is aC^2, which is also the cohesive energy density. This simple relation depends upon the assumption that the ratio of the concentrations of molecules at two different positions relative to a central molecule are independent both of the temperature and of the concentration. The $(\partial p/\partial T)_V$ can be measured directly or it is more often computed as the ratio of the *coefficient of thermal expansion*, $\alpha = (\partial \ln V/\partial T)_p$, to the *coefficient of compressibility*, $\beta = - (\partial \ln V/\partial p)_T$. For real fluids the energy-volume curve looks not unlike Fig. 3–7. Hildebrand has used an empirical equation of the type

$$(\partial E/\partial V)_T = a/V^2 - c/V^{10}, \qquad (D\text{-}27)$$

$$- (E/V) + (E_0/V) = a/V^2 - c/9V^{10} = (\partial E/\partial V) + 8c/9V^{10}. \qquad (D\text{-}28)$$

For many nonpolar liquids at temperatures below the boiling point, the liquid volume at the vapor pressure is just larger than that which

[10]G. Scatchard, *Chem. Rev. 8*, 321 (1931).

[11]J. H. Hildebrand and R. L. Scott, *Solubility of Non-Electrolytes* (Reinhold, New York, 3rd ed., 1950).

corresponds to the maximum in the internal pressure, $V = (5c/a)^{1/8}$, and the internal pressure is 10/11 of the cohesive energy density according to Eqs. (B–23) and (B–24). Mercury is much nearer the minimum in the cohesive energy density curve, and water below 4°C has a negative value of the internal pressure even at the vapor pressure.

The cohesive energy is the enthalpy of evaporation plus the change in enthalpy in expanding from the vapor pressure to the ideal-gas state minus RT. It is thus susceptible to direct measurement, but there are only a few calorimetric measurements of the enthalpy of evaporation that are very accurate. It can be obtained by the Clapeyron relation from the variation in vapor pressure with temperature, for which more good measurements are available. However, there are many substances for which only a single boiling-point measurement is available. Usually we want to know this energy at a temperature below the boiling point.

An approximate answer may be obtained by combining the van der Waals equation for the vapor pressure,

$$\ln p = B - \Delta H/RT, \qquad (D-29)$$

in which ΔH is the enthalpy of evaporation, with Trouton's law that the entropy of vaporization is the same for all substances at the boiling point:

$$\Delta H = RBT_b, \qquad (D-30)$$

in which T_b is the boiling point, if p is given in atmospheres and B is a universal constant.

Hildebrand[12] has shown that the entropy of evaporation at a temperature such that the concentration in the gas phase is the same is much more nearly a constant than the entropy of evaporation when the pressure is the same. Then it is desirable to replace the van der Waals vapor-pressure equation by a relation for $\ln p/T$. I found[10] that low vapor pressures could be represented by the relations

$$\ln (p/T) = (B' - A/RT)/[1 + 0.05(B' - A/RT)], \qquad (D-31)$$

$$\Delta E = \frac{A}{[1 + 0.05(B' - A/RT)]^2}. \qquad (D-32)$$

Eq. (D–30) agrees excellently with the experimental measurements from 3 to 2000 mm Hg with 0.4343 B' equal to 4.7 for "normal liquids," 5.0 for esters, aldehydes, and ketones and 5.7 for water and the alcohols. So a single measurement of a boiling point under any pressure is sufficient to determine the cohesive energy approximately.

---

[12]J. H. Hildebrand, *J. Am. Chem. Soc.* 37, 970 (1915).

The volume of a liquid can be measured easily with high accuracy. It is of considerable theoretical interest. At room temperature the volume of a liquid at atmospheric pressure is about 0.01 percent less than the volume of the liquid under zero external pressure, and it increases about 0.1 percent per Celsius degree rise in temperature.

About 125 years ago Kopp made a correlative study of liquid volumes and found that the volumes at the boiling points were additive functions of the composition. There were, of course, some disturbing exceptions. Kopp's choice of the boiling point was later justified as being approximately a corresponding temperature.

About 75 years ago Traube studied the apparent volumes in solution. The apparent volume of a solute is the volume of the solution minus the volume the solvent in it would occupy at the same temperature and pressure. At the limit of zero concentration, the apparent specific volume of the solute is equal to its partial specific volume. To obtain additive relations, Traube had to add a covolume of about 13 cm^3/mole. His comparisons were at constant temperature.

Sugden[13] eliminated the effect of temperature for pure liquids by utilizing the relation that, at temperatures well below the critical, the surface tension of a liquid is proportional to the fourth power of the density. Then the molal volume multiplied by the fourth root of the surface tension is independent of the temperature. Sugden calls this product the parachor, $P = \bar{W}\gamma^{1/4}/\rho$. He states that the parachor is the volume that 1 mole would occupy if the internal pressure corresponded to unit surface tension. The parachor is an additive function of the composition plus a term for double or triple bond and for ring formation. For a given atom, the parachor decreases with increasing valence. It has therefore been used to determine structure.

It is also of interest to compare simply the volumes of liquids at the same temperature. We shall not expect strict additivity, but the deviations will themselves be of interest.

The volumes at 20°C of the normal aliphatic hydrocarbons differ by about 16 cm^3 per CH$_2$ group, but extrapolation to zero carbon leaves about 34 cm^3 for the two terminal hydrogens and a covolume like that needed by Traube, who also worked at constant temperature. Most of the 18 isomeric octanes have volumes less than that of n-octane. The total spread is less than 10 cm^3 per mole and not related to the structure in a simple way. Derivatives of the n-aliphatic hydrocarbons also show an increment of 16 cm^3 per CH$_2$ group, and this quantity varies only slightly with the temperature, so that the thermal expansion is almost entirely in the end group or covolume.

---

[13]S. Sugden, *J. Chem. Soc. 125,* 32 (1924); *The Parachor and Valency* (Routledge, London, 1930).

SURFACE AND COLLOID CHEMISTRY

Contents

1. Introduction

1–1. Colloids

The word "colloid," meaning gluelike, was devised by Thomas Graham about a century ago to describe substances that do not diffuse through parchment paper or animal membranes through which a "crystalloid," like salt or sugar, diffuses readily. In the absence of a membrane, colloids diffuse much more slowly than crystalloids. Graham thought of colloids and crystalloids as contrasting classes of substances. Since his time, however, many typical crystalloid substances have been prepared as colloids, so now we think of colloidal properties of systems, and consider that different systems may be colloidal to different degrees.

The differences in diffusion, either with or without a membrane, depend largely upon differences in the sizes of the particles. There is also an upper limit of size above which colloidal properties are negligible. We classify as colloidal those particles that will pass through an ordinary filter but not through an ultrafilter—typically a membrane with pores less than 10^{-6} m in diameter.

Visibility depends upon size and upon differences in refractive index. We classify as colloidal by this criterion particles that are not visible in the ordinary microscope but are visible in the ultramicroscope, which is a microscope viewing a light beam at approximately a right angle to its direction. The scattering of light from a beam by dust is familiar to all.

The rate of settling of suspended particles in a gravitational or a centrifugal field and their equilibrium distribution in such a field depend upon their size as well as upon the difference in density between the particles and the suspending medium. A colloid, by this criterion, is a material that does not settle appreciably in the dimensions of

213

ordinary laboratory apparatus in the earth's gravitational field, but that does sediment in a high-speed centrifuge or ultracentrifuge.

These criteria all classify as colloidal particles that have dimensions approximately between 10^{-7} and 10^{-4} cm. The centimeter is obviously an inconveniently large unit. It is customary to use instead the micron ($1\mu = 10^{-6}$ m $= 10^{-4}$ cm), and the millimicron ($1\ m\mu = 10^{-7}$ cm), and the angstrom (1 Å $= 10^{-8}$ cm $= 0.1$ mμ). The colloidal range is then roughly between 1mμ and 1μ. In the globular proteins, such as hemoglobin, all the particles are very nearly if not exactly alike, and they are as stable as smaller organic molecules with similar groups. Other proteins, carbohydrates such as cellulose and starch, and many inorganic colloids such as colloidal gold are stable, but include particles of many sizes within a range. Such particles are often called macro-molecules. Other particles, like the colloidal silver salts and colloidal soap particles, are in rapid dynamic equilibrium with smaller mole-cules, so each particle is continually changing size. These particles are often called micelles. There are particles that show every inter-mediate class of behavior, and the description of them as macro-molecules or as micelles depends as much upon the individual taste of the scientist as upon the characteristics of the system.

1–2. Shape and Surface

Colloid particles may be spherical, but they may also have one di-mension much larger than the other two, to give needlelike particles, or much smaller than the other two, to give plates. If two dimensions are colloidal or subcolloidal but the third is macroscopic, the particles become filaments, which are important in textiles. Plastics are made up of a mass of filaments branched and linked together to give a three-dimensional network. A film has one dimension colloidal or subcolloidal and the other two macroscopic.

Much of the behavior of these particles, filaments, and films may be explained by their very large surfaces and the energy associated with them. The upper limit of colloids may be set as the size below which the surface energy is no longer negligible relative to the co-hesive energy, which may be defined as the energy of evaporation to an ideal gas.

The volume of a parallelepiped with sides of lengths a, b, and c is abc, and its surface area is $2(ab + ac + bc)$. Thus the ratio of surface to volume is $2(1/a + 1/b + 1/c)$. If a is the small dimension in a film, the ratio is practically $2/a$. If a second dimension, b, is reduced to the same size, as in a filament, the ratio is doubled to $4/a$, and if all three dimensions are the same size, as in a cubic colloid particle, the ratio is tripled to $6/a = 6/V^{1/3}$. This is the smallest ratio for a parallepiped, but the smallest value for any solid is that for a sphere of diameter d. The ratio of surface to volume for a sphere is $6/d \cong 4.84/V^{1/3}$.

1-3. Interactions of Monatomic Molecules

One of the earliest scientific generalizations was the statement that two objects cannot be in the same place at the same time. We may modernize this to the Pauli exclusion principle, which applies to electrons, or, less abstractly, to the statement that two molecules repel each other when sufficiently close. On the other hand, the existence of solids or liquids depends upon the fact that molecules attract one another. The precision with which we must know how the energy varies with the distance depends upon the nature of the problem. If the temperature is not much above room temperature, it is often sufficient to say that two molecules cannot get closer than a distance r_1, and that they do not influence each other at all at distances greater than r_2, but that at distances between r_1 and r_2 they have negative mutual energy, which may be regarded as independent of the distance.

Theory tells us, however, that electrically neutral monatomic molecules have an energy of attraction at large distances that is proportional to the inverse sixth power of the distance, and that they are somewhat soft, so that the repulsive energy may better be represented as proportional to the inverse twelfth power of the distance. The Lennard-Jones 6–12 model has the mutual energy

$$u = \epsilon[(r_m/r)^{12} - 2(r_m/r)^6] . \tag{1-1}$$

This gives a minimum energy, $-\epsilon$, when $r = r_m$. The force between the molecules is repulsive when r is less than r_m, and the mutual energy is positive when r is less than $r_m/2^{1/6} \cong 0.89\ r_m$. The force is attractive when r is greater than r_m and reaches a maximum when $r = (4.4)^{1/6}\ r_m \cong 1.28\ r_m$. The energy is about -0.4ϵ at this distance; it decreases to about -0.25ϵ when $r = 2^{1/2}\ r_m$ and to about -0.03ϵ when $r = 2\ r_m$.

1-4. Ions and Dipoles

Before considering the interactions of other types of molecules, we shall consider their behavior in macroscopic electric fields. An ion is a molecule with an electric charge. In an electric field it moves toward a region of lower potential if its charge is positive, and toward a region of higher potential if its charge is negative. Since the ions are charged, this motion constitutes an electric current. Therefore it is very easy to distinguish ions from neutral molecules.

Every molecule has a dielectric constant greater than unity, and therefore tends to move in an electric field toward a region of higher field strength. In general, this dielectric constant is caused by the motion of the electrons relative to the nuclei, to give an induced dipole parallel to the field. Some molecules, however, have a permanent dipole, independent of the field, because the centroid of positive charge and the centroid of negative charge do not coincide. Such molecules therefore orient themselves parallel to an electric field. Since any such orientation is opposed by the random thermal motion, the dielectric constant due to a fixed dipole decreases with increasing temperature. This part of the dielectric constant is proportional to $1/T$, whereas the part due to induced polarization is independent of the temperature. The two parts can also be distinguished by measuring the dielectric constant, the square of the refractive index, at a low frequency and also at so high a frequency that the molecules cannot orient themselves, and so exhibit only their polarization.

1–5. Intermolecular Interactions

Except for repulsion and covalent bonding, the interactions between two molecules may be expressed as the sum of Coulomb energies between charges, each equal to the product of the charges divided by the product of the distance between them and the dielectric constant. Thus, the mutual energy of two ions a large distance apart is inversely proportional to the distance between them. It is repulsive if the two ions have the same sign, attractive if they have opposite signs.

An ion attracts any molecule because it polarizes it to form a dipole which is so aligned that it is attracted by the ion. The moment of the dipole, $e\,l$, is proportional to the field strength, and the mutual energy of the ion and dipole is proportional to the product of the dipole moment and the field strength, or to the square of the field strength. Since the field strength is inversely proportional to the square of the distance, the mutual energy of an ion and a polarizable molecule is proportional to the inverse fourth power of the distance of separation.

The mutual energy of an ion and a fixed dipole depends upon the orientation of the dipole. For each position with attraction there is an antiparallel position with an equal repulsion. But the orientation with attraction will be favored in accordance with the Boltzmann relation, $n_1/n_2 = \exp[-(E_1 - E_2)/kT]$, in which n_1/n_2 is the ratio of the number in the first orientation to the number in the second, E_1

and E_2 are the energies in the two orientations, k is the gas constant per molecule, and T is the absolute temperature.

The two charges on a permanent dipole displace the charges in another molecule and influence the orientation of another dipole differently, so that there is a resultant attraction, which, at large distances, decreases as the inverse sixth power of the distance. Except for substances like the aliphatic amino acids, which exist largely as molecules such as $^+H_3N \cdot CH_2CO_2^-$, and therefore have huge dipole moments, the mutual energy between dipoles becomes important only when the positive end of one dipole and the negative end of another are near the surface. The important negative ends are oxygen or fluorine; the important positive ends are hydrogen attached to oxygen, to nitrogen, to a halogen, or to a carbon atom attached to several halogens. For this reason this type of interaction is often called hydrogen bonding.

The interaction that is responsible for the attraction of neutral monatomic molecules arises from the fact that, although the average position of the electrons has spherical symmetry, the instantaneous position does not. The atom is a rapidly rotating dipole that polarizes another molecule near it and therefore attracts that molecule. The mutual energy varies at large distances as the inverse sixth power of the distance. It is approximately proportional to the five-sixths power of the product of the polarizabilities of the two molecules. These interactions are commonly called van der Waals, London, or dispersion energies.

For polyatomic molecules, the London energy can often be computed as the sum of the effects of the individual atoms, and only atoms very close together need be considered. There are cases, as London himself has warned us, in which electron orbits around several atoms are important, and simple addition can lead to serious errors. The addition of effects due to fixed charges or to polarization is less dangerous.

If two odd electrons with antiparallel spins come close together, they have a very high probability of forming a covalent chemical bond. The energy is usually much larger than for the interactions discussed above, and it varies rapidly with changes in orientation and distance.

1-6. Condensed Systems

Each molecule in a solid or a liquid is in contact with several other molecules, and much of the behavior of a solid or a liquid depends upon how the shapes of the molecules determine their packing. We shall consider here only spherical closest packing, which is the arrangement in many crystals of monatomic molecules and in many ionic crystals with small cations and large anions, such as the oxygen ions. The oxygens form a spherical lattice, and the cations fit into the holes.

A single layer with spherical closest packing is represented by racked pool balls. Each sphere in the interior is surrounded by six others. Let us consider the central ion as 0, and number the adjacent ones from 1 to 6. If we put a ball in the second layer, it will rest on three in the first layer. Three balls may be added touching 0, and they may be added in two ways, 0–1–2, 0–3–4, and 0–5–6, or 0–2–3, 0–4–5, and 0–6–1. Once any one ball in a layer is located, however, it determines the whole layer. The two sets of sites that are possible for the third layer lie either above those occupied by the first layer or above those not chosen for the second. If we call 0, 1, . . . the A sites, 0–1–2, . . . the B sites, and 0–2–3, . . . the C sites, every layer must correspond to one of these sets. The simplest ones are AB AB AB . . . , which gives hexagonal crystals, and ABC ABC . . . , which gives cubic crystals. Some ionic crystals have more complicated forms, such as AB AC AB AC.

Each sphere in the interior is in contact with 12 others, 6 in the same layer, 3 above, and 3 below. A molecule in the surface may be in contact with as many as 9 others or as few as 3. For neutral molecules, the energy of each contact is approximately the $-\epsilon$ of the Lennard-Jones model, and molecules not actually in contact are so far apart that the mutual energy of nearest neighbors furnishes about three-quarters of the total energy. Since each contact is shared by two molecules, the total nearest-neighbor energy is -6ϵ per molecule, and the energy of sublimation about 8ϵ per molecule. The energy necessary to form a surface is about 2ϵ if the surface is a single layer, and about 4ϵ for a surface with the top layer half filled and the second layer complete.

Spherical molecules in a liquid approach the closest-packing lattice, but they are further separated, and often holes develop large enough to hold a molecule. The only deviations from spherical symmetry we need consider here are those in which the interaction of one portion of the molecule involves more energy than that of another. In the liquid those contacts will be broken that have the smallest energy per contact, so the corresponding parts of the molecules will make up the surface, whereas the parts with larger energies will be directed inward.

2. Surface Tension and Interfacial Tension

2-1. Definitions

Surface tension is the differential change of free energy with change of surface area at constant temperature, pressure, and composition. If the bulk phases on both sides of the surface are liquid or solid, it is usually called interfacial tension. If both phases are fluid, this definition is sufficient, for the nature of the surface will not be altered as the area changes, but the surface area of a solid may be changed in more than one way. If the surface area of a liquid is increased, molecules will move from the interior to the surface so that the surface density is unchanged. The surface area of a solid may be increased in the same way by slip along glide planes or by cleavage. If the surface area of an elastic body is increased by stretching, however, the total number of atoms in the surface will remain constant and the surface density will decrease. Moreover, the distances between molecules perpendicular to the direction of stretch may be decreased, though not as much as distances parallel to the stretch are increased. Sometimes this elastic stretching is called surface tension, while the reversible work of extending a surface with constant surface density is called surface free energy. We shall use "surface tension" for the latter process, and describe the other more specifically when necessary.

2-2. Measurement of Surface Tension

Most of the accurate methods for determining surface tension depend upon the relation between the surface tension σ, the mean curvature of the surface $1/b = (1/R_1 + 1/R_2)/2$, and the difference in pressure on the two sides of the surface $p' - p''$:

$$p' - p'' = \sigma(1/R_1 + 1/R_2) = 2\sigma/b, \qquad (2\text{–}1)$$

in which R_1 and R_2 are the principal radii of curvature of the surface. The position of a point in a surface is, of course, a single point. The slope of the surface is determined by the tangent plane or by the normal at that point. The curvature is determined by the osculating paraboloid or by the directions and radii of two circles perpendicular to each other with centers on the normal. The sum $(1/r_1 + 1/r_2)$ is independent of the orientation of these perpendicular surfaces, so the maximum must be paired with the minimum, and this pair is known as the principal radii of curvature, which we have designated by R_1 and R_2. For a sphere or at an axis of rotation of a paraboloid, $R_1 = R_2$; for a rectangular cylinder, one radius is infinite; for a plane, both are infinite; and at a saddle point one radius is positive and one is negative.

Eq. (2–1) is usually derived by equating to zero the sum of the work of extending the surface and the work of volume expansion as the surface is moved parallel to itself so that the new radii are $R_1 + dr$ and $R_2 + dr$. If two infinitesimal arcs at a point in the surface are $\theta_1 R_1$ and $\theta_2 R_2$, they define an area $A = \theta_1\theta_2 R_1 R_2$ (Fig. 2–1). If the surface

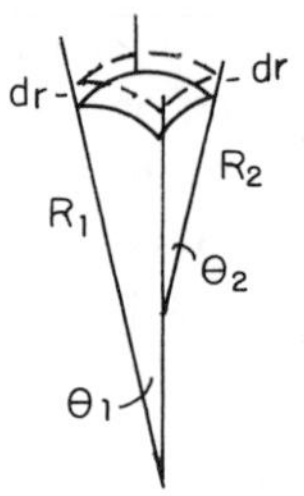

Figure 2-1. Radii of curvature and differential geometry of a surface element.

is moved parallel to itself, the area in the expanded surface is $A + dA$ $= \theta_1\theta_2(R_1 + dr)(R_2 + dr) = \theta_1\theta_2[(R_1 R_2) + (R_1 + R_2)\,dr + (dr)^2]$ and the area $(dr)^2$ may be neglected, so the increase in surface area is $dA = \theta_1\theta_2(R_1 + R_2)dr$. The increase in volume is $dV = \theta_1\theta_2 R_1 R_2 dr$. Then $(p'' - p')\,dV + \sigma dA = 0$ or $(p'' - p')(\theta_1\theta_2 R_1 R_2)dr + \sigma\theta_1\theta_2(R_1 + R_2)\,dr = 0$ or $p' - p'' = \sigma(1/R_1 + 1/R_2)$, which is Eq. (2–1).

Eq. (2–1) may be used to determine the shape of an interface in a gravitational field, in which the difference in pressure is $p' - p''$ $= gh(\rho' - \rho'')$ if g is the acceleration of gravity, h the height above a plane surface, and ρ' and ρ'' are the densities of the two phases. Then Eq. (2–1) may be written

$$bh = 2\sigma/g \, (\rho' - \rho'') = a^2, \qquad (2\text{–}2)$$

which defines a^2. This quantity, which is called the capillary constant or the specific cohesion, is independent of the position of the point in the surface, or of the apparatus.

2–3. Capillary Rise

The most precise measurement of surface tension is made by the capillary-rise method, usually in a vertical tube with circular cross section. At the axis of the tube, h is either maximum or minimum, and all the radii of curvature are equal, b_o.

If the tube is small enough that the curvature is constant across the meniscus, $b_0 = r/\cos \theta$ if θ is the angle of contact of the surface with the tube (Fig. 2–2). This angle depends upon three surface tensions,

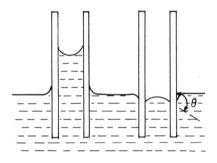

Figure 2–2. Capillary rise and depression.

which we may designate $\sigma = \sigma_{gl}$, σ_{ls}, and σ_{gs}, if g indicates the lighter fluid, l is the heavier fluid, and s is the solid. If $\sigma_{ls} + \sigma_{lg} < \sigma_{gs}$, the heavier fluid is said to wet the solid, θ is less than a right angle, and the meniscus is concave upward. If $\sigma_{gs} + \sigma_{lg} < \sigma_{ls}$, the lighter phase is said to wet the solid if both phases are liquid, or the heavier phase is said not to wet the solid if the lighter phase is a gas, θ is greater than a right angle, and the meniscus is concave downward.

For a liquid that wets the solid and a gas, the angle of contact is usually, perhaps always, zero, provided that the surfaces are clean and the measurements are made with a falling meniscus. This is the case for which precise measurements are possible. Our assumption that

b is constant needs correction. If b_0 is the curvature at the bottom of the meniscus and the tube is small,

$$b_0/r = 1 + 1/3 \ (r/h) - 0.1288 \ (r/h)^2 + 0.1312 \ (r/h)^3, \quad (2\text{--}3)$$

and the surface tension is given by Eq. (2–2) with $b = b_0$. If the tube is large, however, there is a more complicated relation involving $\alpha = \frac{1}{2}(b_0 h)^{1/2} = a/2$:

$$r/\alpha - \ln \alpha/h = 0.8381 + 0.2798 \ \alpha/r + \tfrac{1}{2} \ln \ (r/\alpha). \quad (2\text{--}4)$$

These equations are called after Lord Rayleigh, who added the last term to each. For intermediate tubes, the shape of the surface has been determined by graphic integration. Tables are given in N. K. Adam, *The Physics and Chemistry of Surfaces* (Clarendon Press, Oxford, 3rd ed., 1941). In order to use a moderate amount of liquid, measurements are made with the flatter surface in a tube of moderate radius, and Eq. (2–2) is applied twice.

The case of a finite angle of contact between one right angle and two right angles is very important in pressure-measuring devices that use mercury. Precision demands wide tubes and also very careful consideration of the shape of the meniscus, because mercury has a very high density and a very large surface tension, which is greatly changed by small amounts of impurities.

2–4. Maximum Bubble Pressure

If a capillary tube with an attached manometer is dipped into a liquid and a gas is forced slowly through the tube, the pressure will oscillate as bubbles are emitted (Fig. 2–3). Just after a bubble has broken off, the surface is nearly flat across the mouth of the tube, that is, the curvature is very small and the pressure difference is essentially that

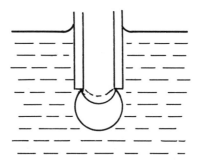

Figure 2–3. Gas bubble forming beneath liquid surface at end of capillary tube.

due to the hydrostatic head. As the bubble builds up, the curvature increases, until the bubble is a hemisphere with radius equal to that of the tube. As the bubble increases further, however, the curvature and the pressure decrease, and the bubble quickly grows so large that it breaks off. The bottom of the bubble is like the bottom of the meniscus in the capillary, and the wetting or not wetting gives an all-or-nothing effect. If the liquid wets the tube, the bubble forms at the inside of the tube; if it does not wet the tube, the bubble forms from the outside. If the tube is very small, the surface tension is given by Eq. (2–2), with $b = b_0$ and h the height corresponding to the maximum pressure. For moderately small tubes, Schrödinger gives the equation

$$b_0/r = 1 - 2r/3h - r^2/6h^2. \qquad (2\text{–}5)$$

Tables for larger tubes are given by Sugden.[1]

For use with small quantities of liquids, a small tube and a large tube are arranged in parallel for equal depths of immersion, with a stopcock in the larger tube so that flow may be diverted to the smaller. Or for comparative results with dilute solutions, carefully matched tubes are inserted in the solution and in the solvent and the depths of immersion at which bubbles alternate from the two tubes are determined. The bubble pressure is probably the simplest accurate method. Obviously, it cannot be used with the gas phase containing only the vapor of the liquid. Although we should expect that the bubbles would have to be formed very slowly, it does not seem necessary in practice.

2–5. Drop Weight

If a liquid flows very slowly from a capillary tube, it will fall as drops. The forming drop will hang from the outer edge of the tube if the liquid wets the tube, from the inner edge if it does not. The surface tension has been found empirically to be given by the relation

$$\sigma = F\, mg/r, \qquad (2\text{–}6)$$

in which F is a function of v/r^3, with v the volume of a drop and m its mass. For $0.86 < v/r^3 < 10.3$, F is 0.26 ± 0.005.

2–6. Sessile Drops, Pendant Drops, and Trapped Bubbles

If a liquid does not wet a surface, the shape of a drop of it placed on that surface is determined by Eq. (2–2). The angle of contact must

---

[1] S. Sugden, *The Parachor and Valency* (Knopf, New York, 1930), chap. XI.

be known and the mathematical difficulties are great. If the liquid does wet the surface, a bubble of gas trapped under a horizontal surface may be studied in the same way. A drop hanging from a capillary tube may be used for liquids that wet or for those that do not wet the solid. Tables have been prepared for using Eq. (2–2) in each of these cases.

2–7. Dipping Ring

If a solid ring is dipped into a liquid that wets it, the force necessary to raise the ring above the surface of the liquid depends upon the surface tension. Since there is a surface on each side of the nearly cylindrical shell trailing the ring, the maximum force is approximately the product of twice the perimeter and the surface tension, or $f = 4\pi r\sigma$. However, there is a meniscus within the ring, the surface is higher on the inside, and the walls of the cylinder are not parallel, so the simple equation may be seriously in error. Corrections have been determined both experimentally and theoretically from Eq. (2–2). The ring must be raised very slowly to avoid more serious errors, which cannot be calculated.

2–8. Fiber Elongation

If a fiber of a very viscous liquid, such as glass, or a soft solid is suspended with a weight at the bottom, the weight will tend to elongate the fiber and the surface tension will tend to shorten it. By using a series of fibers with different weights it is possible to interpolate to that weight which gives no change in length. The fiber may be considered as a cylinder that changes its height at constant volume, $V = \pi h r^2$, but with variable surface area, $A = 2\pi h r$. So $dV/dh = \pi r^2 + 2\pi h r \, dr/dh = 0$, and $dr/dh = -r/2h$. Then $dA/dh = 2\pi r + 2\pi h \, dr/dh = 2\pi r - \pi r = \pi r$, and $\sigma = mg/\pi r$. This method is useful for surfaces that change in area too slowly for use of the other methods.

2–9. Dynamic Methods

If a liquid emerges from an elliptical orifice, the emergent jet has the same elliptical cross section. Surface tension reduces this to a circular cross section, but inertia carries it beyond so that the cross section becomes almost as eccentric as before, but with the axis at right angles to the previous one. Then the jet returns almost to its original shape. The restoring force is the surface tension, and the damping force is the viscosity. The surface tension can be determined from the wavelength and the frequency of flow. If a drop is distorted from its equilibrium shape it will also oscillate about that shape, and the surface tension may be calculated from the oscillation. If ripples are started in a plane surface, gravity and surface tension combine to propagate them, and the surface tension can be calculated from the rate of propagation. A practical application of the qualitative principle is pouring oil on water to reduce waves.

2-10. Interfacial Tension

Interfacial tensions can be measured by the drop method, which may be inverted to "drop" the lighter liquid upward through the heavier. The size of the drops is usually determined from the volume. The dipping-ring and capillary-rise methods, and drop pressure, which is analogous to bubble pressure, may also be used. This latter method has found greater application in the determination of the ratio of pore perimeter to pore area, $2/r$ for a circular pore, with liquids of known surface tension.

2-11. Activity and Particle Size

The potential inside a small drop must be greater than that inside a large one or beneath a plane surface:

$$RT \ln \frac{a}{a_0} = \frac{dG}{dN} = \frac{dG}{dA} \frac{dA/dr}{dN/dr} = \sigma \frac{d\, 4\pi r^2/dr}{d(4\pi r^3 \rho_t/3M)/dr} = \frac{2M\sigma}{r\rho_t}, \tag{2-7}$$

in which M is the molecular weight, ρ_t the density, a the activity, and a_0 the activity beneath a plane surface, for which $1/r_0 = 0$. For water this reduces to $\ln a/a_0 = 1.2 \times 10^{-7}/r$, so the change is only 10 percent for a radius of $10m\mu$. This method has been used to determine the liquid-solid interfacial tensions for slightly soluble solids, which are about ten times the surface tension of water. Some assumption must be made about the shape of the particles, but the chief difficulty arises from the fact that some of the particles are a little larger than the others or become a little larger by chance. Then their lower surface free energy gives a lower solubility and they tend to grow at the expense of the smaller ones. This makes it very difficult to tell the size of the particles that are at equilibrium at the time the solubility is measured.

2–12.   Surface Tension of Pure Liquids. Effect of Temperature

At the critical temperature a liquid and its vapor are identical and the surface tension and total surface energy, like the energy of evaporation, must be zero. At temperatures below the boiling point, which is about two-thirds the critical temperature, the total surface energy and the energy of evaporation are nearly constant. The surface tension then varies almost linearly with the temperature. The surface tensions of several liquids are given in Fig. 2–4. Fused salts have surface tensions of about 50 to 500 dyne/cm and fused metals of 500 to 1000 dyne/cm.

For nonpolar liquids the curves are concave upward and may be expressed approximately by the equation

$$\sigma = \sigma_0(1 - T/T_c)^n. \tag{2-8}$$

Van der Waals derived this equation with $n = 3/2$, but found that experiments indicated that $n = 1.23$. For many liquids it lies between 6/5 and 5/4. Van der Waals also found σ_0 to be proportional to $T_c^{1/3}p_c^{2/3}$, but we shall find it more useful to consider it as an empirical parameter.

The entropy corresponding to Eq. (2–8) is

$$-d\sigma/dT = n\sigma_0(1 - T/T_c)^{n-1}/T_c, \tag{2-9}$$

and the corresponding enthalpy, the total surface energy, is

$$\epsilon = \sigma - T\,d\sigma/dT = \sigma_0(1 - T/T_c)^{n-1}[1 + (n-1)T/T_c]. \tag{2-10}$$

If n is between 6/5 and 5/4, σ is about a quarter of σ_0 at the boiling point, but ϵ is more than nine-tenths of σ_0. So ϵ or σ_0 gives a better comparison of relative surface tensions than does σ itself.

Eötvös proposed the relation

$$\sigma(M/\rho_l)^{2/3} = K(T_c - T), \tag{2-11}$$

with M the molecular weight and ρ_l the density of the liquid, so that M/ρ_l is the molal volume, and $(M/\rho_l)^{2/3}$ would be proportional to the molal surface if all molecules had the same shape. He found that for nonpolar liquids K is about 2.12. For polar molecules it is smaller, and for many chain molecules it is larger. Katayama replaced the density of the liquid by the difference in density of liquid and saturated vapor:

$$\sigma[M/(\rho_l - \rho_g)]^{2/3} = K(T_c - T), \tag{2-12}$$

which gives better agreement at high temperatures.

2–13. The Parachor

If n in Eq. (2–8) is chosen as 6/5 and $(T_c - T)$ is eliminated from this equation and Eq. (2–12), we obtain

$$\sigma = [K^6 T_c^6 / \sigma_0^5][(\rho_l - \rho_g)/M]^4. \qquad (2\text{–}13)$$

The proportionality of the surface tension to the fourth power of the density difference was found by MacLeod. Sugden defined the parachor as

$$P = \sigma^{1/4} M/(\rho_l - \rho_g) = (K^6 T_c^6/\sigma_0^5)^{1/4}. \qquad (2\text{–}14)$$

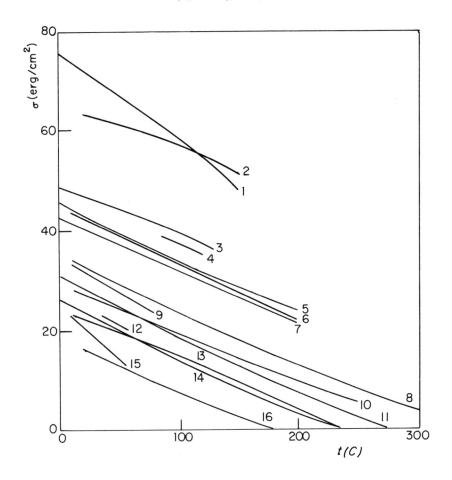

Figure 2–4. Surface tensions vs. temperature: 1, water; 2, glycerol; 3, glycol; 4, acetamide; 5, nitrobenzene; 6, aniline; 7, phenol; 8, chlorobenzene; 9, ethylene dichloride; 10, acetic acid; 11, benzene, acetonitrile (10–90); 12, Ethylidene dichloride; 13, ethyl alcohol; 14, ethyl acetate, ethyl bromide (10–40); 15, ammonia; 16, diethyl ether.

For any one substance the parachor should be independent of the temperature. Sugden also found that it is an additive property of singly bonded atoms plus terms for different types of bonding, such as: H = 17.1, C = 4.8, O = 20.0; 6 ring = 6.1; double bond = 23.2; triple bond = 46.6; ester O = 60; and so forth. He proposed it as a means of determining structure.

2-14. Heat of Surface Formation and of Evaporation

Stefan considered that the energy necessary to bring a molecule to the surface of a liquid should be half the energy necessary to bring it entirely into the gas phase. This corresponds to the closest packing with the top layer half filled and the next layer completely filled which we discussed in Sec. 1-6, that is, to a very dilute gas phase. In practice the comparison is made in enthalpies and the measurements of surface enthalpy ϵ must be made with saturated vapor. The ratio approaches unity as each enthalpy approaches zero at the critical temperature. Comparison has been made at $T/T_c = 0.7$, where the vapor is a nearly perfect gas.

Table 2-1 give some of these results. Substances that have nearly spherical molecules have ratios near $\frac{1}{2}$, but substances with a polar group on one end give a much smaller ratio, which indicates that the molecules are oriented with the nonpolar end toward the gas and the polar end toward the liquid.

2-15. Surface Tension, Total Surface Energy, and Chemical Constitution.

Table 2-2 shows the total surface energy per unit area instead of per molecule for a large number of substances, each of which has a

Table 2-1. Enthalpy of surface formation ϵ_s (10^{-14} erg/molecule) and their ratios to enthalpies of vaporization λ (same units) at a reduced temperature $T/T_c = 0.7$.

Molecule	ϵ_s	ϵ_s/λ	Molecule	ϵ_s	ϵ_s/λ
Mercury	—	0.64	Methyl formate	15.4	0.40
Nitrogen	3.84	.51	Ethyl acetate	18.3	.40
Oxygen	4.6	.50	Acetic acid	11.6	.34
Carbon tetrachloride	18.2	.45	Water	14.4	.28
Benzene	18.4	.44	Ethyl alcohol	11.2	.19
Diethyl ether	15.6	.42	Methyl alcohol	8.5	.16
Chlorobenzene	20.3	.42			

Table 2-2. Total surface energy (ergs/cm²).

Molecule	ϵ_s	Molecule	ϵ_s
n-Octane	51	Isoamyl alcohol	47
Diisoamyl	47	n-Octyl alcohol	49
n-Hexacontane	47	Myricyl alcohol	55
n-Octylene	52	dl-Methyl hexyl carbinol	50
Isoamyl chloride	50	Ethyl ether	52
Isoamyl cyanide	53	Isoamyl ether	47
Isoamyl nitrite	45	Butyric acid	56
Isovaleraldoxime	55	Isocaproic acid	51
Isovaleronitrile	54	Isoamyl acetate	53
Isoamyl amine	52	Isobutyl isovalerate	52
Diisobutyl amine	48	act-Amyl stearate	52
n-Hexyl amine	47	Tripalmitin	52

methyl group at one end. Eleven have methyl groups at both ends, and their average ϵ_s is 50.3. Six have the very polar hydroxyl group at one end, and their average is 51.3, which is not significantly different from 50.3.

Fig. 2–4 includes a few substances that have groups other than methyl at both ends. Ethylidene dichloride ($Cl_2HC \cdot CH_3$) and ethylene dichloride ($ClH_2C \cdot CH_2Cl$) are isomeric, and the difference in surface tension depends upon the fact that there must be a chlorine atom in the surface of the latter. The difference between ethyl alcohol ($HO \cdot H_2C \cdot CH_3$) on one hand and water, glycol ($HOH_2C \cdot CH_2OH$), and glycerol ($HOH_2C \cdot HOHC \cdot CH_2OH$) on the other shows the very large effect when the surface must contain a hydroxyl group.

3. Polycomponent Fluid Surfaces

3-1. Solutions and Mixtures

If a system includes two bulk phases and an interface, a component of that system may be present in all three parts of the system, it may be absent from one bulk phase, or it may even be absent from both bulk phases. We shall consider first two-component systems with both components present in the liquid phase.

When small amounts of one substance are dissolved in another, the surface tension changes as a linear function of the concentration. If the two pure substances have nearly the same surface tension, this linear relation may hold for all mixtures, but even for small differences the curves are often concave upward, giving sometimes a mixture of minimum surface tension. If the pure substances have very different surface tensions, the concavity upward is very marked: the higher surface tension is enormously depressed by small additions, but the smaller one is scarcely raised. Extensive compound formation may lead to complicated curves and even to a maximum surface tension. Illustrative curves are given in Fig. 3–1.

3-2. The Gibbs Adsorption Isotherm

We know from Le Chatelier's law that a substance which reduces the surface tension is more concentrated in the surface than in the bulk, and that one which increases the surface tension is less concentrated in the surface. The fact that a small surface concentration cannot be reduced much but can be increased greatly explains qualitatively the shape of the usual curves. The quantitative relation was derived by Gibbs. Before treating it, we shall consider some general characteristics of extensive properties.

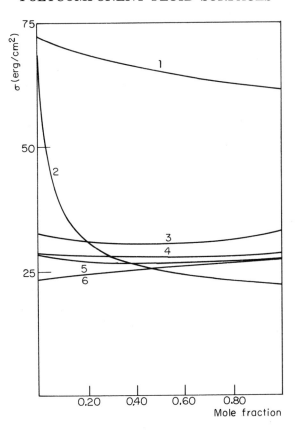

Figure 3-1. Surface tensions of mixtures: 1, water–glycerol; 2, water–ethyl alcohol; 3, carbon disulfide–ethylene dichloride; 4, benzene–ethyl iodide; 5, acetic acid–chloroform; 6, acetone–chloroform.

Let Z represent any extensive property of a system and let $\bar{Z}_k$ $= (\partial Z/\partial N_k)_{T,p,N}$. The subscripts mean that we keep constant during the differentiation the temperature, the pressure, and the quantity of every component not indicated inside the parentheses as varying, in this case every component except k. It may be helpful to consider the volume as a typical extensive property, though the free energy is the one we shall use most often. An extensive property is one that is proportional to the quantity of the system when the temperature, pressure, and composition (mole fraction) are constant. It may therefore be written

$$Z = f(T,p,x)\Sigma_i N_i, \qquad (3\text{–}1)$$

and

$$dZ = f(T,p,x)d\Sigma_i N_i = f(T,p,x)\Sigma_i dN_i. \qquad (3\text{--}2)$$

But

$$dZ = \Sigma_i \bar{Z}_i dN_i. \qquad (3\text{--}3)$$

Integrating Eq. (3–3) at constant composition gives

$$Z = \Sigma_i \bar{Z}_i N_i. \qquad (3\text{--}4)$$

Differentiating Eq. (3–4) totally yields

$$dZ = \Sigma_i d(\bar{Z}_i N_i) = \Sigma_i \bar{Z}_i dN_i + \Sigma_i N_i d\bar{Z}_i. \qquad (3\text{--}5)$$

Subtracting Eq. (3–3) from Eq. (3–5) gives

$$0 = \Sigma_i N_i d\bar{Z}_i. \qquad (3\text{--}6)$$

Eq. (3–6) was derived mathematically from Eq. (3–1) by Euler. The special case for the free energy was derived by Gibbs by a method very similar to that used above, and was exploited by Duhem. It is known as the Gibbs-Duhem equation, and is very useful for determining the chemical potential $\bar{G}_k$ of one component when those of the other components are known.

The surface tension is not usually an extensive property. However, if a body of uniform cross section is extended along one dimension, the increments of the area and of the volume are each proportional to the length, and therefore to each other. This is illustrated in Fig. 3–2. Either of the figures in (a) and either half of (b) or (c) has a volume, $V = 1 \times 1 \times L$ and an outer surface area A' or $A'' = (4 \times L) + 2$.

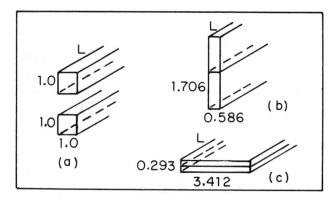

Figure 3–2. Surface areas.

The pair in (a) has no interface A; between the bulk phases in (b), $A = 0.586\,L$, and in (c), $A = 3.412\,L$. If (b) is rotated about an axis parallel to the length L, the interface remaining horizontal, the areas A' and A'' will not be altered, but the area A will increase regularly and reversibly to the maximum in (c). So A is an independent variable.

If the surface A is a small section of a larger surface separating the bulk phases ' and ", the surface areas A' and A'' are both zero, and the free energy of the system is

$$G = \Sigma_i N_i \bar{G}_i + A\sigma$$

and

$$dG = \Sigma_i(\bar{G}_i dN_i + N_i d\bar{G}_i) + \sigma dA + A d\sigma = \Sigma_i \bar{G}_i dN_i + \sigma dA.$$
(3–7)

So

$$0 = \Sigma_i N_i d\bar{G}_i + A d\sigma.$$
(3–8)

No quantity in Eq. (3–8) depends upon the particular method of extending the system. Therefore the equation itself must be general.

Let us call h' the distance from the interface to the upper surface and h'' the distance to the lower surface. Then the volumes of the two bulk phases are $V' = Ah'$ and $V'' = Ah''$ and the total volume is $V = V' + V'' = A(h' + h'')$. We attribute to the surface the difference between the properties of the real system and an imaginary system at the same temperature and pressure composed of a volume Ah' of the first bulk phase and a volume Ah'' of the second bulk phase, with no interface, as in Fig. 3–2(a). We shall designate these surface quantities by a superscript s. The potential of each component is the same in the imaginary system as in the actual one.

By definition, $V^s = V - V' - V'' = 0$, but the corresponding entropy, $S^s = S - S' - S''$, and the number of moles of each component, $N_i{}^s = N_i - N_i' - N_i''$, are not generally zero. If we subtract from Eq. (3–8) the values for the bulk phases in the imaginary system, we obtain

$$0 = A d\sigma + \Sigma_i(N_i - N_i' - N_i'')d\bar{G}_i = A d\sigma + \Sigma_i N_i^s d\bar{G}_i. \quad (3–9)$$

If we define the surface concentration of component i by $\Gamma_i = N_i^s/A$, we obtain

$$d\sigma = -\Sigma_i \Gamma_i d\bar{G}_i, \quad (3–10)$$

which is Gibbs's adsorption isotherm in its most general form. It is still arbitrary, because V' and V'' may have any values consistent with the relation $V' + V'' = V$. Gibbs selects V' in such a way that the surface concentration of the first component is zero, and designates the surface concentration of the second component so determined by $\Gamma_{2(1)}$. He then differentiates the surface tension according to Eq. (3–10) with every $\bar{G}_i$ except $\bar{G}_1$ and $\bar{G}_2$ constant, to give

$$\Gamma_{2(1)} = - \, d\sigma/d\bar{G}_2 = - \, d\sigma/RT \, d \ln a_2 \cong - \, d\sigma/RT \, d \ln c_2. \tag{3–11}$$

Eq. (3–11), rather than the more general Eq. (3–10), is generally known as the Gibbs adsorption isotherm.

Gibbs's method of fixing surface concentrations is analogous to the definition of bulk concentrations relative to the quantity of one component, such as moles per kilogram of water. It has advantages when this component is present in large excess, and it becomes awkward only when this component makes up only a small part of the system. This difficulty is easily avoided in two-component systems.

At constant temperature, equilibrium in the bulk phases requires that

$$V'dp = \Sigma_i N_i' d\bar{G}_i, \qquad V'' \, dp = \Sigma_i N_i'' d\bar{G}_i,$$

or, if $c_i = N_i/V$, that

$$dp = \Sigma_i c_i' d\bar{G}_i = \Sigma_i c_i'' d\bar{G}_i. \tag{3–12}$$

Therefore, for a two-component system,

$$(c_1' - c_1'') \, d\bar{G}_1 = - \, (c_2' - c_2'')d\bar{G}_2. \tag{3–13}$$

Eqs. (3–10) and (3–13) give

$$\frac{d\sigma}{d\bar{G}_2} = - \, \Gamma_2 + \frac{c_2' - c_2''}{c_1' - c_1''} \, \Gamma_1. \tag{3–14}$$

The subscripts are, of course, interchangeable. We find it convenient to express the surface concentrations, not as moles per unit surface area, but as volumes per unit surface area, which makes the angstrom a convenient unit.

3–3. Water-Ethanol Mixtures

Table 3–1 shows the surface concentrations of water-ethanol mixtures at 25°C, corresponding to curve 2 in Fig. 3–1, and the partial vapor pressures of the components. The first column gives the mole fraction of ethanol, $x_2 = N_2/(N_1 + N_2)$; the second gives the volume fraction $\phi_2 = N_2V_2/(N_1V_1 + N_2V_2)$, with V_1 and V_2 the molal volumes of the components. The third column gives the surface tensions, and the fourth and fifth give the surface concentrations as $\Gamma_{2(1)}V_2$ and $- \, \Gamma_{1(2)}V_1$.

Table 3-1. Surface concentrations of water-ethanol mixtures at 25°C.

x_2	ϕ_2	σ (erg/cm²)	$\Gamma_{2(1)}V_2$ Å	$-\Gamma_{1(2)}V_1$ Å	$\Gamma_{2(V)}V_2 = -\Gamma_{1(V)}V_1$ Å
0.00	0.000	72.2	0.0	39.7	0.0
.01	.032	—	1.2	37.5	1.2
.02	.062	—	2.2	33.9	2.1
.05	.146	—	3.4	20.0	3.0
.1	.265	36.4	3.7	10.3	2.8
.2	.448	29.7	3.9	4.6	2.1
.3	.582	27.6	3.5	2.5	1.4
.4	.684	26.35	3.0	1.4	0.9
.5	.765	25.4	2.5	0.8	.6
.6	.830	24.6	2.0	.4	.3
.7	.883	23.85	1.7	.2	.2
.8	.929	23.2	1.5	.1	.1
.9	.967	22.6	1.3	.05	.04
1.0	1.000	22.0	1.2	.0	.0

The sum of these two quantities gives to a fair approximation the distance between the two surfaces located by these two definitions.

The interface may be located in many other ways. The interesting ones probably lie between those fixed by $\Gamma_1 = 0$ and $\Gamma_2 = 0$. Perhaps the most interesting is the one that makes the sum of the volumes of the components going into the surface equal to zero, or

$$\Gamma_{2(v)}V_2 = -\Gamma_{1(v)}V_1, \qquad (3\text{–}15)$$

which with Eq. (3–14) gives

$$d\sigma/\bar{G}_2 = -\Gamma_{2(v)}[1 + V_2(c_2' - c_2'')/V_1(c_1' - c_1'')] \qquad (3\text{–}16)$$

or

$$\Gamma_{2(v)} = \Gamma_{2(1)}[1 + V_2(c_2' - c_2'')/V_1(c_1' - c_1'')]. \qquad (3\text{–}17)$$

The last column of Table 3–1 shows $\Gamma_{2(v)}V_2 = -\Gamma_{1(v)}V_1$. For small alcohol concentrations it approaches $\Gamma_{2(1)}V_2$ and for small water concentrations it approaches $-\Gamma_{1(2)}V_1$. So it avoids the difficulties of $\Gamma_{2(1)}$ and, especially, of $\Gamma_{1(2)}$ in the limits. It is still very asymmetric, because at small concentrations a large positive surface concentration is more easily attained than a large negative one.

3-4. Experimental Tests

There have been many attempts to test experimentally the Gibbs adsorption isotherm, but they have usually turned into tests of

Table 3-2. Surface concentrations.

Substance	c' (gm/kg water)	Γ_1 ($10^{-8} \times$ gm/cm^2)		
		Meas., bubbles	Calc.	Meas., shear
p-Toluidine	2.00	12.2	5.2	6.1
Toluidine	1.79	—	4.9	4.6
Phenol	20.48	14.8	4.8	4.1
Caproic acid	2.59	16.2	6.3	6.8
Caproic acid	3.00	16.9	6.5	5.1
Caproic acid	5.25	20.5	6.3	6.2
Hydrocynnamic acid	1.5	—	5.1	5.6
Hydrocynnamic acid	4.5	—	7.9	5.4
β-Phenylpropionic acid	1.5	—	5.2	5.3

experimental accuracy. Table 3–2 shows some measurements by McBain and his students. The first column gives the substances; the second gives the concentrations in the aqueous phase; the third gives the surface concentrations measured by passing gas bubbles through the solution; the fourth column gives the surface concentrations calculated by the Gibbs adsorption isotherm; and the last column gives the surface concentrations measured with an apparatus that sheared off a large surface very rapidly.

3–5. Electrolyte Solutions

Most salts increase the surface tension of water, and their effects can be described by the equation $\sigma = \sigma_0 + km$ if σ_0 is the surface tension of water, σ is that of a solution of m moles of salt per kilogram of water, and k is usually 1 to 3 dyne/cm. However, k is negative for hydrochloric and nitric acids. The increase of surface tension means a negative adsorption. Langmuir explained this by a surface layer of water one molecule deep. The Gibbs isotherm gives a larger surface concentration, however, and electrostatic theory indicates a repulsion of ions from the surface that should reach to relatively large distances.

    Measurements by Goard on mixed solutions of phenol and salts in water show that the effect of the salts on the surface tension is independent of the phenol concentration, even when the surface concentration of phenol corresponds to a complete unimolecular layer, and that the surface concentration of phenol, at the same potential (or activity) of the phenol, is independent of the salt concentration. This suggests that the two solutes are operating in different regions, the phenol at the surface and the salt well below it.

3-6. Mercury-Vapor Surfaces

The vapor pressure of mercury is very low at room temperatures, and most organic substances are practically insoluble in mercury. So in a system of liquid mercury and organic vapor, each bulk phase is one component, almost pure. The natural location of the surface is such that the surface concentration of mercury is zero. Iredale has measured the effect of benzene on the surface tension of mercury and finds a very rapid decrease that corresponds to an area of about 40 Å² per molecule at one-tenth the saturation pressure, decreasing to 21 Å² per molecule at nine-tenths saturation, and very variable tensions above 99 percent saturation. Other vapors gave the following areas (Å² per molecule): water, 16; ethanol, 29; methyl acetate, 27.

3-7. Surface Films

The interference colors of films of oil on water, like those of soap bubbles, have long fascinated scientists. Benjamin Franklin spread olive oil on a pond and found the film to be 10^{-7} in. thick. Lord Rayleigh used the surface concentration of such films to obtain an approximate value of Avogadro's number. Modern quantitative studies on films of materials not present in either bulk phase follow quite closely the work of Langmuir. He measured the difference in surface tension of a water-air surface and of a water-film–air surface, which is usually called $F = \sigma_0 - \sigma$. This method has the advantage that F varies with the temperature much less than does σ or σ_0. Langmuir's apparatus consisted of a photographic developing tray with waxed sides and an ordinary analytic balance. The tray was completely filled with water. The film was contained between the sides of the tray, a thin metal strip attached to the pointer of the balance perpendicular to the beam, and a movable barrier, the position of which determined the area of the film. The quantity of material in the film was determined by adding a weighed amount of a dilute solution in a volatile solvent. The surface was swept clean with paper strips, and the absence of surface impurities was demonstrated by the absence of any change in F as the area was made very small. Leakage around the ends of the strip on the balance was prevented by small jets of air. Later improvements, mainly due to Adam, are a more precisely shaped tray, a more delicate torsion balance, and light platinum ribbons to replace the air jets.

We should expect that at very low surface concentrations the activity would be proportional to the concentration, so that $d\bar{F}_2 = RT\ d \ln a_2 = RT\ d\ \Gamma_2/\Gamma_2$, and Eq. (3–10) becomes

$$dF = -\ d\sigma = +\ RT\ d\ \Gamma_2, \qquad (3\text{–}18)$$

or

$$F = RT\,\Gamma_2 = kT/A \qquad (3\text{--}19)$$

if A is $1/N\Gamma_2$, the surface area per molecule. If A is given, as is customary, in square angstroms per molecule, FA at 17°C is 400 erg. Most substances give condensed films at tensions less than 1 erg/cm², so the testing is difficult. Diesters of the type $C_2H_5OOC(CH_2)_n$-$COOC_2H_5$ with n about 10 give gaseous films up to about 10 erg/cm². The curves of FA vs. F resemble closely those of pV vs. p for gases just above the critical temperature. They start at the ideal value, decrease to a minimum at about 1 erg/cm², and then rise almost linearly.

The continuity and the fluidity of the films can be shown by sifting with fine powders. When the differential tension rises to that corresponding to the formation of a condensed film, little fluid islands appear, then coalesce at constant tension until the film covers the surface completely. Then the tension rises sharply and the film may become rigid. At large values of the tension, about 60 erg/cm², the film crumples without much change in the tension.

The most important aspect of these films is the evidence that many of them are continuous monomolecular layers with the active groups toward the water and the hydrocarbon chains toward the air. Fig. 3–3 shows Langmuir's original curve for palmitic acid on water at 16°C. The smallest tension Langmuir could measure corresponds to the point Q, the film became rigid at S—which does not happen generally, however—and it crumpled at H.

Fig. 3–4 shows the lower portion of a typical curve with more precise

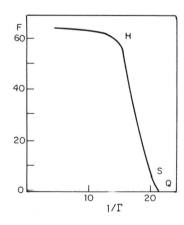

Figure 3–3. Surface film on water: palmitic acid (Langmuir).

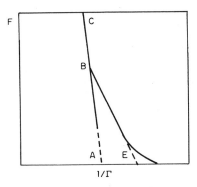

Figure 3–4. Surface film on water: general.

measurements. The line ABC is independent of the chain length and of the active group, the line EB is characteristic of the active group but independent of chain length, and the curved portion varies with the temperature. The different types of compounds are characterized by the intercepts at A and E. The slope of ABC represents the lateral compressibility of the chain; the slope of EB probably represents the compressibility of the shorter active group, but it may represent the work of raising some of these groups from the water. The area at A is 20.7 Å^2. This is somewhat larger than 18.3 Å^2, the perpendicular cross-sectional area of hydrocarbon chains in crystals. It corresponds to a tilt of 63.6°, which is the tilt of the hydrocarbon chains from the basal plane in crystals. Table 3–3 shows the areas of various head groups. That of the carboxyl group may be less than that of the chain. The esters are slightly larger, and the nitriles and isonitriles are larger still. A double bond next the carboxyl group gives a large change in volume. The area of the benzene derivatives corresponds to the benzene ring tilted on edge. The last two values are for the esters of

Table 3-3. Area (Å^2) of active groups.

Group	Area	Group	Area
–COOH	20.7	–CH$_2$–CH:NOH	25.0
–CONH$_2$	21.0	–CH:CH COOH	28.7
–CH$_2$OH	21.7	–CH:CHCOOC$_2$H$_5$	28.7
–COOCH$_3$	22.3	–C$_6$H$_4$NH$_2$	23.7
–COOC$_2$H$_5$	22.3	–C$_6$H$_4$ OH	23.8
–COOCH:CH$_2$	22.3	–C$_6$H$_4$OCH$_3$	23.8
–CN	27.5	–COOCH$_2$CH$_2$OOC–	42.0
–NC	27.5	–COOCH(CH$_2$OOC–)$_2$	63.0

glycol and of glycerol, which have two and three hydrocarbon chains, respectively. Their areas per chain are both 21.0 Å², more than that of a single hydrocarbon chain but less than those of the esters of single-chain acids.

Not everything about these films is simple and well understood. In many cases there is a peculiar effect at higher temperatures, illustrated in Fig. 3–5. At low temperatures the area is nearly independent of

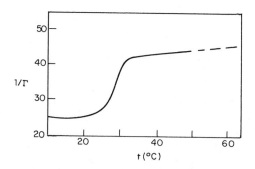

Figure 3–5. Film expansion.

the temperature. Then in a range of about 12°C the area approximately doubles to give an "expanded film" which expands further rather slowly with increasing temperature. The temperature of half expansion varies over the whole measurable range, and there are some films that are expanded over the whole range and some that are always condensed. The expanded films appear continuous, and the area is too small for the molecules to be lying flat in the surface. The current explanation is that the active heads separate but sections of the chains remain in contact.

Katherine Blodgett in Langmuir's laboratory has used acid films containing a small fraction of the barium salt to give permanent films of uniform thickness. If a clean glass or metal strip is pulled upward through the film, the carboxyl groups are turned toward it and adhere to it, whereas the hydrocarbon tails are exposed. If it is immersed again, the hydrocarbon groups of the liquid film adhere to those of the first film, giving a second film with the carboxyl groups exposed. This process can be repeated any number of times, with each of the carboxyl groups of each odd layer pointing inward, those of each even layer pointing outward. The films are stable if there are an odd number of groups. The thickness can be varied in steps of a double layer, about 50 Å, and several steps may be placed on the same strip by stopping the immersion at an appropriate depth each time.

3-8. Soap Bubbles

Ordinary soaps are the alkali salts of long-chain carboxylic acids. They are moderately soluble in water and reduce the surface tension so much that they form condensed monolayers when the bulk concentration is very small. A bubble is a double film with two surfaces, each with the carboxyl ions inward and the hydrocarbon chains outward. Usually there are incomplete double films, which slide readily over the primary film and each other to give the rapidly varying interference colors.

4. Adsorption on Solids

4-1. Vocabulary

The treatment of solid surfaces is more complicated than that of liquid surfaces because it is more difficult to tell what has become of the material removed from another bulk phase, such as a gas. It may be dissolved, or absorbed, in the bulk phase, it may be adsorbed on a moderately smooth interface, or it may be condensed, almost as a new bulk phase, in capillary pores or cracks. McBain has invented the word "sorption" to include all these phenomena. It is also customary to classify adsorption as physical adsorption or chemisorption. The latter obviously involves a chemical action, but the usual criterion is that the heat of evaporation from the surface is much larger than the heat of evaporation, or of sublimation, from the homogeneous bulk phase. The "much" gives latitude for subjective interpretation.

Brunauer has classified adsorption isotherms into five types, illustrated in Fig. 4–1, in which $x = p/p_o$ is the relative vapor pressure

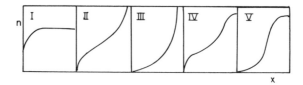

Figure 4–1. The five types of adsorption isotherms.

of the adsorbate and n is the number of molecules adsorbed. Usually the most important problem to be answered by adsorption measurements is the determination of the number of molecules in a monolayer.

Type I includes chemisorption if the initial slope is great enough.

Type III is what is predicted by one simple theory if the energy of adsorption is the same as the energy of condensation. Type II is III superposed on I. Types IV and V illustrate adsorption that is limited because most of the condensation is in capillaries or cracks that have limited volumes; IV is limited II and V is limited III.

The simple explanations of these isotherms depend upon the application to surfaces of the same type of reasoning that leads to the laws of ideal solutions, including Raoult's and Henry's laws and the law of mass action. It also includes theories of unweighted dice and of honestly and efficiently shuffled cards. It may be called the Theory of Independent Probabilities, which states that the probability of an event's happening at one position or time does not depend upon whether another event has or has not happened at any other place or time. We do not believe that any real series of events meets exactly the criteria of this theory, but we do know that very often the results calculated by assuming this theory give useful approximations of real systems.

4-2. The Langmuir Adsorption Isotherm

Langmuir studied adsorption at pressures so low compared to saturation pressure that he obtained curves of type I. He explained the curves by assuming that the surface is made up of a number of equivalent sites that react independently. If there are s such sites in the surface, s_1 of them are occupied at equilibrium and $s_o \, [= s - s_1]$ are free. The rate of evaporation is proportional to s_1 and the rate of adsorption to s_o and to the partial pressure p_A. At equilibrium the two rates must be equal:

$$k's_1 = k''s_o p_A$$

or

$$s_1/s_o = k''p_A/k' = k_1 p_A. \tag{4-1}$$

We may measure the adsorption as the fraction θ of sites occupied; $\theta = s_1/s = v/v_m$ if v is the volume of gas (under standard conditions) adsorbed at pressure p_A and v_m is the volume adsorbed to form a monolayer. Then $s_o/s = 1 - \theta = 1 - v/v_m$. So Eq. (4-1) may be written

$$s_1/s = k_1 p_A/(1 + k_1 p_A). \tag{4-2}$$

This corresponds to a curve of type I. It will be recognized as the law of mass action for a reaction of the type $A + B = AB$, in which $s_1/s_o = (AB)/(B)$. If the probabilities are independent, it makes no difference whether each site is on a separate molecule, several sites are on each large colloidal molecule, or all of them are on a single surface.

The desired quantities k_1 and s are best obtained from the measured values of s_1 and p_A by rearranging Eq. (4–2) to give a straight line, such as

$$p_A/s_1 = 1/k_1 s + p_A/s \qquad (4–3)$$

or

$$s_1/p_A = k_1(s - s_1). \qquad (4–4)$$

If p_A/s_1 is plotted against p_A, the slope of the straight line is $1/s$, and the slope divided by the intercept is k_1. If s_1/p_A is plotted against s_1, the intercept on the s_1-axis is s and that on the other axis is $k_1 s$, or the slope is $- k_1$.

We should expect s to be independent of the temperature and k_1 to be given approximately by the expression

$$k_1 = k_{10}\, e^{H_1/RT}, \qquad (4–5)$$

in which H_1 is the heat per mole of evaporation from the surface; the factor k_{10} is related to the standard entropy of evaporation by the equation $k_{10} = \exp(- S_1^0/R)$, and is expected to be virtually independent of temperature. Experiments often give good agreement with Eqs. (4–2) and (4–5), with H_1 considerably larger than the heat of evaporation from the bulk phase and s corresponding to only about one-twentieth of the surface being covered. The large heat and small number of sites both indicate chemisorption. The latter explains the independence of the probabilities. Usually the number of sites appears to increase with increasing temperature, as though more surfaces were being opened up as the temperature increases.

If more than one substance is being adsorbed on the same sites, we may still let s_0 be the number of vacant sites, s_A the number occupied by A, s_j the number occupied by J, and so on. Then Eq. (4–1) becomes

$$s_A = s_0 k_A p_A = s k_A p_A/(1 + \Sigma_j k_j p_j), \qquad (4–6)$$

which reduces to Eq. (4–2) if A is the only adsorbate. If there are sites of different kinds, characterized by different constants, k_A' for s' sites, k_A'' for s'' sites, and so forth, the total number of sites occupied by A may be determined by adding Eqs. (4–6) for the different types of sites.

4–3.   The Brunauer-Emmett-Teller (BET) Isotherm

Brunauer, Emmett, and Teller studied multilayer adsorption of a single adsorbate up to the vapor pressure of the liquid. In their simplest theory they consider unlimited adsorption, such as types II and III. We shall let s, s_0, s_1, and k_1 have the same meaning as in Eqs. (4–1) and (4–2), and let s_2 be the number of sites containing two molecules, s_i the number containing i molecules, and so forth, with

$$s_i/s_{i-1} = k_i p_A. \tag{4-7}$$

The total number of sites is

$$s = s_0 + \Sigma_i s_i, \tag{4-8}$$

and the number of molecules adsorbed is

$$n = \Sigma_i i s_i; \tag{4-9}$$

all sums are taken for $i = 1$ to ∞. The average number of molecules per site, θ, which can now be greater than unity, is

$$\theta = n/s = \Sigma_i i s_i/(s_0 + \Sigma_i s_i). \tag{4-10}$$

If we assume that the specific effect of the substrate surface extends only to the first layer, so that $k_2 = k_3 = k_i = k$, the sums may be solved readily. Then, if $c = k_1/k$ and $x = kp_A$,

$$s_1 = k_1 p_A s_0 = c\, s_0 x,$$
$$s_2 = kp_A s_1 = c\, s_0 x^2,$$
$$s_i = k_1 p s_0 (kp)^{i-1} = cs_0 x^i,$$

and

$$\theta = \frac{cs_0 \Sigma_i i x^i}{s_0(1 + c\Sigma x^i)}. \tag{4-11}$$

However, $\Sigma x^i = x/(1 - x)$ and $\Sigma i x^i = xd(\Sigma x^i)/dx = x/(1 - x)^2$. So

$$\theta = \frac{cx}{(1 - x)[1 - x + cx]}$$

$$= \frac{cx}{(1 - x)[1 + (c - 1)x]} \tag{4-12}$$

$$= \frac{(c - 1)x}{[1 + (c - 1)x]} + \frac{x}{1 - x}. \tag{4-13}$$

When $p_A = p_A^0$, the vapor pressure of liquid A, θ becomes infinite and x becomes unity. So x must equal p/p_0 and $k = 1/p_A^0$. The second form of Eq. (4–12) is the BET isotherm usually seen. Eq. (4–13) shows that it may be split into a Langmuir term with $(c - 1) = k_1 p_{A0}$ and a term independent of k_1. The parameters are most easily determined from the form

$$\frac{x}{n(1 - x)} = \frac{1}{cs} + \frac{c - 1}{cs} x. \tag{4-14}$$

If $x/n(1 - x)$ is plotted against x, the intercept is $1/cs$ and the slope is $(c - 1)/cs$. These determine c and s. It is claimed that good straight lines are obtained from $x = 0.05$ to 0.3. The deviations at low pressures are attributed to a few very active spots, and those at high pressures to a finite limit to adsorption.

If we express the constants as in Eq. (4–5),

$$k_1 = k_{10}\, e^{H_1/RT}, \qquad k = k_0\, e^{H_L/RT}.$$

Since $k = 1/p_A^0$, H_L must be the heat of evaporation of the liquid. It is very probable that k_{10} differs but little from k_0. If the two are the same,

$$c = k_1/k = e^{(H_1 - H_L)/RT},$$

in which $(H_1 - H_L)$ is the heat of wetting of the surface. Because of the irregularities at low pressures, this should be the heat of wetting of a surface with the most active spots covered.

Type II curves correspond to Eq. (4–12) or (4–13) with $c > 1$, and Type III to $c \leq 1$. If $c = 1$, $H_1 = H_L$ and the substrate corresponds to a fixed layer of the adsorbate. From Eq. (4–13) we see that $\theta = 1$ when $x = 0.5$, but that the first layer is only half filled, the second a quarter filled, and so on. This is probably very unrealistic because the mutual energy of molecules in the same layer is neglected. The real adsorbed layers are undoubtedly much denser. Still smaller values of c correspond to a surface that is not wetted by the liquid. The BET picture corresponds to a series of filaments one molecule in cross section instead of the drops that are formed at such surfaces.

Brunauer, Teller, and their associates have calculated equations that correspond to types IV and V. The equations become very intricate, and anyone interested is referred to their original papers.

The usual way of determining the surface area of solids is to multiply s found from Eq. (4–14) by an area per molecule of the adsorbate. The BET method is to calculate this area for nitrogen on the assumption that the molecules are spheres and that the density is the same as the density of the bulk liquid or of the bulk solid. A precision of about 15 percent is claimed. The earlier method was to replace s by n at "point B" defined as the low-pressure end of the "straight portion" of a graph such as that of Fig. 4–1. The difference from the BET value was seldom as great as 15 percent.

4–4.   The Harkins-Jura Isotherm

Harkins and Jura start with the empirical generalization for condensed films on liquids, which corresponds to the fact that the lines of Fig. 3–4 are straight and that the intercept at A depends upon the substrate but the slope does not. Analytically this is expressed as

$$F = \sigma_0 - \sigma = b - aA/N_2 = b - a/\Gamma. \tag{4-15}$$

When the substrate is varied, b changes but a does not. Harkins and Jura assume that this generalization may be extended to solid substrates. Rearranging Eq. (4–15) and substituting the Gibbs isotherm, Eq. (3–11), leads to

$$\Gamma = a/(b + \sigma - \sigma_0) = - (d\sigma/d\bar{G}_2)_A. \tag{4-16}$$

Integration gives

$$G_2 - G_{20} = k - \frac{(b + \sigma - \sigma_0)^2}{2a} = k - \frac{a}{2\Gamma^2}. \tag{4-17}$$

Then, to the approximation to which the vapor may be considered a perfect gas,

$$RT \ln \frac{p}{p_0} = k - \frac{a}{2\Gamma^2} = k - \frac{aA^2}{2N_2^2} = k - \frac{a'A^2}{2V^2}. \tag{4-18}$$

A plot of log p/p_0 against $1/V^2$ should give a straight line with slope proportional to A^2. The authors say that the straight portion does not go to such small pressures as those of the BET plots because low pressures do not give a condensed film, but that the straight portion does go to higher pressures.

To determine the constant for each gas, they determine the slope for the adsorption of that gas upon a solid whose area has been determined by an absolute method. This method is only applicable to particles without pores or cracks, that is, to fine crystals. They allow the particles to come to equilibrium with saturated water vapor and then measure the heat of immersion in water. This heat is the surface enthalpy of water multiplied by the surface area of the water-coated particle. Since the enthalpy is known, the area is determined. This is not exactly the surface area of the solid, but the small correction can be determined to a good approximation. The volume of water adsorbed indicates the thickness of the film, δ. If the particle were a sphere of radius r, the ratio of the area of the wetted particle to that of the dry particle would be $(1 + \delta/r)^2$. If it were a very long circular cylinder, the ratio would be $(1 + \delta/r)$. For a very flat plate, it would be 1. Sharp edges and corners give effects hard to figure, but always small. The agreement between the determination of surface area by the Harkins and Jura method with that by the BET method is excellent.

4-5. The Freundlich Isotherm

The Freundlich isotherm is much older than the others we have discussed. It was derived empirically by Freundlich and may be expressed as $V/m = Kp^n$ or

$$\log (V/m) = \log K + n \log p, \qquad (4\text{–}19)$$

with V the volume of adsorbed gas (standard conditions), m mass of adsorbate, and p the pressure of the gas; K and n are characteristic parameters, with n less than unity and sometimes as small as 0.2.

The Freundlich isotherm differs from the Langmuir isotherm in that it gives an infinite slope in the limit of zero pressure and gives no asymptotic limit. It can give agreement over a limited range of pressure. It differs from the BET type II isotherm in that there is no inflection.

The Freundlich isotherm has also been found to correspond approximately to a continuous distribution of energies of adsorption, if the number of sites with any given energy decreases rapidly as the energy increases, perhaps about as the inverse square. A truly constant n less than unity would require that this distribution go to infinite energy, but a cutoff at a large finite value would make n constant to pressures as low as can be studied.

There is much discussion whether a typical surface, particularly one important in technical adsorption or catalysis, is represented better by a constant energy of condensation or by a wide distribution of energies, and whether the mutual energy of molecules in the same layer can be ignored as in the Langmuir and BET treatments. There seem to be all varieties of surfaces, from nearly uniform to very heterogeneous. The interactions within a layer are probably negligible in a monolayer with widely separated sites. They may be as much as half the energy of evaporation for the distant layers, so the surface should be much more compact than is pictured in the BET treatment.

5. Physical Properties of Colloidal Solutions

5-1.   Introduction

It is convenient to classify colloidal solutions as: monodisperse if all the particles are of the same kind, particularly of the same size; paucidisperse if they fall into a few groups, each of the same kind; and polydisperse if there is a wide and almost continuous distribution of sizes. Thus a solution of hemoglobin is monodisperse; blood plasma is paucidisperse since it contains albumin, several kinds of globulins, and fibrinogen; and gelatin, the molecules of which contain from 1 to at least 500 amino acid residues, is polydisperse. These are all macromolecules or intrinsic colloids. At the other extreme we have colloid particles such as soap micelles or some colloidal salts which are continually exchanging small molecules with the ambient solution. These are called extrinsic colloids, and they are generally polydisperse. There are all gradations. In some cases primary colloid particles that are relatively stable may combine to form less tightly bound secondary aggregates.

Important characteristics of a colloidal solution are the average molecular weight and the distribution of molecular weights about this average. Three sorts of average molecular weight are of interest in different experiments: the number-average, or first-moment average, molecular weight, $\Sigma_i N_i \bar{W}_i / \Sigma_i N_i = \Sigma_i w_i / \Sigma_i w_i \bar{W}_i$; the weight-average, or second-moment average, molecular weight, $\Sigma_i N_i \bar{W}_i^2 / \Sigma_i N_i \bar{W}_i = \Sigma_i w_i \bar{W}_i / \Sigma_i w_i$; and the Z-average, or third-moment average, molecular weight, $\Sigma_i N_i \bar{W}_i^3 / \Sigma_i N_i \bar{W}_i^2 = \Sigma_i w_i \bar{W}_i^2 / \Sigma_i w_i \bar{W}_i$, in which N_i is the number of molecules with molecular weight $\bar{W}_i$ and w_i is the weight of material with molecular weight $\bar{W}_i$. In a monodisperse system these averages are all the same. Their ratios give measures of the variations in molecular weights.

5–2. Osmotic Pressure

Graham distinguished colloids from crystalloids by their inability to
pass through membranes. If there is a membrane that will permit the
passage of solvent but not that of some or all of the solutes, it is
called semipermeable. Many natural plant and animal membranes
are permeable to water but not to colloid particles. Early workers used
these and also membranes of cupric ferricyanide precipitated in the
pores of an earthenware cup, which would stand more than 200 atm
pressure difference. Modern studies, which are limited to low pressures,
usually use collodion or cellophane membranes. The equilibrium pressure difference across a semipermeable membrane between a solvent
and a solution containing a colloid solute is called the osmotic pressure.
Van't Hoff based his theory of solutions on the fact that the osmotic
pressure is RT times the concentration of solute, or the same pressure
that the solute would exert as a gas at the same concentration. This
does not mean, however, that the pressure is caused by bombardment
of the walls by solute molecules.

Fig. 5–1 is a diagram of an osmometer containing solutions A and

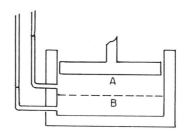

Figure 5–1. Diagram of osmometer.

B, separated by a membrane, represented by the broken line. There is
a piston by which the pressure on solution A may be altered and the
volume of A fixed. The difference in height in the two side tubes
measures the difference in pressure and the rate of rise or fall in the
B tube measures the rate of net flow through the membrane. Origi-
nally A and B are both pure solvent. Solvent molecules are continually
diffusing in each direction, but if the pressure is the same on the two
sides, there will be no net flow through the membrane. If the pressure
on A is increased, the potential of the solvent on that side is increased,
and the net flow will be from A to B.

If a colloid is added to solution A, say one particle to each million molecules of solvent, there is now one molecule per million that cannot pass from A to B, so the rate at equal pressure will be less by one part per million from A to B than from B to A. This can be compensated by a pressure that increases the activity of each solvent molecule in A by one part per million. This pressure does not depend upon the rate, but is a function only of the fractional decrease in rate due to the colloid.

At equilibrium, the temperature and the activity, or potential, of each component that can diffuse through the membrane must be the same on the two sides. We shall consider first the case in which only the solvent can cross the membrane. On the solution side, the potential of the solvent is reduced by the presence of the solute and may be increased by increasing the pressure. If we let $RT \ln a_1$ be the potential of the solvent in the solution at the initial pressure minus the potential of pure water at the same temperature and pressure,

$$d\bar{G}_1 = RT \, d \ln a_1 + \bar{V}_1 dP = 0, \tag{5-1}$$

where P is the osmotic pressure. For small pressures and solute concentrations, we may let $\bar{V}_1 = V_1$, the molal volume of the pure solvent, which is constant; then integration gives

$$PV_1 = - RT \ln a_1. \tag{5-2}$$

It is convenient to define several measures of concentration: $x_i = N_i/\Sigma_j N_j$ is the mole fraction; $\phi_i = N_i V_i/\Sigma_j N_j V_j = N_i V_i/V$ is the volume fraction; $C_i = w_i/V = \phi_i \rho_i$ is the weight-volume concentration; and $C_i' = w_i/N_1 V_1 = \phi_i \rho_i/\phi_1$ is the weight-solvent volume concentration. Here N_i is the number of moles of component i, V_i is the partial volume of component i in very dilute solution in component 1, the solvent, V is the total volume, and ρ_i is the density $\bar{W}_i/V_i$ of component i. We also define the free energy of mixing, ΔG^M, as the difference in free energy between the solution and the unmixed components at the same temperature and pressure.

For ideal solutions,

$$\Delta G^M = RT \, \Sigma_i N_i \ln x_i \tag{5-3}$$

and

$$\ln a_1 = \ln x_1. \tag{5-4}$$

Flory and Huggins found that for many high polymers, linear and branched, in solvents that give no heat of mixing, Eqs. (5-3) and (5-4) should be replaced by

$$\Delta G^M = RT \, \Sigma_i N_i \ln \phi_i \tag{5-5}$$

and

$$\ln a_1 = \ln \phi_1 + \Sigma_i \phi_i (1 - V_1/V_i), \qquad (5\text{--}6)$$

which reduce to Eqs. (5–3) and (5–4) if the volumes are all equal. For a two-component system, we expand the logarithm in Eq. (5–6) and add the term $h\phi_2^2$, which should represent the heat of dilution in simple cases, to give

$$\ln a_1 = - \phi_2 V_1/V_2 + \phi_2^2 (h - {}^1\!/_2) + \cdots. \qquad (5\text{--}7)$$

Combining Eqs. (5–2) and (5–7) gives

$$\frac{P}{RT} = \frac{\phi_2}{V_2} - \frac{\phi_2^2(h - {}^1\!/_2)}{V_1} + \cdots \qquad (5\text{--}8)$$

$$= \frac{C_2}{\bar{W}_2} - \frac{C_2^2(h - {}^1\!/_2)}{V_1 \rho_2^2} + \cdots \qquad (5\text{--}9)$$

$$= \frac{C_2'}{\bar{W}_2} - \frac{C_2'^2(h - {}^1\!/_2 + V_1\rho_2/\bar{W}_2)}{V_1 \rho_2^2} + \cdots. \qquad (5\text{--}10)$$

To determine the molecular weight and the interaction, it is customary to plot P/RTC_2 against C_2 or P/RTC_2' against C_2'. In either case, the lines are nearly straight, so the determination of the intercept, $1/\bar{W}_2$, and the initial slope, the coefficient of C_2^2 or $C_2'^2$ in Eq. (5–9) or (5–10), is not difficult. Since $\bar{W}_2 = w_2/N_2$, it is the number-average molecular weight for a mixed solute.

Fig. 5–2 shows the results of osmotic-pressure measurements by Flory for solutions of some isobutylenes in cyclohexane and in benzene. The intercepts of the two largest polymers are the same in benzene as in cyclohexane, for the molecular weight is independent

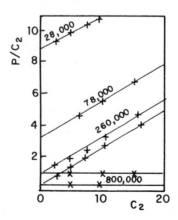

Figure 5-2. Osmotic pressures of isobutylenes of indicated molecular weights: $+$ in C_6H_{12}; $\times$ in C_6H_6.

of the solvent. The slopes are very different for the two solvents, but they are independent of the size of the polymer. For cyclohexane, h is nearly zero, so ρ_2 must be independent of the polymer size, as it should be. For benzene, h is very nearly $1/2$, and it must also be independent of polymer size, as we should expect.

5-3.   The Donnan Equilibrium

If the colloid particles are electrically charged, the colloid component must contain enough small ions to make it neutral. Most membranes are permeable to small ions as well as to solvent, and the solutions on both sides must remain electrically neutral. If the solvent can ionize, this will lead to solvolysis, which is called membrane hydrolysis if water is the solvent.

Let us assume that the colloid is negatively charged with an equivalent weight $1/z_2$, let it be present as the sodium salt, and let parentheses indicate concentrations per liter of solvent. Then at equilibrium,

$$(Na^+) + (H^+) = z_2 C_2' + (OH^-),$$
$$(Na^+)_0 + (H^+)_0 = (OH^-)_0,$$
$$(Na^+)(OH^-) \, \gamma_{NaOH}^2 = (Na^+)_0(OH^-)_0 \, \gamma_0{}^2{}_{NaOH},$$
$$(H^+)(OH^-) \, \gamma_{HOH}^2 = (H^+)_0(OH^-)_0 \, \gamma_0{}^2{}_{HOH},$$

in which the subscript zero represents the outside solution, without colloid; γ is the activity coefficient.

If the solutions are ideal,

$$P = RT[C_2'/\bar{W}_2 + (Na^+) + (H^+) + (OH^-) \\ - (Na^+)_0 - (H^+)_0 - (OH^-)_0].$$

If we can neglect (OH^-) and $(H^+)_0$, the osmotic pressure becomes

$$P = RT[C_2'(1/\bar{W}_2 + z_2) - 2(Na^+)_0]$$

if the colloid is a strong acid, and

$$P = RT[C_2'/\bar{W}_2 + (Na^+) - 2(Na^+)_0]$$

if the colloid is a weak acid. The uncertainties due to accidental variation in pH make the determination of molecular weight quite impossible for such solutions.

The hydrolysis can be suppressed and the molecular weight determined readily by adding a large excess of a salt with small ions, such as sodium chloride. If the (H^+) and (OH^-) are ignored, which is permissible,

$$(Na^+) = z_2 C_2' + (Cl^-),$$
$$(Na^+)_0 = (Cl^-)_0,$$
$$(Na^+)(Cl^-) \, \gamma_{NaCl}^2 = (Na^+)_0(Cl^-)_0 \, \gamma_0{}^2{}_{NaCl}.$$

If the solutions are ideal,

$$(\text{Na}^+) = (\text{Cl}^-)_0\left[1 + \left(\frac{z_2 C_2'}{2(\text{Cl}^-)_0}\right)^2\right]^{1/2} + \frac{z_2 C_2'}{2}, \tag{5–11}$$

$$(\text{Cl}^-) = (\text{Cl}^-)_0\left[1 + \left(\frac{z_2 C_2'}{2(\text{Cl}^-)_0}\right)^2\right]^{1/2} - \frac{z_2 C_2'}{2}, \tag{5–12}$$

and

$$
\begin{aligned}
P &= RT\left(\frac{C_2'}{\bar{W}_2} + (\text{Na}^+) + (\text{Cl}^-) - (\text{Na}^+)_0 - (\text{Cl}^-)_0\right) \\
&= RT\left[\frac{C_2'}{\bar{W}_2} + 2(\text{Cl}^-)_0\left\{\left[1 + \left(\frac{z_2 C_2'}{2(\text{Cl}^-)_0}\right)^2\right]^{1/2} - 1\right\}\right] \tag{5–13} \\
&\cong RT\left(\frac{C_2'}{\bar{W}_2} + \frac{C_2'^2 z_2^2}{4(\text{Cl}^-)_0}\right). \tag{5–14}
\end{aligned}
$$

The approximation in Eq. (5–14) is very good because the next term in the expansion is proportional to $C_2'^4$. The equilibrium expressed in Eqs. (5–11) and (5–12) is called the Donnan equilibrium, and the corresponding pressure from Eq. (5–13) or (5–14) is called the Donnan pressure. It is proportional to the square of the colloid concentration and is inversely proportional to the outside salt concentration. Although the Donnan pressure at moderate concentrations may be greater than the pressure that is proportional to the colloid concentration, it is not difficult to extrapolate P/RTC_2' to zero concentration.

The Donnan pressure is easy to understand from the effect of the ions on the activity of the water, but quite incompatible with pressure due to bombardment by solute that cannot pass through the membrane.

Fig. 5–3 shows the coefficient b in the equation

$$P\bar{W}_2/RTC_2' = 1 + bC_2'/\bar{W}_2 + \cdots \tag{5–15}$$

as a function of $z_2\bar{W}_2$ for bovine serum albumin in 0.15 M sodium chloride solution. The broken curve is $(z_2\bar{W}_2)^2/4(\text{Cl}^-)_0$, which is the value for an ideal solution. The facts that the minimum value is not zero and that the wing is less steep than the broken line show that the solution is not ideal. The horizontal displacement of the minimum from $z_2 = 0$ may be due to an error in calculating z_2 when it is positive.

5–4.   Light Scattering

The scattering of light by particles much smaller than the wavelength of light is called Rayleigh scattering. The blue of the sky and the

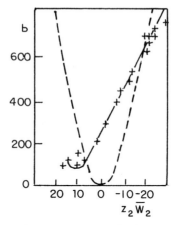

Figure 5–3. Slope of osmotic-pressure curve for bovine serum albumin in 0.15M NaCl (pH increasing from left to right).

related red of the sunset are due to such scattering by the molecules of the earth's atmosphere, as first explained by Rayleigh in 1871. The Rayleigh theory is not easily extended to dense fluids of interacting particles, but Einstein and later Smoluchowski showed that the phenomenon could be treated as due to fluctuations in the refractive index. In pure fluids these are directly related to fluctuations in density, but solutions show an additional scattering due to fluctuations in concentration. For a two-component system, the relation is

$$d(P/RT)/dC_2' = (1/V_1)\, d\ln a_1/dC_2' = HC_2'/\tau, \qquad (5\text{–}16)$$

in which $H = 32\pi^3 n^2 (dn/dC_2')^2/3N_A\lambda^4$, n is the refractive index of the solution, N_A is Avogadro's number, λ is the wavelength of the light, and τ is the turbidity. Combining Eqs. (5–15) and (5–16) gives

$$HC_2'/\tau = 1/\bar{W}_2 + 2bC_2'/\bar{W}_2'^2 + \cdots. \qquad (5\text{–}17)$$

The ratio of the initial slope to the intercept is twice as great as for the osmotic pressure. The light-scattering measurements are complementary to osmotic-pressure measurements in that the turbidity increases with the molecular weight. For most high polymers, dn/dC_2' is independent of the molecular weight, and light-scattering measurements give the weight-average molecular weight. In general, the average depends also on dn/dC_2'.

For more than two components, the slope is a complicated function involving $\bar{W}_2^2(dn/dC_2')^2$, $\bar{W}_3^2(dn/dC_3')^2$, $\bar{W}_2\bar{W}_3(dn/dC_2')(dn/dC_3')$, and so forth. In a mixture of a colloid and a crystalloid solute, however, all

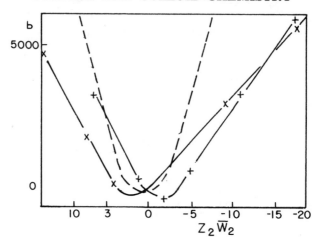

Figure 5–4. Slope of light-scattering curves of bovine serum albumin in 0.003M salts: $\times$ NaCl; $+$ NaSCN (pH increasing from left to right).

but the first term may be unimportant. Fig. 5–4 shows the values of b for bovine serum albumin in 0.003 M NaCl or NaSCN. It should be compared with Fig. 5–3. The minima are much nearer zero charge, and are negative rather than positive. Since the salt concentration is 1/50 as great, the theoretical curve of Fig. 5–4 is 50 times as large as that of Fig. 5–3.

For these small particles the light scattering is symmetric about the plane perpendicular to the beam. As one or more dimensions approach the wavelength of the light, Eq. (5–16) no longer describes the total turbidity, and the light scattering is no longer symmetric. From the variation of the intensity of the scattered light with the angle it is possible to learn much about the dimensions of these larger molecules. The methods and the interpretation are too complicated for reproduction here.

5–5. Equilibrium Distribution in a Gravitational or Centrifugal Field

The treatment of the distribution of a colloid, or other material, in a gravitational or centrifugal field is analogous to the treatment of osmotic pressure in that we let the change in activity be the change due to composition at constant gravitational or centrifugal potential. Then

$$d\bar{G}_i = RT \, d \ln a_i + \bar{V}_i(\rho_i - \rho_0)gdh = 0 \qquad (5\text{–}18)$$

or

$$d\bar{G}_i = RT \, d \ln a_i - \bar{V}_i(\rho_i - \rho_0)\omega^2 x dx = 0, \qquad (5\text{–}19)$$

in which g is the acceleration of gravity, h is the height, ω is the angular velocity of the centrifuge, and x is the distance from the axis. For ideal solutions of constant density, we may integrate to obtain

$$RT \ln C/C_0 = -g\bar{V}_i(\rho_i - \rho_0)(h - h_0), \qquad (5\text{–}20)$$

$$RT \ln C/C_0 = \omega^2 \bar{V}_i(\rho_i - \rho_0)(x^2 - x_0^2)/2. \qquad (5\text{–}21)$$

Each solute distributes itself independently, except as ρ_0 is affected by the concentrations.

In the study of colloidal solutions it has become customary to determine dC'/dx directly from dn/dx by measuring the distortion of a linear scale seen through the cell, or by one of the schlieren methods. Then $d \ln a_i/dx$ in Eq. (5–19) is replaced by $[1/C'_i + (d \ln \gamma_i)/dC'_i]$ $(dn/dx)/(dn/dC'_i)$. For a mixed solute, integration gives the Z, or third-moment, average molecular weight. The measurement of dn/dx throughout the cell gives in principle a method of determining all the moments from a single measurement. However, the limitations of the constancy of the temperature and the angular velocity for periods of 10 to 30 days at about 15,000 rev/min prevent too great exploitation of these possibilities. The determination of the number-average molecular weight from osmotic pressure and the weight-average molecular weight from light scattering increases the precision to which the equilibrium centrifuge results may be interpreted.

Perrin determined Avogadro's number from the distribution of gamboge particles in the earth's gravitational field, using Eq. (5–20). His values of N_A by this method were accurate to about 10 percent. Other measurements are described in the next section.

5–6. Fluctuations

The visibility of large individual colloid particles gives a convenient means of verifying the assumptions of the kinetic theory. We consider first the fluctuations in the number n of particles in a given volume. If the average number is n_0, the probability $P(n)$ of finding any specified n is

$$P(n) = n_0{}^n e^{-n_0}/n!. \qquad (5\text{–}22)$$

(Recall that $0! = 1$.) Enough observations must be made to smooth out the effect of a run of luck. Table 5–1 gives the results of measurements on a gold sol by Westgren.

Table 5-1. Fluctuations in concentration: gold sol, 250 Å; $n_0 = 1.43$.

Relative frequency				n				
	0	1	2	3	4	5	6	7
Observed	381	568	357	175	67	28	5	2
Calculated	380	542	384	184	66	19	5	2

More complicated expressions give the frequency of any sequence, the average duration of any number, and the average time between repetitions of a number.

If the solution is so dilute that a single particle can be followed, it is found to be moving rapidly in what is called Brownian movement, which is the thermal motion of kinetic theory for which the average kinetic energy is $3kT/2$ for any particle, regardless of its size or surroundings. It is not possible to follow the exact path of any molecule with a velocity corresponding to this energy because the velocity and direction change about 10^{20} times a second.

Einstein developed a method of studying such motion, which is now called a "random walk," by measuring the change in position in one direction, say along the x-axis, at equal intervals of time. He found the average of the square of the displacement, $\overline{x^2}$, to be

$$\overline{x^2} = 2\,Dt, \tag{5-23}$$

if D is the diffusion coefficient and t the time interval. If Stokes's law is obeyed,

$$D = kT/6\pi\eta r, \tag{5-24}$$

in which $k \ [= R/N_A]$ is Boltzmann's constant, T the absolute temperature, η the viscosity, and r the radius of the spherical particles. Since R is known, this gives a means of calculating Avogadro's number N_A. Perrin studied the fluctuations of gamboge particles, which were nicely spherical. After the experiment he acidified the solution and the particles adhered to the walls so that he could count the number corresponding to a given weight. Often they formed rows long enough that he could measure the radius directly, as well as compute it from the number and the density of the bulk gamboge. He obtained Avogadro's number to within about 40 percent before more precise methods were developed. Later observers have obtained much better agreement with the accepted methods.

The average kinetic energy of rotation around any axis is equal to its average kinetic energy of translation parallel to that axis, unless the rotation is so rapid that the energy is quantized. Einstein has derived the equation

$$\overline{\theta^2} = kTt/4\pi\eta r^3, \qquad (5\text{--}25)$$

in which $\overline{\theta^2}$ is the average of the square of the angular displacement about any axis. Perrin was able to test this equation by studying particles with a scar at one point. The agreement was within his experimental error, which was much larger than for his other methods.

5-7. Diffusion

Fick's laws of diffusion were derived in analogy with Fourier's laws for the flow of heat. They do not, therefore, depend upon the assumption of the existence of molecules. For the case in which the concentration varies only in the x-direction, these laws are

$$dn/dt = -DA\ \partial C/\partial x, \qquad (5\text{--}26)$$

$$\partial C/\partial t = D\ \partial^2 C/\partial x^2, \qquad (5\text{--}27)$$

in which n is the net number of units that pass in the direction of increasing x through a plane of area A perpendicular to the x-axis, x is distance, t is time, C is concentration, and D is the diffusion coefficient.

Einstein related the diffusion coefficient to the frictional coefficient by the equation

$$D = kT/f, \qquad (5\text{--}28)$$

in which f is the frictional coefficient or the force necessary to give a particle unit velocity, by equating the force that leads to diffusion to $kT(d \ln C)/dx$. This obviously assumes ideal solutions. For nonideal solutions, Eq. (5–26) must be replaced by

$$dn/dt = -DAC(\partial \ln a)/\partial x = -DA(1 + C(d \ln \gamma)/dC)\partial C/\partial x, \qquad (5\text{--}29)$$

and Eq. (5–27) should be changed accordingly. For spherical particles, which obey Stokes's law,

$$f = 6\pi\eta r. \qquad (5\text{--}30)$$

The usual application of Fick's equation is to apply the proper boundary conditions to Eq. (5–26) to fit the experimental conditions. We shall consider only the case of a practically infinitely high column with initial concentrations C_0 below a plane at zero height and zero above that plane. An exceedingly sharp boundary can be formed by removing liquid at the boundary through a capillary. The chief experimental difficulties are to minimize disturbances from convection and from vibration. The resulting equation is

$$C = \frac{C_0}{2}\left(1 - \frac{2}{\pi^{1/2}}\int_0^y e^{-n^2}dn\right),$$
(5–31)

in which C is the concentration at point x and time t and $y = x/2(Dt)^{1/2}$. The function in parentheses is 1 minus the probability integral, which may be found in tables. Eq. (5–31) gives S-shaped curves, with $C = C_0/2$ at $x = 0$, and maximum slope at that point.

It is customary to measure dC/dx directly, by various methods of determining the change of refractive index with distance (Sec. 5–5). Differentiation of Eq. (5–31) gives

$$-dC/dx = (C_0/2(\pi Dt)^{1/2})e^{-x^2/4Dt}.$$
(5–32)

This curve is symmetric about the maximum at the original boundary. The height of this maximum is a linear function of $t^{-1/2}$.

Diffusion coefficients are often measured by the rate of diffusion through a porous diaphragm, the two solutions being stirred to keep their concentrations uniform. Eq. (5–26) then becomes

$$dn/dt = -D(C_2 - C_1)A/(x_2 - x_1).$$
(5–33)

The effective ratio $A/(x_2 - x_1)$ of area to thickness, which is not given by the external dimensions of the diaphragm, is determined from measurements with a material with known diffusion coefficient.

The presence of different components with different velocities is shown by the shape of the curves of Eq. (5–32) and by a variation in time of $d \ln (C_2 - C_1)/dt$ in Eq. (5–33).

5–8. Sedimentation

A particle moves in an external field at an average rate that is equal to the force F divided by the frictional coefficient f. In a gravitational field with constant g, the motion of a particle of mass m, volume v, and density ρ in a solution of density ρ_0 is

$$dx/dt = gv(\rho - \rho_0)/f = gm(1 - \rho_0/\rho)/f.$$
(5–34)

Since the Brownian movement is superimposed upon this imposed motion, identical particles in a suspension will not all fall with the same speed, but the upper boundary will spread, just as a stationary diffusion boundary spreads. For large particles this spreading will be negligible. If each particle that reaches the bottom of the column remains there, the rate of deposition of uniform particles is constant until the upper boundary reaches the bottom. For concentrated suspensions this rate may be measured by the difference in height of a small side tube filled with the suspending medium alone and the column containing the suspension. It may be measured more accurately by a balance pan at the bottom of the column.

In a polydisperse system, each component will have its own rate of deposition until it is almost completely deposited. This method therefore gives a method of determining how much of each component is present, classified by $m(1 - \rho_0/\rho)/f$.

Smaller particles will not remain at the bottom but will begin to diffuse upward as soon as they reach the bottom. Their sedimentation may be studied by the rate of fall of the upper boundary, which is shown by the maximum in dC/dx. In a monodisperse system there will be a single boundary, in a paucidisperse system there may be several, but in a polydisperse system there is again a single boundary, which spreads more rapidly than that of a monodisperse system if there is no interaction between the particles, but which may in practice be abnormally sharp.

For these smaller particles it is customary to replace the earth's gravitational field by the much more powerful centrifugal field of an ultracentrifuge, which may be 20,000 to 350,000 times that of gravity. The cell is cut as a small sector of the rotor, so that the walls will not interfere with motion in the radial field. In the centrifugal field, the acceleration of gravity g in Eq. (5–13) is replaced by $\omega^2 x$, in which ω is the angular velocity and x is measured from the axis of rotation. It is customary to determine the sedimentation constant s from the equation

$$s = (dx/dt)/\omega^2 x = m(1 - \rho_0/\rho)/f. \qquad (5\text{–}35)$$

Often the sedimentation is expressed as s_{20}^0, which is s multiplied by $\eta/(1 - \rho_0/\rho)$ and divided by the same quantity for water at 20°C; the unit is the Svedberg (S), and $1\ S = 1 \times 10^{-13}$ cm/sec per unit field.

The frictional coefficient may be eliminated by combining Eqs. (5–28) and (5–35) to give

$$m = kTs/D(1 - \rho_0/\rho), \qquad (5\text{–}36)$$

or, if k is replaced by R, the molecular weight is determined by measurements of sedimentation velocity and of diffusion. It is customary to extrapolate to the values at zero concentration s_0 and D_0 before eliminating f_0.

The interpretation of the kind of average molecular weight that is determined by this method for a polydisperse system is not at all simple. It is intermediate between the number average and the weight average.

5-9.   Viscosity

The viscosity coefficient of a fluid, η, often called merely the viscosity, is defined as the ratio of the shearing stress per unit area to the velocity gradient. To picture this definition, it is convenient to consider two parallel infinite planes separated by the fluid to a distance of 1 cm. Then the viscosity coefficient is the force per unit area of one of these planes necessary to maintain a relative velocity of 1 cm/sec parallel to the planes. The fluidity is the reciprocal of the viscosity.

For a Newtonian fluid the viscosity coefficient is independent of the rate of shear. For such fluids it is convenient to measure the viscosity in a capillary viscometer, that is, to measure the rate of flow through a tube of radius a and length L with a pressure difference P between the two ends. The annular section next to the wall, $r = a$, will not move at all, and the maximum velocity is at the axis, $r = 0$. The force per unit area acting on the cylinder of radius r is $P\pi r^2/2\pi rL$, which is equal to $- \eta dv/dr$. So

$$dv = - (P/2\eta L)rdr, \qquad\qquad (5\text{–}37)$$

and

$$v = (a^2 - r^2)\ P/4\eta L. \qquad\qquad (5\text{–}38)$$

Integrating over the tube,

$$V/t = \int_0^a 2\pi rvdr = \pi a^4 P/8\eta L$$

or

$$\eta = \pi a^4 Pt/8LV, \qquad\qquad (5\text{–}39)$$

in which V/t is the volume flowing per unit time. This is known as Poiseuille's law. It obviously depends upon the flow's being laminar. At high velocities, the flow becomes turbulent, and V/t is no longer proportional to P. Long before this happens it becomes necessary to make a correction for the kinetic energy with which the fluid leaves the capillary. This correction can be reduced by increasing the length of the capillary, decreasing the pressure, or decreasing the radius a. As a is made small, however, difficulties due to dust particles increase. Ordinarily, the pressure is the difference h in height of the inlet and outlet times the density, and the constants of the apparatus are obtained by measuring the time of flow of a liquid of known viscosity.

The unit of viscosity is the poise ($gm\ cm^{-1}\ sec^{-1}$). The viscosity of water is 0.01 poise at 20°C and it decreases by about 2 percent per degree decrease in temperature. Usually the viscosity is approximately an exponential function of the temperature.

Einstein found that for very dilute solutions of spheres that are large compared with the solvent molecules,

$$\eta/\eta_0 = 1 + 2.5\phi_2, \tag{5-40}$$

in which η_0 is the viscosity of the solvent and ϕ_2 is the volume fraction of the solute, $C_2/1000\rho_2$. If the particles are solvated, ϕ_2 must correspond to the solvated particles. If the particles are not spherical, the coefficient is increased, and it is increased very greatly for long slender molecules.

Convenient ways of handling the viscosities are to plot

$$(\eta - \eta_0)/\eta_0 c_2 = E + FC_2 + \cdots \tag{5-41}$$

or

$$(\ln \eta/\eta_0)/C_2 = E + F'C_2 + \cdots. \tag{5-42}$$

Usually Eq. (5–42) is fitted by a straight line better than Eq. (5–41), which permits a surer extrapolation to E. Eq. (5–40) corresponds to $E = 2.5/1000\rho$.

5-10. Plastic Flow. Thixotropy

If the viscosity coefficient varies with the rate of shear, the capillary viscometer is not at all suitable since the rate of shear varies from zero to a maximum. Such systems are studied in viscometers that have two concentric cylinders, one of which is fixed and the other rotates. For this case,

$$\frac{dv}{dx} = \frac{2\omega}{r^2(1/R_1^2 - 1/R_2^2)}, \tag{5-43}$$

in which R_1 is the radius of the inner cylinder and R_2 that of the outer cylinder. The maximum variation is from $1/R_2^2$ to $1/R_1^2$. The end effects may be reduced by having the cylinders long and the distances between the ends large.

In the Couette type for precise laboratory work and in the Mac-Michael commercial viscometer, the outer cylinder is rotated at a constant rate and the torque on the inner cylinder is measured. In the commercial Stormer instrument, the outer cylinder is fixed, the inner cylinder is rotated with constant torque, and its speed is determined. The Stormer instrument is more convenient, but is apparently not so precise. In either instrument it is easily possible to run a series of experiments at different stresses and rates of shear.

If the stress is plotted against the rate of shear, a Newtonian fluid gives a straight line through the origin like curve A in Fig. 5–5, and

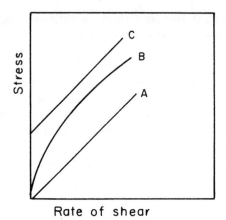

Figure 5-5. Plastic flow.

the slope is the coefficient of viscosity. Some liquids give curves like *B*, in which the viscosity, defined as the ratio of the ordinate to the abscissa, decreases as the stress increases. The slope of the curve is called the plasticity. Often the curve appears more like *C*. The plasticity is nearly constant down to very low shears, and the viscosity approaches infinity. The apparent intercept on the stress axis is called the yield point.

Plastic flow is very important in paints, which should flow easily under the large stress of the brush, but with difficulty under the small stress of gravity.

In many suspensions the viscosity is a function of the time of flow. In some cases the viscosity increases with time of flow to an asymptotic limit, whereas in other cases it decreases very rapidly to the asymptote. Sometimes there is a combination of the two behaviors, first a very rapid decrease to a minimum and then a rise to the asymptote. The rapid decrease is called thixotropy. It occurs with particles that are very slender needles or very thin plates, and is attributed to the breakdown of a sort of brush-pile structure to one in which the molecules are parallel and therefore interfere less with each other's flow. Thixotropy is a desirable property for paints, and is very important in the muds used to line drilled wells.

Very asymmetric small colloid molecules may be oriented by a shearing force and may show their orientation by double refraction, called double refraction of flow. A Couette-type viscometer with very small separation between the cylinders and very high rates of flow is used, with means of examining the fluid by a light beam between the cylinders and parallel to their axes.

6. Electric Effects at Surfaces and in Colloidal Solutions

6-1. Introduction

The discussion of electric potential is complicated by the facts that there are several possible levels of rigor and that it is convenient, and almost necessary, to shift frequently from one level to another. The first aspect is a general one in the use of thermodynamics in real systems. Rigorous thermodynamics requires a system at equilibrium, but we believe that in a system not very far from equilibrium there are a pressure, a temperature, a chemical potential, or activity, of each component, an electric potential, and so on, that are related to one another and to the concentrations of the components at that point as though the system were in equilibrium, and that do not depend upon the gradients of any of these quantities. This is the first axiom of the thermodynamics of irreversible processes. The second axiom is that for small gradients of a potential the corresponding flux is proportional to the gradient.

There is a special difficulty with electric potential because we cannot define exactly, even in principle, the electric potential difference between two different media, such as two metals, a metal and a solution, or even two aqueous solutions of different concentrations. The electric potential difference is the work of transferring unit charge of electricity without any transfer of matter, and it is not possible to separate electricity from matter. In rigorous thermodynamic treatments we consider the potential difference between two pieces of the same metal, usually the copper terminals of a potentiometer. So we can measure exactly only the electric potential difference of a complete cell, which we can relate to the changes of state that occur in the cell. We can often locate these changes of state at an electrode, a membrane, or a liquid junction, and it is sometimes convenient to talk of the potential difference as though it also were located at the same boundary as the corresponding change in state. This corresponds to considering the ions as components, although the chemical potential of a single ion species cannot have rigorous meaning.

Theoretical studies by the use of physical models usually involve the electric potential, but often this potential does not appear in the final results. However, many results in colloid chemistry are expressed in terms of the electrokinetic, or zeta, potential ζ, which is defined in terms of a model that is certainly inadequate. Many workers believe that ζ is no longer a useful concept. So much of the earlier work has been expressed in terms of ζ, however, that it is necessary to become familiar with the concept (see Sec. 6–6).

6–2. Electric Double Layers

When there is no electric current, a homogeneous phase will have a uniform electric potential ϕ. Two neighboring phases, however, will not have the same potential. The change in potential will not be infinitely steep, but at a plane surface may be like that represented in Fig. 6–1. Then the electric force is the negative of the potential gradient $- d\phi/dx$, and the electric density is $- (d^2\phi/dx^2)D/4\pi$, in which D is the dielectric constant. This leads to a positive charge on the left of the boundary and a negative charge on the right. These charges comprise an electric double layer. There must be such a layer if the electric potential changes at the boundary but remains constant in each phase.

Helmholtz, who first noted the necessity for a double layer, assumed that these charges may be considered as concentrated in parallel planes. Such a distribution is called a Helmholtz double layer. Gouy applied Boltzmann's distribution law as $c_i = c_{i0} \exp (\phi \epsilon z_i/kT)$, in which c_i is the concentration of ions of species i, with valence z_i, at a point where the potential is ϕ, and noted that this gives a diffuse exponential double layer, which is often called a Gouy double layer. Stern is usually credited with a compromise between the two types, such as would be obtained if mechanical forces, which should be included in the Boltzmann relation, keep all ions away from the region between the two planes. Fig. 6–1–b shows an infinite repulsion between the two planes, with a diffuse distribution outside. This is the type of distribution assumed by Debye and Hückel around a spherical ion.

6–3. Liquid-Junction Potentials. Mixture Boundaries

In the cell

$$\text{H}_2(p), \ \text{H}^+\text{Cl}^-(m), \ \text{Hg}_2\text{Cl}_2(s) + \text{Hg}(l),$$

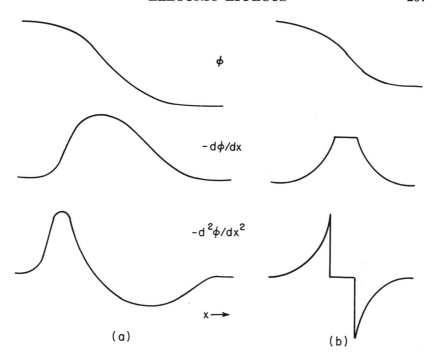

Figure 6–1. Electric double layers at a plane surface: (a) diffuse or Gouy double layer, (b) Stern double layer (charges excluded from central region). The three curves are the electric potential and its first and second derivatives in a direction normal to the surface.

the changes of state from the reactions at the two electrodes per faraday of electricity are

$$\tfrac{1}{2}H_2(p) = H^+(m) + \epsilon^-$$

and

$$\tfrac{1}{2}Hg_2Cl_2(s) + \epsilon^- = Cl^-(m) + Hg(l),$$

and the total change of state in the cell is

$$\tfrac{1}{2}H_2(p) + \tfrac{1}{2}Hg_2Cl_2(s) = H^+Cl^-(m) + Hg(l).$$

Except for the fact that the anode solution, saturated with hydrogen, is not exactly the same as the cathode solution, saturated with calomel, the potential difference of the whole cell may be calculated by rigorous thermodynamics. The potential of either electrode, however, requires a knowledge of a single-ion activity.

We define t_i, the transference number of species i, as the number of moles of species i transferred per faraday, $\mathfrak{F}$, in the direction of the positive current. Thus the transference number of an anion is negative, and a neutral species may have a transference number. The usual transport number T_i $[= t_i z_i]$ is positive for all ions and zero for all neutral species. However, it is possible to consider motion relative to any species, that is, to assume that the transference number of that species is zero. Usually the solvent is chosen as the stationary species. Then the electric potential difference E_L arising from this change of state is given by

$$- dE_L \, \mathfrak{F}/RT \;=\; \Sigma_i t_i d \ln a_i. \tag{6-1}$$

Since $\Sigma_i t_i z_i = 1$, we may add and subtract $(1/z_S)d \ln a_S$, if S is a standard ion, to give

$$\frac{-dE_L \mathfrak{F}}{RT} = \frac{1}{z_s} d \ln a_s + \Sigma_i t_i \!\left(d \ln a_i - \frac{z_i}{z_s} d \ln a_s \right)\!. \tag{6-2}$$

The terms in the summation in Eq. (6-2) involve no single-ion activities, and the term $(1/z_s)d \ln a_s$, integrated from solution α to solution β, combines with the terms for the two electrodes to eliminate all single-ion activities. Of course, diffusion is occurring across the boundary, so the first axiom of nonequilibrium thermodynamics is involved.

If there is a single solute, its concentration fixes both its activity and the transference numbers of the ions, so the electromotive force of a concentration cell with transference is fixed by the terminal concentrations and the electrodes, if only the temperature and pressure are constant through the boundary. If we have two cells, such as

E_I: Hg(l) + Hg$_2$Cl$_2(s)$ HCl(m_0), H$_2(p)$, HCl(m), Hg$_2$Cl$_2(s)$ + Hg(l),
E_II: Hg(l) + Hg$_2$Cl$_2(s)$ HCl(m_0), HCl(m), Hg$_2$Cl$_2(s)$ + Hg(l),

we have the relation

$$t_\mathrm{H} = dE_\mathrm{II}/dE_\mathrm{I}, \tag{6-3}$$

which may be used to calculate any one of the three quantities t_H, E_I, or E_II if the other two are known as functions of the concentration m.

If there are two or more solutes, it is necessary to know how the solutions are mixed in order to integrate Eq. (6-1) or (6-2). A mixture boundary or Henderson boundary is obtained by any mechanical mixing, whether by shear of a stopcock or other similar device, by flowing the two solutions together with laminar flow, or by turbulent flow of one liquid into the other. In the mixture boundary the concentration of component i at any point is given by $xm_{i\beta} + (1 - x)\, m_{i\alpha}$,

$dm_i = (m_{i\beta} - m_{i\alpha})dx$, and x varies from zero in α to unity in β. If the mobility u_i is given the same sign as t_i, we have

$$t_i = u_i m_i / \Sigma_j z_j u_j m_j,$$

and for ideal solutions

$$d \ln a_i = dm_i / m_i.$$

Then

$$\frac{-dE_L\mathfrak{F}}{RT} = \Sigma_i t_i d \ln a_i = \frac{\Sigma_i u_i dm_i}{\Sigma_j z_j u_j m_j} = \frac{\Sigma_i u_i (m_{i\beta} - m_{i\alpha})dx}{\Sigma_j z_j u_j m_{j\alpha} + x\Sigma_j z_j u_j (m_{j\beta} - m_{j\alpha})x}$$

(6–4)

and

$$\frac{-E_L\mathfrak{F}}{RT} = \frac{\Sigma_i u_i (m_{i\beta} - m_{i\alpha})}{\Sigma_j z_j u_j (m_{j\beta} - m_{j\alpha})} \ln \frac{\Sigma_j z_j u_j m_{j\beta}}{\Sigma_j z_j u_j m_{j\alpha}}.$$

(6–5)

The assumption that the solutions are ideal is not as drastic as it appears at first sight, for there is compensation between the assumption of constant mobility and that of constant activity coefficients, as well as compensation between the deviations for different ions. The assumption that, for each ion species i, $d(u_i \gamma_S / \gamma_i u_S) = 0$, with S any arbitrarily selected standard ion, is almost, though not quite, sufficient for Eq. (6–5) to hold.

Eq. (6–5) explains the use of the saturated potassium chloride bridge to reduce liquid-junction potentials. We note that $\Sigma_j z_j u_j m_j$ is proportional to the conductivity L, and that $|\Sigma_i u_i m_i|$ cannot be greater than $|\Sigma_j z_j u_j m_j|$. If the mobilities of potassium and chloride ion were exactly equal, $(\Sigma_i u_i m_i)_\beta$ would be zero, and we should have

$$\frac{E_L\mathfrak{F}}{RT} < \frac{L_\alpha / L_\beta}{1 - L_\alpha / L_\beta} \ln (L_\alpha / L_\beta),$$

which approaches zero as L_α / L_β approaches zero.

6–4. Planck Boundaries. Membranes

Max Planck calculated the electromotive force for a boundary at which the concentration in solution α is kept constant up to a plane A, that in solution β is kept constant down to another parallel plane B, and diffusion gives a steady-state gradient from A to B. His answer is an implicit expression in $\int_\alpha^\beta \Sigma_i t_i d \ln a_i$, which looks very different from Henderson's expression but gives identical values for the potential difference in some important simple cases, and usually gives values not very different from Henderson's simpler expression, Eq. (6–5).

In practice the Planck boundary requires a membrane, which should be considered another component. It is usually simpler to consider motion relative to the membrane, so that u_M is zero and the mobility of the water u_W is not zero but the negative of the mobility the membrane would have if the water were fixed. With uncharged membranes and holes large relative to the ions, this makes very little difference, but if the surface of the pores is charged by ionization of the walls or by adsorption of ions from the solution, the effect may be very large.

A typical ion-exchange resin is a sulfonated polystyrene, in which there are six equivalents of sulfonate ions per kilogram of water, or three equivalents per liter of wet resin. Electric neutrality requires that the concentration of the mobile cation in such a resin be equal to the sum of the concentrations of the fixed sulfonate ion and the mobile anion. The Donnan equilibrium (Sec. 5–3) is approximately satisfied. So, if the solutions outside are less than one molal with only one cation species, the concentration of mobile anion inside the resin at equilibrium or in the steady state is so small that the transference number of the cation is nearly unity and the membrane between solutions α and β gives almost the same potential as a pair of electrodes reversible to the cation, one in each solution. An anion exchanger under these conditions behaves as a pair of electrodes reversible to the anion. There is no oxidation or reduction at these membrane "electrodes."

The ion-exchanger membranes show but little selectivity among different ions of the same sign, except that they screen out colloidal ions. A glass membrane, however, is so much more easily permeable to protons than to any other ions that it serves as a very satisfactory pair of hydrogen electrodes until, at very high ratios of alkali ion to hydrogen ion, alkali ions are also transferred.

The equilibrium across a semipermeable membrane with colloid on one side applies only to neutral components, and not to the ions. This means that there must be an electric potential difference across the membrane, which is called the Donnan potential, E_D. In the ideal case it is given by

$$E_D \, \mathfrak{F}/RT = (1/z_i) \ln c_{i\alpha}/c_{i\beta}. \tag{6–6}$$

The Donnan equilibrium for ideal solutions demands that this be the same for any ion type.

6–5. Electric Endosmosis and Streaming Potentials

The customary electric potential measurements described above are made by imposing very small electric currents upon a system in a steady state of flow without current or pressure difference across the membrane, and interpolating to find the potential difference that gives no current. There are several other interesting steady-state relations between the flux of electricity, or electric current, I, the flux of matter, treated as a volume flux, J, the electric potential difference ΔE, the pressure difference ΔP, and the differences in chemical potential due to differences in chemical composition, $\Delta \bar{G}_i$, ..., which correspond to our use of $RT\, d \ln a_i$. We shall consider only the simple case in which ΔT and every $\Delta \bar{G}_i$ are zero, and abbreviate ΔE and ΔP to E and P. Then the fundamental relations are

$$I = L_E E \ + L_{EP} P, \qquad (6\text{--}7)$$

$$J = L_{EP} E + L_P P. \qquad (6\text{--}8)$$

The fact that L_{EP} is the same in the two equations is one of the Onsager reciprocal relations, which are the fundamental laws of the thermodynamics of irreversible processes. If we use subscripts to indicate which of these quantities are zero, we have from Eqs. (6–7) and (6–8) 24 relations of the type $(I/E)_P = L_E$, of which 12 are reciprocals of the other 12, such as $(I/E)_P$ and $(E/I)_P$, and 4 of the pairs are equal to 4 others by the reciprocal relations. It is convenient to group them in pairs that differ only in subscripts:

$$(I/E)_P = L_E, \qquad (6\text{--}9)$$

$$(I/E)_J = L_E - L_{EP}^2/L_P; \qquad (6\text{--}10)$$

$$(J/P)_E = L_P, \qquad (6\text{--}11)$$

$$(J/P)_I = L_P - L_{EP}^2/L_E; \qquad (6\text{--}12)$$

$$(I/P)_E = (J/E)_P = L_{EP}, \qquad (6\text{--}13)$$

$$(I/P)_J = (J/E)_I = L_{EP} - L_E L_P/L_{EP}; \qquad (6\text{--}14)$$

$$(E/P)_I = -(J/I)_P = -L_{EP}/L_E, \qquad (6\text{--}15)$$

$$(E/P)_J = -(J/I)_E = -L_P/L_{EP}. \qquad (6\text{--}16)$$

Obviously only three of these relations are independent, since they are all expressed in terms of the three coefficients L_E, L_P, and L_{EP}. These may be measured most directly by Eqs. (6–9), (6–11), and (6–13) as the electric conductivity at zero pressure difference, the permeability at zero electric potential difference, and the electric permeability at zero pressure difference. It may be more convenient, however, to measure the electric conductivity at zero mass flow, the permeability at zero electric current, the electric permeability at zero current or its reciprocal the streaming potential, as in Eqs. (6–10), (6–12), or (6–14), or to measure the endosmotic ratio $(J/I)_P$ in Eq. (6–15) or any one of the remaining ratios.

Electric endosmosis is the flow of matter through a membrane caused by a potential difference. It is usually measured as $(J/I)_P$. Electrophoresis is the related flow of colloid particles through a liquid caused by a potential gradient.

6-6.   The Electrokinetic, or Zeta, Potential

Consider two parallel surfaces separated by a small distance δ. The surfaces have zero curvature in the x direction, and the other principal radius of curvature is everywhere large relative to δ. One of the surfaces has a positive charge of uniform density ρ, and the other an equal negative charge. The rest of the system is electrically neutral, so the two surfaces constitute a Helmholtz double layer and the electric potential between them is ζ. The space between the surfaces is filled with a fluid with viscosity η and dielectric constant D. If an electric potential is applied in the x direction, the surfaces will move relative to each other and all the volume behind a surface will move at the same velocity as the surface, so that the velocity gradient is concentrated in the region between the surfaces.

The electric force per unit area of either surface is $\rho\, dE/dx$ and the viscous force per unit area is $v\eta/\delta$, if v is the relative velocity. In the steady state,

$$v = (\rho\delta/\eta)\, dE/dx. \tag{6-17}$$

In general, δ will not be known, so we eliminate it by the equation for a parallel-plate condenser, $\delta\rho = D\zeta/4\pi$, to give

$$v = (D\zeta/4\pi\eta)\, dE/dx. \tag{6-18}$$

If the surfaces are closed to give a cylinder of solid surrounded by a continuous liquid, v is the electrophoretic velocity of the solid. If the surfaces are closed to give a cylinder of liquid surrounded by solid, v is the velocity of endosmosis. It has been found that for flow along parallel walls the assumption of a Helmholtz double layer is unnecessary if ζ is defined as the potential difference between the surface of shear and the liquid a large distance away.

In endosmosis it is usually not possible to determine the velocity v, but we determine instead the material flux $J = vA$, in which A is the cross-sectional area of liquid. We eliminate A by determining the steady-state current $I = \gamma A\, dE/dx$, in which γ is the conductivity, to give

$$(J/I)_P = D\zeta/4\pi\eta\gamma. \tag{6-19}$$

It is generally accepted that the assumption of zero curvature in the x direction is unnecessary but that Eq. (6–19) is valid for any membrane, and that the chief difficulty is the assumption that the conductance, and particularly the dielectric constant and the viscosity, are the same as in the bulk fluid. The conductance should be an average over the cross section of the pore, but the dielectric constant and the viscosity are particularly important in the vicinity of the surface of shear.

Smoluchowski, who first applied Eq. (6–18) to electrophoresis, believed that here also it is independent of the shape of the particles. However, the force on an isolated sphere in a nonconducting medium is $\epsilon dE/dx$ if ϵ is the charge, the potential at the surface of radius a is $\zeta = \epsilon/Da$, and the resistance according to Stokes's law is $6\pi\eta va$. Equating gives

$$v = (D\zeta/6\pi\eta)\, dE/dx, \tag{6–20}$$

which is only two-thirds the value given by Eq. (6–18). In this case the other half of the double layer is at an infinite distance. The best modern theory is that of Henry, who finds that each of these theories is true within limits and that the transition depends upon κa, if κ is the Debye function

$$\kappa^2 = 4\pi N_A^2 \epsilon^2 \Sigma_i c_i z_i^2 / 1000\, RTD, \tag{6–21}$$

in which N_A is Avogadro's number, ϵ the protonic charge, c_i (mol/lit) the concentration of species i, and z_i its valence (see Sec. 6–7). The reciprocal $1/\kappa$ is a measure of the thickness of the ion atmosphere. For water at room temperature, $\kappa \cong (\Sigma_i c_i z_i^2/2)^{1/2}/3 \times 10^{-8}$ cm^{-1}. If the ratio of v to that given by Eq. (6–21) is plotted against $\log \kappa a$, the curve looks like the curve for ϕ in Fig. 6–2, with the right-hand asymptote at 1, the left-hand asymptote at 2/3, and the value 5/6 at $\log \kappa a = 1$. The value 2/3 is correct to within 1 percent if κa is less than 1/2, and the value 1 is correct to within 1 percent if κa is greater than 300.

6–7. The Ion Atmosphere

The Debye-Hückel treatment of the relation between the electric potential ϕ and the distribution of ions is based on the combination of Boltzmann's law,

$$n_i = n_{i0}e^{-\phi\epsilon z_i/kT}, \tag{6–22}$$

in which n_{i0} is the number of ions per cubic centimeter, and Poisson's equation,

$$\nabla^2\phi = \frac{\partial^2\phi}{\partial x^2} + \frac{\partial^2\phi}{\partial y^2} + \frac{\partial^2\phi}{\partial z^2} = -\frac{4\pi\epsilon}{D}\Sigma_i n_i z_i, \tag{6–23}$$

to give

$$\nabla^2\phi = (-4\pi\epsilon\,\Sigma_i n_{i0}z_i/D)e^{-\phi\epsilon z_i/kT}. \tag{6–24}$$

This equation cannot be solved analytically, so Debye and Hückel expand the exponential to give

$$\nabla^2\phi = -4\pi\epsilon\Sigma_i n_{i0}z_i/D + (4\pi\epsilon^2\Sigma_i n_{i0}z_i^2/DkT)\,\phi + \cdots \cong \kappa^2\phi.$$
(6–25)

The first term is zero because the solution as a whole must be electrically neutral, and the coefficient of ϕ in the second term becomes identical with κ^2 of Eq. (6–21) when the change is made from molecules per cubic centimeter to moles per liter and from k to R.

For the one-dimensional plane-surface case, the solution is

$$\phi = ce^{-\kappa x} + c'e^{\kappa x}.$$
(6–26)

In the case of a spherically symmetric distribution, with r the distance from the center, the solution is

$$\phi = ce^{-\kappa r}/r + c'e^{+\kappa r}/r.$$
(6–27)

If there is an ion of valence z_k that repels other ions as a rigid sphere, so that only coulombic forces are operative when $r > a$ but there is infinite repulsion when $r \leq a$, the solution becomes

$$\phi = \frac{\epsilon z_k e^{\kappa-(r-a)}}{D(1 + \kappa a)r},$$
(6–28)

$$\Sigma_i n_i z_i = -\frac{z_k \epsilon^2 \Sigma_i n_{i0}z_i^2 e^{\kappa a}}{kTD(1 + \kappa a)}\,\frac{e^{-\kappa r}}{r} = -\frac{z_k \kappa^2}{4\pi}\,\frac{e^{\kappa a}}{1 + \kappa a}\,\frac{e^{-\kappa r}}{r}.$$
(6–29)

The number of ions in a spherical shell between r and $r + dr$, divided by κdr, is

$$4\pi r^2 \Sigma_i n_i z_i = -\frac{z_k e^{\kappa a}}{1 + \kappa a}\,\kappa r e^{-\kappa r},$$
(6–30)

and the number at a distance greater than r is $-z_k e^{\kappa a}(1 + \kappa r)e^{-\kappa r}/(1 + \kappa a)$. Fig. 6–2 shows $\kappa r e^{-\kappa r}$ and $(1 + \kappa r)e^{-\kappa r}$ plotted against κr. To obtain the fraction a of the charge around an ion of finite size, it is necessary to normalize by dividing each ordinate by that of curve B at $\kappa r = \kappa a$. Curve A gives the best justification for considering $1/\kappa$ as the "thickness of the ion atmosphere," for more of the atmosphere is at this distance than at any other. However, there is more than 70 percent of the atmosphere at a greater distance.

In electrophoresis it is usually assumed that $\zeta = \phi_{r=a} = ez_k/Da$ $(1 + \kappa a)$. Thus electrophoresis gives a method of separating colloids approximately according to the ratio of charge to radius, although there are complications for nonspherical molecules. Tiselius revolu-

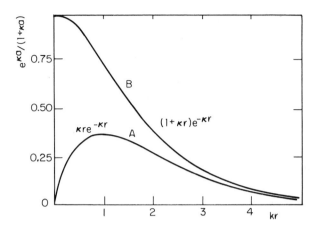

Figure 6–2. Distribution of charge about a central ion: (A) fraction of charge of ion atmosphere between (κr and $\kappa r + d\kappa r$) divided by $d\kappa r$; (B) fraction of charge outside κr. The ordinate of either curve is to be divided by the ordinate of curve B at $\kappa r = \kappa a$ to normalize.

tionized the method by using cells of rectangular, instead of circular, cross section, and by operating at 4°C, which is near the temperature of maximum density, to avoid stirring by convection due to heating from the electric current. It is customary to work in buffer solutions to fix the charge and the potential gradient. The boundaries are determined as for diffusion or sedimentation.

The quantity ϵ^2/DkT is about 7×10^{-8} cm for water at room temperature. So it appears that neglect of the higher terms in the expansion of the Boltzmann integral is justified only when the ratio $z_k z_i / a$ is small. There have been many attempts to improve the original treatment. One of the first and most enlightening is that of Bjerrum, who applied the Boltzmann distribution law at zero concentration, in the range where curve A of Fig. (6–12) is a straight line, to give

$$4\pi r^2 n_i = 4\pi r^2 n_{i0} e^{-z_k z_i \epsilon^2 / DkTr}. \tag{6–31}$$

The curves for $z_k z_i = 0$, and for $z_k z_i$ positive and negative are shown in Fig. 6–3, together with the last two calculated by the Debye-Hückel approximation. The minimum in the curve for $z_k z_i$ negative is at $r = z_k z_i \epsilon^2 / DkT = a'$, and the abscissas are given as r/a'. Bjerrum notes that the minimum is so low and so flat that there will be very few ions near $r = a'$, and that the mutual energy of attraction is $2kT$ at this distance, so that the ions within the minimum may be considered as bound to the central particle; those outside are treated by

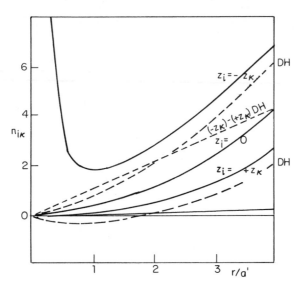

Figure 6–3. Distribution of charge close to central ion. The broken curves give the Debye-Hückel approximations.

the Debye-Hückel approximation. Bjerrum calculates an electrostatic association constant, which may be approximately expressed as

$$K = 0.001 \, N_A (\epsilon^2 z_k z_i / DkT)^3 \exp[(\epsilon^2 z_k z_i / DkTa) - 1]. \quad (6\text{--}32)$$

We should not expect very good quantitative agreement, for most of these distances of approach are so small that the repulsion should not be considered that of rigid spheres and the dielectric constant should not be taken as equal to that of the solvent. One of the most important results is the rather surprising one that the correction to the simple Debye theory is small until a/a' becomes less than one-half, that is, the distance of closest approach less than one-quarter, of the distance at which the Debye-Hückel approximation gives negative values of the concentration when $z_k z_i$ is positive. We note also, however, that when this correction is important the determining factor is no longer the ionic strength but only the concentration of ions of the opposite charge.

6–8.   Lyophobic Colloids

Colloids are sometimes classed as lyophilic if they dissolve or disperse readily and lyophobic if they do not. Usually lyophobic colloids are unstable unless they carry a rather large charge so that the particles are kept apart by electrostatic repulsion. These colloids are extremely sensitive to electrolytes, particularly those having polyvalent ions of sign opposite to those of the colloid, which should be tightly bound in accordance with Eq. (6–32). Precipitation by electrolytes is called coagulation. An early empirical generalization was that coagulation is produced by bivalent ions at one-thirtieth of the equivalent concentration necessary for univalent ions, and trivalent ions require a concentration only one-thirtieth that for bivalent ions. For other colloids the ratios are somewhat different, however, and there are specific effects of ions of the same charge.

A slight excess of trivalent or tetravalent ions may redisperse a coagulated lyophobic colloid. This is doubtless related to the fact that these ions may reverse the direction of endosmotic flow. The most probable explanation is that enough high-valence mobile ions are firmly attached to univalent fixed charges to change the sign of the surface charge.

A lyophilic colloid is usually heavily solvated. If the surface of a lyophobic colloid is covered by a lyophilic colloid, the behavior to surrounding molecules is that of the surface. For example, argyrol is a concentrated dispersion of silver stabilized by an organic lyophilic colloid, which is used medicinally.

6–9. Electrocapillary Effects

If two mercury reservoirs are connected electrically and if mercury drops from one to the other through a solution of a mercurous salt, there is a positive current in the wire from the lower to the upper reservoir. If the outside circuit is broken, the current cannot continue, but the mercurous salt is concentrated at the lower reservoir. These effects may be explained by the adsorption of mercurous ions at the expanding surface as the drops form and their release as the drops coalesce, giving a higher concentration at the bottom and thus a concentration cell. If the electric circuit is closed, the ions will be formed from the mercury at the upper surface and will form mercury at the lower surface. If the circuit is open, the anions will be carried down by electrostatic attraction.

The adsorption of ions at a mercury surface is also responsible for the action of the capillary electrometer, which was used at the beginning of the century to determine the null point in potentiometric measurements. The capillary electrometer is a cell consisting of a large mercury anode and a thread of mercury in a glass capillary as cathode, with an electrolyte whose anion combines with the mercurous ion to form an insoluble salt, and with excess of that salt at the anode. When a potential is imposed, the adsorption of ions at the cathode is changed. This changes the interfacial tension, and therefore the level of mercury in the capillary changes. When the electrode is short-circuited, the mercury returns to its original position. If the potential is balanced, so that there is no difference in potential when the circuit is closed, there will be no motion of the surface.

The concentration of mercurous ions in the capillary is always small, so that they contribute but little to the transference, but they alone are electrolyzed. As current passes, all those near the cathode are electrolyzed, the electrode becomes polarized, and the current stops. The surface tension in this static condition can be measured for each value of the imposed electromotive force. The curve so determined passes through a maximum and is nearly parabolic. A parabola would have the equation

$$\sigma_0 - \sigma = k(E - E_0)^2, \tag{6-33}$$

with a maximum $\sigma = \sigma_0$ when $E = E_0$.

At the anode the current density is very much less than at the cathode, and the solid mercury salt prevents much change in concentration, so practically all the change is at the cathode. When the potential drop across the cathode is zero, the electric double layer has disappeared, and the orientation of the double layer must be reversed as the potential passes through zero. When the double layer has disappeared, the ions must be adsorbed in equivalent amounts. In the classical treatment of the capillary electrometer, it is assumed that this adsorption is zero and that the disappearance of the double layer corresponds to the maximum in the surface tension, so that E_0 is the potential of the rest of the cell. Under these conditions the potential drop at the dropping mercury electrode would also be zero. This would then give two methods of measuring the absolute values of single-electrode potentials. Different solutions give different values for the potentials. If they are approximately correct, the absolute value of the molal hydrogen electrode at 25°C is about -0.2 V.

This is one of the aspects of electrosurface chemistry that have been revived in recent years and studied from the standpoint of the thermodynamics of irreversible processes much more carefully than in the early days. We can only indicate the direction that this work is taking. Another very important field is the extension of the theories of Debye to highly charged colloids of irregular and variable shapes.

6–10. Dielectric Properties

In Sec. 1–4 we discussed the contributions to the dielectric constant of the oscillation of electrons and of the rotation of permanent dipoles, and we noted that the two could be distinguished because the first is the square of the refractive index at high frequencies and the second is the difference between the squares of the refractive index at frequencies so low that the molecule can follow the field and that at high frequencies. If we plot the square of the refractive index against the logarithm of the frequency, we obtain a curve that looks again like the upper curve in Fig. 6–2. The inflection occurs at the characteristic frequency. If there are two or more characteristic frequencies, they are sometimes resolved to give a curve with steps, or they may give an unresolved curve stretched out over a larger frequency range. The reciprocal of the characteristic frequency, or the relaxation time, of liquid water is about 2×10^{-11} sec. The relaxation times of proteins in water solutions are 1,000 to 10,000 times this value. At intermediate frequencies, the proteins lower the dielectric constants of aqueous solutions; at lower frequencies, they raise them. The characteristic frequencies and the change of dielectric constant at each step help to determine the shape of the colloidal particles and the distribution of charges within them.

Bibliography of Publications
by George Scatchard

Index

Bibliography of Publications
by George Scatchard

1. M. T. Bogert and G. Scatchard, "Researches on Quinazolines. XXXIII. A New and Sensitive Indicator for Acidimetry and Alkalimetry, and for the Determination of Hydrogen-Ion Concentrations Between the Limits of 6 and 8 on the Sorensen Scale," *J. Am. Chem. Soc. 38*, 1606–1615 (1916).
2. G. Scatchard, "2-Uraminobenzoic Acid, Benzoylene Urea and Some of Their Derivatives: and the Use of Dinitrobenzoylene Urea as an Indicator." Dissertation (Eschenbach Printing Co., Easton, Pa. 1916).
3. A. Smith, H. Eastlack, and G. Scatchard, "The Transition of Dry Ammonium Chloride," *J. Am. Chem. Soc. 41*, 1961–1969 (1919).
4. M. Bogert and G. Scatchard, "Researches on Quinazolines. XXXIV. The Synthesis of Certain Nitro and Amino Benzoylene Ureas and Some Compounds Related Thereto," *J. Am. Chem. Soc. 41*, 2052–2068 (1919).
5. V. Grignard, G. Rivat, and G. Scatchard, "Sur le Sulfure d'Éthyle β-β'-Bi-iodé et son Application à la Détection et au Dosage de l'Ypérite," *Ann. Chim., 9ᵉ série, 15*, 5–18 (1920).
6. G. Scatchard, "The Speed of Reaction in Concentrated Solutions and the Mechanism of the Inversion of Sucrose. I," *J. Am. Chem. Soc. 43*, 2387–2406 (1921).
7. G. Scatchard, "The Hydration of Sucrose in Water Solution as Calculated from Vapor-Pressure Measurements," *J. Am. Chem. Soc. 43*, 2406–2418 (1921).
8. G. Scatchard, "The Speed of Reaction in Concentrated Solution and the Mechanism of the Inversion of Sucrose. II," *J. Am. Chem. Soc. 45*, 1580–1592 (1923).
9. G. Scatchard, "Electromotive-Force Measurements with a Saturated Potassium Chloride Bridge or with Concentration Cells with a Liquid Junction," *J. Am. Chem. Soc. 45*, 1716–1723 (1923).
10. G. Scatchard, "The Influence of Gelatin on Transference Numbers," *J. Am. Chem. Soc. 46*, 2353–2357 (1924).

11. G. Scatchard, "The Activities of Strong Electrolytes. I. The Activity of Hydrochloric Acid Derived from the Electromotive Force of Hydrogen-Silver-Chloride Cells," *J. Am. Chem. Soc. 47*, 641–648 (1925).

12. G. Scatchard, "The Activities of Strong Electrolytes. II. A Revision of the Activity Coefficients of Potassium, Sodium, and Lithium Chlorides, and Potassium Hydroxide," *J. Am. Chem. Soc. 47*, 648–661 (1925).

13. G. Scatchard, "The Activities of Strong Electrolytes. III. The Use of the Flowing Junction to Study the Liquid-Junction Potential between Dilute Hydrochloric Acid and Saturated Potassium Chloride Solutions; and the Revision of Some Single-Electrode Potentials," *J. Am. Chem. Soc. 47*, 696–709 (1925).

14. G. Scatchard, "The Activity of Strong Electrolytes. IV. The Application of the Debye-Hückel Equation to Alcoholic Solutions," *J. Am. Chem. Soc. 47*, 2098–2111 (1925).

15. G. Scatchard, "Electromotive-Force Measurements in Aqueous Solutions of Hydrochloric Acid Containing Sucrose," *J. Am. Chem. Soc. 48*, 2026–2035 (1926).

16. G. Scatchard, "The Unimolecularity of the Inversion Process," *J. Am. Chem. Soc. 48*, 2259–2263 (1926).

17. G. Scatchard, "The Milner and Debye Theories of Strong Electrolytes," *Phil. Mag. 2*, 577–586 (1926).

18. G. Scatchard, "A Revision of Some Activities in Water-Alcohol Mixtures," *J. Am. Chem. Soc. 49*, 217–218 (1927).

19. G. Scatchard, "The Interaction of Electrolytes with Non-electrolytes," *Chem. Rev. 3*, 383–402 (1927).

20. G. Scatchard, "Mixed Solutions of Electrolytes and Non-electrolytes," *Trans. Faraday Soc. 23*, 454–462 (1927).

21. J. A. Beattie, L. J. Gillespie, G. Scatchard, W. C. Schumb, and R. F. Tefft, "Density (Specific Gravity) and Thermal Expansion (under Atmospheric Pressure) of Aqueous Solutions of Inorganic Substances and of Strong Electrolytes," *International Critical Tables* McGraw-Hill, New York, vol. 3, p. 51 (1928).

22. G. Scatchard, "The Moore Laboratory of Chemistry, Amherst College," *Nucleus* (Dec. 1929).

23. G. Scatchard, "The Rate of Reaction in a Changing Environment," *J. Am. Chem. Soc. 52*, 52–61 (1930).

24. G. Scatchard and R. F. Tefft, "Some Electromotive Force Measurements with Calcium Chloride Solutions," *J. Am. Chem. Soc. 52*, 2265–2271 (1930).

25. G. Scatchard and R. F. Tefft, "Electromotive Force Measurements on Cells Containing Zinc Chloride. The Activity Coefficients of the Chlorides of the Bivalent Metals," *J. Am. Chem. Soc. 52*, 2272–2281 (1930).

26. G. Scatchard, "Note on the Equation of State Explicit in the Volume," *Proc. Nat. Acad. Sci. 16*, 811–813 (1930).

27. G. Scatchard and T. F. Buehrer, "An Effect of the Breadth of Junction on the Electromotive Force of a Simple Concentration Cell," *J. Am. Chem. Soc. 53*, 574–578 (1931).

28. G. Scatchard, "Equilibria in Non-electrolyte Solutions in Relation to the Vapor Pressures and Densities of the Components," *Chem. Rev. 8*, 321–333 (1931).

29. G. Scatchard, "Thermal Expansion and the Debye-Hückel Heat of Dilution," *J. Am. Chem. Soc. 53*, 2037–2039 (1931).

30. G. Scatchard, "Interatomic Forces in Binary Alloys," *J. Am. Chem. Soc. 53*, 3186–3188 (1931).

31. G. Scatchard, "Die Anwendung der Debyeschen Elektrolyttheorie auf konzentrierte Lösungen," *Physik. Z. 33*, 22–32 (1932).

32. G. Scatchard, "Statistical Mechanics and Reaction Rates in Liquid Solutions," *Chem. Rev. 10*, 229–240 (1932).

33. G. Scatchard and J. G. Kirkwood, "Das Verhalten von Zwitterionen und von mehrwertigen Ionen mit weit entfernten Ladungen in Elektrolytlösungen," *Physik. Z. 33*, 297–300 (1932).

34. G. Scatchard, P. T. Jones, and S. S. Prentiss, "The Freezing Points of Aqueous Solutions. I. A Freezing Point Apparatus," *J. Am. Chem. Soc. 54*, 2676–2690 (1932).

35. G. Scatchard, P. T. Jones, and S. S. Prentiss, "The Freezing Points of Aqueous Solutions. II. Potassium, Sodium, and Lithium Nitrates," *J. Am. Chem. Soc. 54*, 2690–2695 (1932).

36. G. Scatchard and S. S. Prentiss, "The Freezing Points of Aqueous Solutions. III. Ammonium Chloride, Bromide, Iodide, Nitrate, and Sulfate," *J. Am. Chem. Soc. 54*, 2696–2705 (1932).

37. G. Scatchard, "The Effect of the Forces Between Solvent Molecules on the Properties of Electrolyte Solutions," Communication from the Research Laboratory of Physical Chemistry, Massachusetts Institute of Technology, No. 276, *Chemistry at the Centenary Meeting of the British Association, 1931*, 70–72 (1932).

38. G. Scatchard, "Ligevaegt i Blandiger af Ikke-Elektrolyter" (Abstract), *Kemisk Maanedsblad* (Copenhagen) *13*, 77–78 (1932).

39. G. Scatchard, "The Coming of Age of the Interionic-Attraction Theory," *Chem. Rev. 13*, 7–27 (1933).

40. G. Scatchard and S. S. Prentiss, "An Objective Study of Dilute Aqueous Solutions of Uni-univalent Electrolytes," *Chem. Rev. 13*, 139–146 (1933).

41. G. Scatchard and S. S. Prentiss, "The Freezing Points of Aqueous Solutions. IV. Potassium, Sodium and Lithium Chlorides and Bromides," *J. Am. Chem. Soc. 55*, 4355–4362 (1933).

42. G. Scatchard, S. S. Prentiss, and P. T. Jones, "The Freezing Points of Aqueous Solutions. V. Potassium, Sodium and Lithium Chlorates and Perchlorates," *J. Am. Chem. Soc. 56*, 805–807 (1934).

43. G. Scatchard and S. S. Prentiss, "The Freezing Points of Aqueous Solutions. VI. Potassium, Sodium and Lithium Formates and Acetates," *J. Am. Chem. Soc. 56*, 807–811 (1934).

44. G. Scatchard, "Non-electrolyte Solutions," *J. Am. Chem. Soc. 56*, 995–996 (1934).

45. G. Scatchard and S. S. Prentiss, "Freezing Points of Aqueous Solutions. VII. Ethyl Alcohol, Glycine and Their Mixtures," *J. Am. Chem. Soc. 56*, 1486–1492 (1934).

46. G. Scatchard and S. S. Prentiss, "Freezing Points of Aqueous Solutions. VIII. Mixtures of Sodium Chloride with Glycine and Ethyl Alcohol," *J. Am. Chem. Soc. 56*, 2314–2319 (1934).

47. G. Scatchard and S. S. Prentiss, "The Freezing Points of Aqueous Solutions. IX. Mixtures of the Reciprocal Salt Pair: Potassium Nitrate–Lithium Chloride," *J. Am. Chem. Soc. 56*, 2320–2326 (1934).

48. G. Scatchard and W. J. Hamer, "The Application of Equations for the Chemical Potentials to Partially Miscible Solutions," *J. Am. Chem. Soc. 57*, 1805–1809 (1935).

49. G. Scatchard and W. J. Hamer, "The Application of Equations for the Chemical Potentials to Equilibria Between Solid Solution and Liquid Solution," *J. Am. Chem. Soc. 57*, 1809–1811 (1935).

50. G. Scatchard and Marjorie A. Benedict, "The Freezing Points of Aqueous Solutions. X. Dioxane and Its Mixtures with Lithium, Sodium and Potassium Chlorides," *J. Am. Chem. Soc. 58*, 837–842 (1936).

51. G. Scatchard, "Concentrated Solutions of Strong Electrolytes," *Chem. Rev. 19*, 309–327 (1936).

52. G. Scatchard, "Change of Volume on Mixing and the Equations for Non-electrolyte Mixtures," *Trans. Faraday Soc. 33*, 160–166 (1937).

53. G. Scatchard, review of *Ions in Solution* by R. W. Gurney, *J. Am. Chem. Soc. 59*, 2751 (1937).

54. G. Scatchard, C. L. Raymond, and H. H. Gilmann, "Vapor-Liquid Equilibrium. I. Apparatus for the Study of Systems with Volatile Components," *J. Am. Chem. Soc. 60*, 1275–1278 (1938).

55. G. Scatchard and C. L. Raymond, "Vapor-Liquid Equilibrium. II. Chloroform-Ethanol Mixtures at 35, 45, and 55°," *J. Am. Chem. Soc. 60*, 1278–1287 (1938).

56. G. Scatchard, W. J. Hamer, and S. E. Wood, "Isotonic Solutions. I. The Chemical Potential of Water in Aqueous Solutions of Sodium Chloride, Potassium Chloride, Sulfuric Acid, Sucrose, Urea and Glycerol at 25°," *J. Am. Chem. Soc. 60*, 3061–3070 (1938).

57. G. Scatchard, Symposium on Intermolecular Action: introduction; discussion, *J. Phys. Chem. 43*, 1–3, 281–296 (1939).

58. G. Scatchard, S. E. Wood, and J. M. Mochel, "Vapor-Liquid Equilibrium. III. Benzene-Cyclohexane Mixtures," *J. Phys. Chem. 43*, 119–130 (1939).

59. G. Scatchard, "The Nature of the Critical Complex and the Effect of Changing Medium on the Rate of Reaction," *J. Chem. Phys. 7*, 657–663 (1939).

60. G. Scatchard, S. E. Wood, and J. M. Mochel, "Vapor-Liquid Equilib-

rium. IV. Carbon Tetrachloride–Cyclohexane Mixtures," *J. Am. Chem. Soc. 61*, 3206–3210 (1939).

61. G. Scatchard, S. E. Wood, and J. M. Mochel, "Vapor-Liquid Equilibrium. V. Carbon Tetrachloride–Benzene Mixtures," *J. Am. Chem. Soc. 62*, 712–716 (1940).

62. G. Scatchard, "The Effect of Solvents on Reaction Rates" (discussion of paper by Laidler and Eyring) *Ann. N. Y. Acad. Sci. 39*, 341–344 (1940).

63. G. Scatchard, "The Calculation of the Compositions of Phases in Equilibrium," *J. Am. Chem. Soc. 62*, 2426–2429 (1940).

64. G. Scatchard, "The Sorting of Mixed Solvents by Ions," *J. Chem. Phys. 9*, 34–41 (1941).

65. G. Scatchard and L. F. Epstein, "The Calculation of the Thermodynamic Properties and the Association of Electrolyte Solutions," *Chem. Rev. 30*, 211–223 (1942).

66. G. Scatchard, "Equilibrium Thermodynamics and Biological Chemistry," *Science, 95*, 27–32 (1942).

67. G. Scatchard, review of *Elementary Physical Chemistry* by H. S. Taylor and H. A. Taylor, 3rd ed., *Chem. Eng. News 21*, 196, 198 (1943).

68. G. Scatchard, "Constants of the Debye-Hückel Theory," *J. Am. Chem. Soc. 65*, 1249 (1943).

69. G. Scatchard, sections in *Proteins, Amino Acids and Peptides* by E. J. Cohn and J. T. Edsall (Reinhold, New York, 1943): chap. 3, "Thermodynamics and Simple Electrostatic Theory"; chap. 8 with J. T. Edsall, "Solubility of Amino Acids, Peptides and Related Substances in Water and Organic Solvents"; chap. 24, sec. 10, "Interactions in Protein Solutions Calculated from Electromotive Force and Osmotic Measurements."

70. G. Scatchard, S. T. Gibson, L. M. Woodruff, A. C. Batchelder, and A. Brown. "Chemical, Clinical, and Immunological Studies on the Products of Human Plasma Fractionation. IV. A Study of the Thermal Stability of Human Serum Albumin," *J. Clin. Invest. 23*, 445–453 (1944).

71. G. Scatchard, A. C. Batchelder, and A. Brown, "Chemical, Clinical, and Immunological Studies on the Products of Human Plasma Fractionation. VI. The Osmotic Pressure of Plasma and of Serum Albumin," *J. Clin. Invest. 23*, 458–464 (1944).

72. G. Scatchard, J. L. Oncley, J. W. Williams, and A. Brown, "Size Distribution in Gelatin Solutions. Preliminary Report," *J. Am. Chem. Soc. 66*, 1980–1981 (1944).

73. G. Scatchard, review of *The Physical Chemistry of Electrolytic Solutions* by H. S. Harned and B. B. Owen, *J. Am. Chem. Soc. 66*, 1043 (1944).

74. G. Scatchard, "Louis John Gillespie (1886–1941)," *Proc. Am. Acad. Arts Sci. 75*, 164 (1944).

75. G. Scatchard, L. E. Strong, W. L. Hughes, Jr., J. N. Ashworth, and

A. H. Sparrow, "Chemical, Clinical, and Immunological Studies on the Products of Human Plasma Fractionation. XXVI. The Properties of Solutions of Human Serum Albumin of Low Salt Content," *J. Clin. Invest. 24*, 671–679 (1945).

76. G. Scatchard, S. E. Wood, and J. M. Mochel, "Vapor-Liquid Equilibrium. VI. Benzene–Methanol Mixtures," *J. Am. Chem. Soc. 68*, 1957–1960 (1946).

77. G. Scatchard, S. E. Wood, and J. M. Mochel, "Vapor-Liquid Equilibrium. VII. Carbon Tetrachloride–Methanol Mixtures," *J. Am. Chem. Soc. 68*, 1960–1963 (1946).

78. G. Scatchard, "Physical Chemistry of Protein Solutions. I. Derivation of the Equations for the Osmotic Pressure," *J. Am. Chem. Soc. 68*, 2315–2319 (1946).

79. G. Scatchard, A. C. Batchelder, and A. Brown, "Preparation and Properties of Serum and Plasma Proteins. VI. Osmotic Equilibria in Solutions of Serum Albumin and Sodium Chloride," *J. Am. Chem. Soc. 68*, 2320–2329 (1946).

80. G. Scatchard, A. C. Batchelder, A. Brown, and Mary Zosa, "Preparation and Properties of Serum and Plasma Proteins. VII. Osmotic Equilibria in Concentrated Solutions of Serum Albumin," *J. Am. Chem. Soc. 68*, 2610–2612 (1946).

81. J. L. Oncley, G. Scatchard, and A. Brown, "Physical-Chemical Characteristics of Certain of the Proteins of Normal Human Plasma," *J. Phys. Colloid Chem. 51*, 184–198 (1947).

82. G. Scatchard, L. F. Epstein, J. Warburton, Jr., and P. J. Cody, "Thermodynamic Properties of Saturated Liquid and Vapor of Ammonia-Water Mixtures," *Refrig. Eng. 53*, 413–421 (1947).

83. G. Scatchard, review of *Thermodynamics for Chemists* by S. Glasstone, *J. Am. Chem. Soc. 70*, 886 (1948).

84. S. S. Gellis, J. R. Neefe, J. Stokes, Jr., L. E. Strong, C. A. Janeway, and G. Scatchard, "Chemical, Clinical, and Immunological Studies on the Products of Human Plasma Fractionation. XXXVI. Inactivation of the Virus of Homologous Serum Hepatitis in Solutions of Normal Human Serum Albumin by Means of Heat," *J. Clin. Invest. 27*, 239–244 (1948).

85. G. Scatchard, "The Scientific Work of Edwin Joseph Cohn," address on Cohn's receiving the Richards Medal, *Nucleus 25*, 263–276 (1948).

86. G. Scatchard and Elizabeth S. Black, "The Effect of Salts on the Isoionic and Isoelectric Points of Proteins," *J. Phys. Colloid Chem. 53*, 88–99 (1949).

87. G. Scatchard, "Equilibrium in Non-electrolyte Mixtures," *Chem. Rev. 44*, 7–35 (1949).

88. G. Scatchard, review of *Outlines of Physical Chemistry* by Farrington Daniels, *J. Chem. Ed. 26*, 120 (1949).

89. G. Scatchard, "The Attractions of Proteins for Small Molecules and Ions," *Ann. New York Acad. Sci. 51*, 660–672 (1949).

90. G. Scatchard, I. H. Scheinberg, and S. H. Armstrong, Jr., "Physical Chemistry of Protein Solutions. IV. The Combination of Human Serum Albumin with Chloride Ion," *J. Am. Chem. Soc.* 72, 535–540 (1950).

91. G. Scatchard, I. H. Scheinberg, and S. H. Armstrong, Jr., "Physical Chemistry of Protein Solutions. V. The Combination of Human Serum Albumin with Thiocyanate Ion," *J. Am. Chem. Soc.* 72, 540–546 (1950).

92. G. Scatchard, review of *Thermodynamics. An Advanced Treatment for Chemists and Physicists* by E. A. Guggenheim, *J. Chem. Ed.* 27, 291 (1950).

93. G. Scatchard, "The Social Behavior of Molecules," *Am. Scientist* 38, 437–442 (1950).

94. G. Scatchard, "The Colloid Osmotic Pressure of Serum," *Science* 113, 201–202 (1951).

95. G. Scatchard, "Molecular Interactions in Protein Solutions," chap. 10 (pp. 239–266) in *Science in Progress*, Sigma Xi Lecture, 1951.

96. G. Scatchard, G. M. Kavanaugh, and L. B. Ticknor, "Vapor-Liquid Equilibrium. VIII. Hydrogen Peroxide–Water Mixtures," *J. Am. Chem. Soc.* 74, 3715–3720 (1952).

97. G. Scatchard, L. B. Ticknor, J. R. Goates, and E. R. McCartney, "Heats of Mixing in Some Non-electrolyte Solutions," *J. Am. Chem. Soc.* 74, 3721–3724 (1952).

98. G. Scatchard and L. B. Ticknor, "Vapor-Liquid Equilibrium. IX. The Methanol–Carbon Tetrachloride–Benzene System," *J. Am. Chem. Soc.* 74, 3724–3729 (1952).

99. G. Scatchard, "Solutions of Nonelectrolytes," *Annual Rev. Phys. Chem.* 3, 259–274 (1952).

100. G. Scatchard, "Some Physical Chemical Aspects of 'Plasma Extenders'," *Ann. New York Acad. Sci.* 55, 455–464 (1952).

101. G. Scatchard, "Molecular Interactions in Protein Solutions," *Am. Scientist 40*, 61–83 (1952).

102. G. Scatchard, "Some Electrochemical Properties of Ion Exchange Membranes," *Nucleus 30*, 76–79 (1952).

103. G. Scatchard, "Ion Exchanger Electrodes," *J. Am. Chem. Soc.* 75, 2883–2887 (1953).

104. G. Scatchard, "Equilibria and Reaction Rates in Dilute Electrolyte Solutions," *Nat'l Bur. Standards Circ.* 524, 185–192 (1953).

105. G. Scatchard and R. A. Westlund, Jr., "Equilibrium of Solid α-Silver-Zinc Alloys with Zinc Vapor," *J. Am. Chem. Soc.* 75, 4189–4193 (1953).

106. G. Scatchard, "Transport of Ions Across Charged Membranes" (pp. 128–143) in *Ion Transport Across Membranes*, H. T. Clarke, Ed. (Academic Press, New York, 1954).

107. G. Scatchard, W. L. Hughes, Jr., F. R. N. Gurd, and P. E. Wilcox, "The Interaction of Proteins with Small Molecules and Ions,"

chap. 11 (pp. 193–219) in *Chemical Specificity in Biological Interactions*, F. R. N. Gurd, Ed. (Academic Press, New York, 1954).

108. G. Scatchard, A. Gee, and Jeanette Weeks, "Physical Chemistry of Protein Solutions. VI. The Osmotic Pressures of Mixtures of Human Serum Albumin and gamma-Globulins in Aqueous Sodium Chloride," *J. Phys. Chem. 58*, 783–787 (1954).

109. G. Scatchard, A. Brown, R. M. Bridgforth, Jeanette Weeks, and A. Gee, "A Precision Modification of the Hepp Type Osmometer," mimeographed report from Department of Chemistry, M.I.T. (undated).

110. G. Scatchard and R. G. Breckenridge, "Isotonic Solutions. II. The Chemical Potential of Water in Aqueous Solutions of Potassium and Sodium Phosphates and Arsenates at 25°C," *J. Phys. Chem. 58*, 596–602 (1954); corr. *59*, 1234 (1955).

111. J. S. Johnson, K. A. Kraus, and G. Scatchard, "Distribution of Charged Polymers at Equilibrium in a Centrifugal Field," *J. Phys. Chem. 58*, 1034–1039 (1954).

112. G. Scatchard, "Some Aspects of the Physical Chemistry of Protein Solutions," Richards Medal Address, *Nucleus 31*, 211–217 (1954).

113. G. Scatchard, "Transport of Ions Across Charged Membranes," chap. 3 (pp. 18–32) in *Electrochemistry in Biology and Medicine*, T. Shedlovsky, Ed. (Wiley, New York, 1955).

114. G. Scatchard, review of *Electrolyte Solutions. The Measurement and Interpretation of Conductance, Chemical Potential and Diffusion*, by R. A. Robinson and R. H. Stokes, *J. Colloid Sci. 11*, 289–291 (1956).

115. G. Scatchard and R. H. Boyd, "The Equilibrium of alpha-Silver-Cadmium Alloys with Cadmium Vapor," *J. Am. Chem. Soc. 78*, 3889–3893 (1956).

116. G. Scatchard, "General Introduction to Discussions of Membrane Phenomena," *Faraday Soc. Discussions 21*, 27–30 (1956).

117. G. Scatchard and F. Helfferich, "The Effect of Stirring on Cells with Cation Exchanger Membranes," *Faraday Soc. Discussions 21*, 70–82 (1956).

118. G. Scatchard, review of *The Molecular Theory of Solutions* by I. Prigogine, *Chem. Eng. News* (Sept. 30, 1957), p. 84.

119. G. Scatchard, J. S. Coleman, and Amy L. Shen, "Physical Chemistry of Protein Solutions. VII. The Binding of Some Small Anions to Serum Albumin," *J. Am. Chem. Soc. 79*, 12–20 (1957).

120. G. Scatchard, notes for course: Equilibrium in Solutions.*

121. G. Scatchard, notes for course: Surface and Colloid Chemistry.*

122. G. Scatchard, "The Interpretation of Activity and Osmotic Coefficients, chap. 2 (pp. 9–18) in *The Structure of Electrolytic Solutions*, W. J. Hamer, Ed. (Wiley, New York, 1959).

123. B. A. Soldano, R. W. Stoughton, R. J. Fox, and G. Scatchard,

---

*The present volume is the published version of references 120 to 121.

"A High Temperature Isopiestic Unit," chap. 14 (pp. 224–235) in *The Structure of Electrolytic Solutions*, W. J. Hamer, Ed. (Wiley, New York, 1959).

124. J. S. Johnson, G. Scatchard, and K. A. Kraus, "The Use of Interference Optics in Equilibrium Ultracentrifugations of Charged Systems," *J. Phys. Chem. 63*, 787–793 (1959).

125. G. Scatchard and Judith Bregman, "Physical Chemistry of Protein Solutions. VIII. The Effect of Temperature on the Light Scattering of Serum Albumin Solutions," *J. Am. Chem. Soc. 81*, 6095–6100 (1959).

126. G. Scatchard and S. Zaromb, "Physical Chemistry of Protein Solutions. IX. A Light Scattering Study of the Binding of Trichloroacetate Ion to Serum Albumin," *J. Am. Chem. Soc. 81*, 6100–6104 (1959).

127. G. Scatchard, Y. V. Wu, and Amy L. Shen, "Physical Chemistry of Protein Solutions. X. The Binding of Small Anions by Serum Albumin," *J. Am. Chem. Soc. 81*, 6104–6109 (1959).

128. G. Scatchard, "John Gamble Kirkwood, 1907–1959," *J. Chem. Phys. 33*, 1279–1281 (1960).

129. G. Scatchard, B. Vonnegut, and D. W. Beaumont, "New Type of Freezing-Point Apparatus. Freezing Points of Dilute Lanthanum Chloride Solutions," *J. Chem. Phys. 33*, 1292–1298 (1960).

130. J. S. Johnson, K. A. Kraus, and G. Scatchard, "Activity Coefficients of Silicotungstic Acid; Ultracentrifugation and Light Scattering," *J. Phys. Chem. 64*, 1867–1873 (1960).

131. G. Scatchard, "Osmotic Coefficients and Activity Coefficients in Mixed Electrolyte Solutions," *J. Am. Chem. Soc. 83*, 2636–2642 (1961).

132. G. Scatchard and N. J. Anderson, "The Determination of the Equilibrium Water Content of Ion Exchange Resins," *J. Phys. Chem. 65*, 1536–1539 (1961).

133. G. Scatchard, review of *Electrolytic Dissociation* by C. B. Monk, *J. Am. Chem. Soc. 83*, 5043 (1961).

134. R. M. Rush and G. Scatchard, "Molal Volumes and Refractive Index Increments of $BaCl_2$–HCl Solutions. Mixture Rules," *J. Phys. Chem. 65*, 2240–2242 (1961).

135. G. Scatchard and T. P. Lin, "The Equilibrium of alpha-Silver-Zinc-Cadmium Alloys with Zinc and Cadmium Vapors," *J. Am. Chem. Soc. 84*, 28–34 (1962).

136. G. Scatchard and J. Pigliacampi, "Physical Chemistry of Protein Solutions. XI. The Osmotic Pressures of Serum Albumin, Carbonylhemoglobin and their Mixtures in Aqueous Sodium Chloride at 25°," *J. Am. Chem. Soc. 84*, 127–134 (1962).

137. G. Scatchard, "The Gibbs Adsorption Isotherm," *J. Phys. Chem. 66*, 618–620 (1962).

138. G. Scatchard, review of *States of Matter* by E. A. Moelwyn-Hughes, *J. Am. Chem. Soc. 84*, 1518 (1962).

139. G. Scatchard, review of *Thermodynamics* by G. N. Lewis and M. Randall, revised by K. S. Pitzer and L. Brewer, *J. Am. Chem. Soc.* *84*, 3791 (1962).

140. G. Scatchard, "Half a Century as Part-Time Colloid Chemist," Kendall Company Award in Colloid Chemistry Address, American Chemical Society, March 26, 1962. Later published in "Twenty Years of Colloid and Surface Chemistry: The Kendall Award Addresses" (K. J. Mysels, C. M. Samour and J. M. Hollister, editors) American Chemical Society, Washington, D.C. (1973).

141. G. Scatchard, "Solutions of Electrolytes," *Annual Rev. Phys. Chem.* *14*, 161–176 (1963).

142. G. Scatchard, "Basic Equilibrium Equations," pp. 105–117 in *Conference on the Ultracentrifuge* (Academic Press, New York, 1963).

143. G. Scatchard, G. M. Wilson, and F. G. Satkiewicz, "Vapor-Liquid Equilibrium. X. An Apparatus for Static Equilibrium Measurements," *J. Am. Chem. Soc. 86*, 125–127 (1964).

144. G. Scatchard and F. G. Satkiewicz, "Vapor-Liquid Equilibrium. XII. The System Ethanol–Cyclohexane from 5 to 65°," *J. Am. Chem. Soc. 86*, 130–133 (1964).

145. G. Scatchard and G. M. Wilson, "Vapor-Liquid Equilibrium. XIII. The System Water–Butyl Glycol from 5 to 85°," *J. Am. Chem. Soc. 86*, 133–137 (1964).

146. D. H. Freeman and G. Scatchard, "Electrokinetic Behavior of Ion-Exchange Resin," *Science 144*, 411–412 (1964).

147. G. Scatchard, "The Effect of Dielectric Constant Difference on Hyperfiltration of Salt Solutions," *J. Phys. Chem. 68*, 1056–1061 (1964).

148. G. Scatchard and W. T. Yap, "The Physical Chemistry of Protein Solutions. XII. The Effects of Temperature and Hydroxide Ion on the Binding of Small Anions to Human Serum Albumin," *J. Am. Chem. Soc. 86*, 3434–3438 (1964).

149. D. H. Freeman and G. Scatchard, "Volumetric Studies of Ion-Exchange Resin Particles Using Microscopy," *J. Phys. Chem. 69*, 70–74 (1965).

150. G. Scatchard, "The Computation of the Activity Coefficients of Small Ions from the Loeb, Overbeek and Wiersema Tables," *Z. physik. Chem. 228*, 354–363 (1965).

151. G. Scatchard, "The Osmotic Pressure, Light Scattering and Ultracentrifuge Equilibrium of Polyelectrolyte Solutions," chap. 16 (pp. 347–360) in *Chemical Physics of Ionic Solutions*, B. E. Conway and R. G. Barradas, Eds. (Wiley, New York, 1966).

152. G. Scatchard, "Water; A Review," *Fed. Proc. 25*, 954–957 (1966).

153. G. Scatchard, "Remarks on Osmotic Pressure," *Fed. Proc. 25*, 1115–1117 (1966).

154. G. Scatchard and W. H. Orttung, "The Electromotive Forces of NH₄Br and KBr Concentration Cells. Are the Abnormalities in the

Freezing Points of Ammonium Salts Due to Surface Effects?" *J. Colloid Sci. 22*, 12–18 (1966).

155. G. Scatchard and W. H. Orttung, "Diffusion and the Bi-ionic Potential with an Ion Exchanger Membrane," *J. Colloid Sci. 22*, 19–22 (1966).

156. S. Y. Tyree, Jr., R. L. Angstadt, F. C. Hentz, Jr., R. L. Yoest, and G. Scatchard, "The Osmotic Coefficients and Other Related Properties of Aqueous 12-Tungstosilicic Acid ($H_4W_{12}SiO_{40}$) at 25°," *J. Phys. Chem. 70*, 3917–3921 (1966).

157. G. Scatchard, "The Excess Free Energy and Related Properties of Solutions Containing Electrolytes," *J. Am. Chem. Soc. 90*, 3124–3127 (1968).

158. G. Scatchard, corrections to "The Excess Free Energy and Related Properties of Solutions Containing Electrolytes," *J. Am. Chem. Soc. 91*, 90 (1969).

159. G. Scatchard, Edwin J. Cohn Lecture, "Edwin J. Cohn and Protein Chemistry," *Vox Sanguinis 17*, 37–44 (1969).

160. Y. C. Wu, R. M. Rush, and G. Scatchard, "Osmotic and Activity Coefficients for Binary Mixtures of Sodium Chloride, Sodium Sulfate, Magnesium Sulfate, and Magnesium Chloride in Water at 25°. I. Isopiestic Measurements on the Four Systems with Common Ions," *J. Phys. Chem. 72*, 4048–4053 (1968).

161. Y. C. Wu, R. M. Rush, and G. Scatchard, "Osmotic and Activity Coefficients for Binary Mixtures of Sodium Chloride, Sodium Sulfate, Magnesium Sulfate, and Magnesium Chloride in Water at 25°. II. Isopiestic and Electromotive Force Measurements on the Two Systems without Common Ions," *J. Phys. Chem. 73*, 2047–2053 (1969).

162. G. Scatchard, R. M. Rush, and J. S. Johnson, "Osmotic and Activity Coefficients for Binary Mixtures of Sodium Chloride, Sodium Sulfate, Magnesium Sulfate, and Magnesium Chloride in Water at 25°. III. Treatment with Ions as Components," *J. Phys. Chem. 74*, 3786–3796 (1970).

163. H. F. Gibbard, Jr., and G. Scatchard, "The Vapor-Liquid Equilibria of Synthetic Seawater Solutions from 25–100°C," *J. Chem. Eng. Data 17*, 498–501 (1972).

164. H. F. Gibbard, Jr., and G. Scatchard, "Liquid-Vapor Equilibrium of Aqueous Lithium Chloride, from 25° to 100°C and from 1.0 to 18.5 Molal, and Related Properties," *J. Chem. Eng. Data 18*, 293–297 (1973).

165. H. F. Gibbard, Jr., G. Scatchard, R. A. Rousseau and J. L. Creek, "Liquid-Vapor Equilibrium of Aqueous Sodium Chloride from 298 to 373K and from 1 to 6 mol kg^{-1}, and Related Properties," *J. Chem. Eng. Data 19*, 281–288 (1974).

Index